AF621661

THE NEW AVENUES IN BIOINFORMATICS

Cellular Origin and Life in Extreme Habitats and Astrobiology

Volume 8

Series Editor:

Joseph Seckbach
Hebrew University of Jerusalem, Israel

The New Avenues in Bioinformatics

Edited by

Joseph Seckbach

The Hebrew University of Jerusalem, Israel

and

Eitan Rubin

Harvard University, Cambridge, MA, U.S.A.

KLUWER ACADEMIC PUBLISHERS

DORDRECHT / BOSTON / LONDON

A C.I.P Catalogue record for this book is available from the Library of Congress.

ISBN 1-4020-2639-0 (HB)
ISBN 1-4020-2834-2 (e-book)

Published by Kluwer Academic Publishers,
P.O. Box 17, 3300 AA Dordrecht, The Netherlands.

Sold and distributed in North, Central and South America
by Kluwer Academic Publishers,
101 Philip Drive, Norwell, MA 02061, U.S.A.

In all other countries, sold and distributed
by Kluwer Academic Publishers,
P.O. Box 322, 3300 AH Dordrecht, The Netherlands.

Printed on acid-free paper

Modeling and simulating the action of different anti-viral drugs using a SADT actigram. This system formalize the medical knowledge for data management in the expert systems Genesyx, "Expert Immuno" and "Expert SIDA" (developed by the Société de Bio-Informatique et de Biotechnologie, Tours, France). These expert systems can be apply to drug discovery, education, diagnostic and therapeutic purpose.

Actigram made by S. Villeret, P. Bobola, C. Gaudeau and F. Aboli.

All Rights Reserved
© 2004 Kluwer Academic Publishers
No part of this work may be reproduced, stored in a retrieval system, or transmitted in any form or by any means, electronic, mechanical, photocopying, microfilming, recording or otherwise, without written permission from the Publisher, with the exception of any material supplied specifically for the purpose of being entered and executed on a computer system, for exclusive use by the purchaser of the work.

Printed in the Netherlands.

This book is dedicated to:
Professor Tova Arzi (Emeritus Professor, Tel-Aviv University, Israel), who was the senior editor's (JS) first teacher of plant anatomy and revealed to him many secrets of the Plant World, with all the best wishes for health and happiness.

TABLE OF CONTENTS

ACKNOWLEDGEMENTS

This bioinformatics book was inspired a couple of years ago when the editors attended the Netherlands-Israel Bioinformatics Conference, organized by OPTIN in The Hague. We thank Ms. Jennifer Peersmann, the OPTIN director, for inviting us to participate in that conference. This book is the eight-volume in the series "***Cellular Origins, Life in Extreme Habitats and Astrobiology (COLE)***," edited by J. Seckbach.

We acknowledge all the contributors for their chapters and our reviewers who examined the chapters. Specific appreciation goes to Fern Seckbach who proofread several sections of this book, and to all people who assisted us in this project. We thank Professor François Raulin (University of Paris 12 & Paris 7 Faculty of Sciences and Technology) for suggesting the title for this volume. Last but not least, we are grateful and deeply appreciative of the kindness of Dr. Frans van Dunne and Ms. Claire van Heukelom, our Kluwer representatives, for their constant faithful handling of this volume, as they have done with all the other books in this COLE series.

April 30, 2004

Joseph Seckbach
Eitan Rubin

LIST OF AUTHORS FOR *THE NEW AVENUES IN BIOINFORMATICS*

All senior authors are underlined

ANXOLABÉHÈRE DOMINIQUE.
LABORATOIRE DE DYNAMIQUE DU GENOME ET EVOLUTION INSTITUT JACQUES MONOD, 2 PLACE JUSSIEU, 75251 PARIS CEDEX 05, FRANCE
E-mail: hq@ccr.jssieu.fr

ARKIN ISAIAH T.
DEPARTMENT OF BIOLOGICAL CHEMISTRY, THE ALEXANDER SILBERMAN INSTITUTE OF LIFE SCIENCES. THE HEBREW UNIVERSITY OF JERUSALEM, GIVAT-RAM, JERUSALEM, 91904, ISRAEL
E-mail: arkin@cc.huji.cc.il

BENCZE LAJOS
MÜLLER LABORATORY, INSTITUTE OF ORGANIC CHEMISTRY, UNIVERSITY OF VESZPREM, EGYETEM U. 6. H-8200 VESZPREM, HUNGARY

BOBOLA PHILIPPE
SOCIÉTÉ DE BIO-INFORMATIQUE ET DE BIOTECHNOLOGIE, TOURS, FRANCE
E-mail: lbibi@netcourier.com

BUJNICKI JANUSZ M.
BIOINFORMATICS LABORATORY, INTERNATIONAL INSTITUTE OF MOLECULAR AND CELL BIOLOGY, KS. TROJDENA, 4 02-109 WARSAW, POLAND
E-mail: iamb@genesilico.pl

CAGLIOTI LUCIANO
DEPARTMENT OF CHEMISTRY AND TECHNOLOGY OF BIOLOGICALLY, ACTIVE COMPOUNDS, UNIVERSITY "LA SAPIENZA" OF ROME, P. LE A. MORO, 5, I-00185 ROMA, ITALY

DE BAETS BERNARD
DEPARTMENT OF APPLIED MATHEMATICS, BIOMETRICS AND PROCESS CONTROL, GHENT UNIVERSITY, COUPURE LINKS 653, 9000 GHENT, BELGIUM
E-mail: Bernard.DeBaets@rug.ac.be

DEGROEVE SVEN
DEPARTMENT OF PLANT SYSTEMS BIOLOGY, FLANDERS INTERUNIVERSITY INSTITUTE FOR BIOTECHNOLOGY (VIB), K.L.LEDEGANCKSTRAAT 35, 9000 GHENT, BELGIUM
E-mail: svgro@gengenp.rug.ac.be

EINERHAND MARK P.W.
VEREENIGDE. NIEUWE PARKLAAN 97, 2587 BN THE HAGUE, THE NETHERLANDS E-mail: m.einerhand@vereenigde.nl

FISCHER DANIEL
BIOINFORMATICS, DEPARTMENT COMPUTER SCIENCE, BEN GURION UNIVERSITY, BEER-SHEVA 84015, ISRAEL
E-mail: dfischer@cs.bgu.ac.il

GAUDEAU CLAUDE
LABORATOIRE DE BIO-INFORMATIQUE ET DE BIOTECH-NOLOGIE, BIO ESPAS, TOURS, FRANCE
E-mail: sbibi@netcourrier.com

GILL-MORE RAVEH
COMPUGEN, LTD, 72 PINCHAS ROSEN ST. TEL-AVIV 69512, ISRAEL
E-mail: raveh@compugen.co.il

KOLTAI, HINANIT
DEPARTMENT OF GENOMICS AND BIOINFORMATICS, THE AGRICULTURAL RESEARCH ORGANIZATION, THE VOLCANI CENTER, BET DAGAN 50250, ISRAEL

KREISBERG-ZAKARIN RACHELI
BIOINFORMATICS UNIT, GEORGE S. WISE FACULTY OF LIFE SCIENCES, TEL-AVIV UNIVERSITY, ISRAEL
E-mail: racheli@post.tau.ac.il

LEONOV HADAS
DEPARTMENT OF BIOLOGICAL CHEMISTRY, THE ALEXANDER SILBERMAN INSTITUTE OF LIFE SCIENCES. THE HEBREW UNIVERSITY OF JERUSALEM, GIVAT-RAM, JERUSALEM, 91904, ISRAEL
E-mail: hleonov@cs.huji.ac.il

LINIAL MICHAL
THE LIFE SCIENCE INSTITUTE, THE HEBREW UNIVERSITY OF JERUSALEM, ISRAEL
E-mail: michall@cc.huji.ac.il

LORENZ RALPH D.
LUNAR AND PLANETARY LABORATORY, UNIVERSITY OF ARIZONA, AZ. 85721 USA
E-mail: rlorenz@lpl.arizona.edu

LUCAS YVES
LABORATOIRE VISION ET ROBOTIQUE, IUT DE BOURGES, UNIVERSITÉ D' ORLÉANS, FRANCE
E-mail: yves.lucas@libertysurf.fr

MAGALI MORIN
SOCIÉTÉ DE BIO-INFORMATIQUE ET DE BIOTECHNOLOGIE, TOURS, FRANCE

MARKMAN OFER
PROCOGNIA LTD. PROCOGNIA (ISRAEL) LTD., 3 HABOSEM ST., ASHDOD, 77610, ISRAEL
E-mail: ofer.markman@procognia.com

PALYI GYULA,
DEPARTMENT OF CHEMISTRY, UNIVERSITY OF MODENA AND REGGIO EMILIA, VIA CAMPI, 183, I-41100 MODENA, ITALY
E-mail: palyi@unimo.it

PRUESS MANUELA
EMBL OUTSTATION, THE EUROPEAN BIOINFORMATICS INSTITUTE (EBI) WELLCOME TRUST GENOME CAMPUS, HINXTON, CAMBRIDGE, CB10 1SD, UK
E-mail: mpr@ebi.ac.uk

QUESNEVILLE HADI
LABORATOIRE DE DYNAMIQUE DU GENOME ET EVOLUTION INSTITUT JACQUES MONOD, 2 PLACE JUSSIEU, 75251 PARIS CEDEX 05, FRANCE
E-mail: hq@ccr.jssieu.fr

ROEDERER JUAN G.
GIOPHYSICAL INSTITUTUTE, UNIVERSITY OF ALASKA-FAIRBANKS, FAIRBANKS, AK 99775, USA
E-mail: jgr@gi.alaska.edu

ROUZE PIERRE
LABORATOIRE ASSOCIÉ DE L'INRA (FRANCE), K.L. LEDEGANCKSTRAAT 35, 9000 GHENT, BELGIUM
E-mail: Pierre.Rouze@gengenp.rug.ac.be

RUBIN EITAN
HARVARD UNIVERSITY, THE BAUER CENTER FOR GENOMIC RESEARCH, CAMBRIDGE, MA. USA
E-mail: erubin@cgr.harvard.edu

RYCHLEWSKI LESZEK
BIOINFOBANK INSTITUTE, UL. LIMANOWSKIEGO 24A, 60-744 POZNAN, POLAND
E-mail: leszek@bioinfo.pl

SAEYS YVAN,
DEPARTMENT OF PLANT SYSTEMS BIOLOGY, FLANDERS INTERUNIVERSITY INSTITUTE FOR BIOTECHNOLOGY (VIB), K.L.LEDEGANCKSTRAAT 35, 9000 GHENT, BELGIUM
E-mail: yvsae@gengenp.rug.ac.be

SASSON ORI
THE SCHOOL OF COMPUTER SCIENCE AND ENGEENIRING, THE HEBREW UNIVERSITY OF JERUSALEM, ISRAEL
E-mail: Ori@osasson.ocm

SECKBACH JOSEPH
THE HEBREW UNIVERSITY OF JERUSALEM, HOME: P.O.BOX 1132, EFRAT, 90435, ISRAEL
E-mail: seckbach@huji.ac.il

THEVOT FREDERIC
SOCIÉTÉ DE BIO-INFORMATIQUE ET DE BIOTECHNOLOGIE, TOURS, FRANCE
E-mail: fredericpat@yahoo.fr

UNGER RON
FACULTY OF LIFE SCIENCE, BAR-ILAN UNIVERSITY, RAMAT GAN, 52900, ISRAEL
E-mail: ron@biocom1.ls.biu.ac.il

VAN DE PEER YVES
DEPARTMENT OF PLANT SYSTEMS BIOLOGY, FLANDERS INTER-UNIVERSITY INSTITUTE FOR BIOTECHNOLOGY (VIB), K.L. LEDEGANCK STRAAT 35, 9000 GHENT, BELGIUM
E-mail: yves.vandepeer@gengenp.rug.ac.be

VAN MELLE JOHANNES
VEREENIGDE. NIEUWE PARKLAAN 97, 2587 BN THE HAGUE, THE NETHERLANDS
E-mail: j.vanmelle@vereenigde.nl

VOLPIN HANNE
DEPARTMENT OF GENOMICS AND BIOINFORMATICS, THE AGRICULTURAL RESEARCH ORGANIZATION, THE VOLCANI CENTER, BET DAGAN 50250, ISRAEL
E-mail: hanne@agri.gov.il

XIE HANQING
COMPUGEN, INC., 7 CENTER DRIVE, SUITE 9, JAMESBURG, NJ 08831, USA
E-mail: han@cgen.com

YAKIR BENNY
DEPARTMENT OF STATISTICS, THE HEBREW UNIVERSITY OF JERUSALEM, ISRAEL
E-mail: msby@mscc.huji.ac.il

ZUCCHI CLAUDIA
DEPARTMENT OF CHEMISTRY, UNIVERSITY OF MODENA AND REGGIO EMILIA, VIA CAMPI, 183, I-41100 MODENA, ITALY
E-mail: Zucchi@unimore.it

GLOSSARY*

Active site—the truedimensional location in an enzyme or in a functional protein where the actual activity takes place (usually where the enzymatic reaction happens—see *site*)

Algorithm—An algorithm is a procedure for solving a problem. The word derives from the name of the Persian mathematician, Al-Khowarizmi (825 AD). A computer program can be viewed as an elaborate algorithm.

Alignment—the identification of similar regions in two or more sequences (nucleotides or amino acids).

Alpha Helix—a helical structure of a peptide, one of the most common structural forms in proteins, key in the secondary structure, characteristic to soluble as well as membrane proteins.

Analogue, analogous, analogy—similar or similarity in structure (e.g. base analogue), most often used for proteins similar in structure but of different evolutionary origin, one of the most common bioinformatics methodology to deduce or guess a gene or protein function.

Aneuploidy—It is the abnormality where the number of chromosomes is not an exact multiple of the haploid number.

Annotation—the action of improving databases, and the most valued outcome of bioinforatics to databases. Often includes referencing, error corrections, notations, comments, direction and links to other resources.

Anticodon—the triplet of bases on the tRNA that decodes the amino acid (see codon, redundancy).

Base pair—a definition noting the two matching nucleic acids on the two opposite strands of the double helix, most often a measure for length or size of a gene plasmid, or DNA segment

Beta sheet—a sheet like structure in a protein, mediated by inter-peptide hydrogen bonds, key in the secondary structure of proteins, common in soluble proteins.

Biochips—common definitions to all experimental systems in which arrays of biomolecules are constructed on a single surface. (most common—DNA chips, protein chips, peptide chips, glycan chips)

Biological system—A natural (not artificial) system exhibiting purpose-directed interactions that are controlled by information.

---

*Composed and edited by Drs. **Ofer Markman**[1+2], **Sophia Kossida**[2] and **Eitan Rubin**[3] with entries from the chapters' authors: Professor **Juan Roederer**, Dr. **Ron Unger** and also from **R. D. Lorenz,** and **Dr. Hanqing Xie.**

[1]Procognia (Israel) Ltd. 3 Habosem St., Ashdod 77601 Israel, http://www.procognia.com, [2]Internet Biologists, http://www.internetbiologists.org, and [3]The Editor of this volume.

cDNA*—a common molecular biology tool in which the single stranded mRNA molecule is translated to the identical in content sense, anti-sense double stranded DNA molecule. (e.g. cDNA libraries).

Cluster –ing—a statistical tool in which a set of data is segregated to groups of the most similar data items within the group according to predefined criteria

Complementary—see sense anti-sense, a strand of nucleic acid that matches a certain nucleic acid strand according to the A-T G-C pairing rules.

Consciousness—Coherent, cooperative and synchronous interaction between cortical and limbic structures of an animal brain during information -processing.

Consensus Sequence—a bioinformatic tool, a virtual sequence, defining the characteristic sequence of a multiply alligned similar sequence. May use nucleic acid or amino acid codes and symbols defining non physical peptide or nucleic acid units such as "aromatic amino acid", "purine", "hydrogen donor" or "bulky amino acid" as well as newly defined notations by the user.

Convergence—the end point of an iterative algorithm run, the point in which the difference between each iteration is smaller that a predefined threshold.

C-terminal—an orientation in the peptide chain arbitrarily close to the COOH end of the peptide chain. In a gene or in a DNA fragment, the orientation in the DNA arbitrarily close to the COOH of the "would be translated" peptide

Deconvolution—algorithmic or mathematical conversion of interpreted signals or output data into a meaningful pattern, by treating overlapping signals and information redundancy.

Dendrogram—a "tree like" graphics to present the result of a binary clustering algorithm. Often the tree arbitrarily portrays cluster distances in the height of the tree nodes.

Disulphide bond—a long range, covalent binding, maybe intra-chain but often inter-chain, cystein—cystein bond in a protein or peptide

DNA—a nucleic acid, based on the deoxyribose phosphodiester backbone. The most common genetic material.

DNA chip—see biochip, a DNA arraying technology and platform

DNA fingerprinting—a technique to identify individual molecule or individual organism, pathogen, person or animal on the basis of tandem repeats of DNA sequences that are characteristic to each individual.

Domain—(1) a section of a protein with strong similarities to other proteins or other regions of the same protein, with all the copies functioning in a similar way; (2) a region of the protein that structurally self-organizes into a functional unit. The two views of a domain often but not always overlap.

Downstream—(1) for DNA: toward the 3' end; (2) for metabolic/signaling pathways: in subsequenct reactions

Endemic (disease or species)—Native to a specific region

Endogenous—Produced from within the organism

Entropy—A formal measure of information content. In a physical, chemical or other system, or a message, the entropy of a particular configuration of the system relates to the improbability of that configuration.

Epidemic—Disease affecting simultaneously large number of individulas

Epitope—small and undefined structural element, a ligand to antibody

EST (Expressed Sequence Tag)—The result of single pass sequencing of cDNA. Much cheaper to produce than full-length sequencing, these low-quality sequences are useful for gaining insight into gene structure and expression level.

Exon—the untranslated part of a gene that result in a protein (see intron)

Expression—the processes that cause a gene to express its phenotype—most often used as a synonym to *transcription and translation.*

Fold—a schematic and ionized representation of the three dimensional structure of a protein.

Frameshift mutation—A mutation which causes a change in the amino acid.

Genetic code—The three letter code that translates nucleotides into amino acids.

Genome—The DNA code that makes up the whole genetic composition of an organism.

Genomics—The study of the full gene repertoire of an organism.

Glycans—glycoprotein) or a lipid (as in glycolipid), in proteins glycans are the most abundant post translational modification (PTM). Chains of sugars which are connected to a protein (as in ?

Glycogenomics—is the connection of genomics to the field of glycobiology.

Glycoinformatics—is the general technology that stores, processes and analyzes the information on glycomolecules.

Glycome—the general glycan phenotype of an organism.

Glycomics—is the interface of wet-bench glycoanalytical technology and glycoinformatics to make sense of the glycome (the general glycan phenotype of an organism).

Glycomolecules—a general term describing sugars, oligosaccharides, polysaccharides and glyco-conjugates (glycoproteins, glycolipids and glycoseaminoglycans); the molecules that bind them (lectins); and the molecule that processes them—e.g. glycosyltransferases and glycosidases.

Glyco-proteomics—is the analysis of glycans on distinct proteins, and by thus is a subdivision of glycomics.

Information—That which represents a consistent and reproducible correspondence between a spatial and/or temporal pattern at a “sender” and a pattern-specific change elicited at a “recipient”.

In-silico—in the computer, modeled by computer or computer simulated.

In-situ—at the site, related to data collected from tissue or cells or to experiments done directly on the unprocessed biological material.

Intron—a part of a gene the product of which is a protein and is part of the code for the translated protein.

In-vivo—on or in live organisms, refers to experiments performed or data collected, except in human in which the adjective "clinical-" is used.

ligand—the molecule that binds to a receptor, used also as an arbitrary name for a binding partner

Linkage—a measure of distance on the DNA molecule, relates to the probability of two DNA fragments to be inherited independently.

Map—annotated sequence, often with low resolution, maps are often rich in positional information and order or binding sites, cleavage sites and functional sites such as genes or gene regulation sites.

Mapping—the process of defining the sequence, positioning sites and annotating the sequence.

Model organism—Organisms that depict advantages for answering scientific questions. Drsophila melanogaster, the fruit fly, for example is a model organism in genetics. The life cycle is relative short, there are many progenies, it is easy, inexpensive to maintain and needs little place to store.

Modeling—the process of predicting the complex structure or behavior of a complex system only on the basis of prior knowledge and not on the basis of direct measurements. Often referred to the process of predicting the behavior or structure on the basis of incomplete or insufficient direct measurements.

Motif—see short structural motifs

mRNA—the molecule that carry the message to the ribosome for the translation of a protein from a gene.

Multimerity—dimer, trimer, teteramer etc—composision of a molecule by several identical, similar or related protein units—(e.g the IgG dimer, Jun-Fos dimer, the G-protein trimer)

N-Glycosylation—in a glycoprotein, the most common type of glycosylation, in which the glycan binds to an aspargine (N) side chain.

N-terminal see C-terminal—an orientation in the peptide chain arbitrarily close to the the NH_2 end of the peptide chain. In a gene or in a DNA fragment, the orientation in the DNA arbitrarily close to the NH end of the "would be translated" peptide.

O-glycosylation—in a glycoprotein, the second most common glycosylation, in which the glycan is bound to the –OH group of Serine or threonine.

Open reading frame (ORF)—A stretch of DNA without stop codon which codes a protein be it the full the length of it or partial length of it.

Orthologous—see *Paralogous* for the full citation" ... Where the homology is the result of speciation so that the history of the gene reflects the history of the species (for example alpha hemoglobin in man and mouse) the genes should be called **orthologous** (ortho = exact).

Paralogous—The original quotation is by Walter Fitch (1970, Systematic Zoology 19:99–113): "Where the homology is the result of gene duplication so that both copies have descended side by side during the history of an organism, (for example, alpha and beta hemoglobin) the genes should be called paralogous (para = in parallel). Where the homology is the result of speciation so that the history of the gene reflects the history of the species (for example alpha hemoglobin in man and mouse) the genes should be called **orthologous** (ortho = exact)."

Peptide—the basic chemical entity of proteins, a chain of amino acids

Phylogenetic tree—Graphical representation of the evolutionary relationships of the different opreational taxonomic units (OTU) which could be species, genes.

Phylogenetics—The study of the evolutionarly relationships among the different life forms.

Primary sequence or structure—the most schematic and basic presentation of a polymer structure, describing only the covalent chemistry of the molecules, i.e. the order of building units.

Protein—peptide, often used for the long peptides with defined 3D structures, a common name for the unmodified and post translationaly modified ones

Protein chip—see biochip

Protein Structure Prediction—The three dimensional structure of proteins is determined by their linear amino acid sequence. Understanding the process by which proteins fold, and in particular being able to predict, computationally, the three dimensional structure of a protein is one of the grand challenges of computational biology and bioinformatics

Proteome/mics—The full complement of peptide, proteins and their variants and modifications (e.g. glycoproteins, phosphorylated peptides and proteins), in a certain cell, tissue or organ at a certain time. The qualitative and quantative study of the proteins in an organism is termed proteomics.

Purpose (of a process)—An expected specific change that would occur naturally only with a very low probability or not at all.

Query—the input for a search in a database according to predetermined rules and an automated algorithm.

Redundancy—a definition in information analysis that refers to the extent in which different strings or words convey the exact meaning and function, often refered to the extent in which different codons translate to the same amino-acid.

Repeat—a part of the biopolymer that appears along the length of the polymer in several very similar or identical repeats.

Replication—The exact duplication of the DNA molecule, by synthesis for each of the DNA strands its complementary strand.

RNA—a family of nucleic acid biopolymers the backbone of which is on the basis of deoxyribose phosphodiester. (include mRNA, tRNA, rRNA)

Secondary structure—a schematic description of the structure of a biopolymer (often Protein or RNA) according to hydrogen bonding patterns.

Self-consciousness—Capability of the human brain to make a single, coherent real-time representation of its own state of information-processing.

Sense, Antisense—an arbitrarily defined direction of sequences, often if sense is the mRNA sequence, then antisense is exactly the sequence complementary to it according to the A-T G-C pairing rules.

Sequencing—the determination of the sequence of the biopolymer (RNA, DNA, glycan, peptie or protein) by experimental procedure

Short structural motifs—It was noted that short (up to 10 residues) fragments of proteins tend to form structural motifs that re-occur many times in the database of known protein structure. These motifs correspond to the well known secondary structure elements of proteins, helices, beta-sheets, and turns, (but with finer details) and also to typical structural ways in which secondary structure elements are connected.

Signal sequence—a part of the protein molecule the function of which is to direct the traslated protein to a certain cellular organelle, and which is cleaved during the protein maturation process.

Site—a definition of a position on a biopolymer. Often refered to a position on the sequence in which a function, binding or an enzyme cleavage is potential. Alternatively, a small structural element in a molecule in which a significant function, such as binding or an enzymatic reaction take place.

SNP (single nucleotide polymorphism)—Single nucleotide polymorphisms (SNPs) are common DNA sequence variations among individuals. They show up with higher frequency than mutations.

Speciation—The formation of one or more species from an old one. The new species are no longer capable of interbreeding.

Structural building blocks—It was found that a canonical set of short structural motifs, often called structural building blocks, can be extracted such that proteins can be reconstructed by using these "standard" building blocks. Many studies have shown that the size of such a set is about hundred building blocks.

Taxonomy—Study of scientific classification

Thinking (human)—Capability of the human brain to recall stored information, process it and re-store it in modified form without any concurrent sensory and somatic input.

Transcription and translation—the processes that most often cause a gene to express its phenotype in the form of protein *expression*, often conceived as a two stage process: transcription in which a *DNA* is converted into an mRNA and translation in which the mRNA is translated to a peptide in the ribosome.

Transcriptome—The full complement of activated genes, mRNAs, or transcripts in a particular tissue or cell at a particular time is refered to as transcriptome. See whole transcriptome

tRNA—transfer RNA, the molecule carrying the amino acid and mediates its insertion in the correct place in the translation process in the ribosome.

Upstream—a term used to define position on the DNA strand in relation to other position, towards the 3' end of the DNA molecule

Whole transcriptome—the complete set of RNA molecules derived from the genome during the lifetime of an organism. See transcriptome.

Xenologous—Xenologous are sequences which arise from introduction of DNA by lateral gene transfer.

MORE TERMS[1]

No glossary is ever full. Terms of interest can be searched in any of the internet glossaries. It is important to know that terms are often right for their context. Therefore, terms in these sites are most suitable for the context they were defined for.

The bioinformatic glossary of **California State University**
http://www.bscbioinformatics.com/Fac/Tools/glossary.html

The bioinformatic glossary of **The European Bioinformatics Institute (EBI)**
http://www.ebi.ac.uk/2can/glossary/index.php

The bioinformatics glossary of **Cambridge HealthTech Institute (edited by Mary Chity)** one of few glossaries relevant for the modern lifescientist at
http://www.genomicglossaries.com/content/Bioinformatics_gloss.asp

The Bioinformatics Glossary at **the City Univ. of NY, College of Staten Island** (Courseware) http://www.library.csi.cuny.edu/~davis/Bio_326/bioinfo_glossary.html

The glossary **of www.SequenceAnalysis.com** by **A. S. Louka** at
http://www.sequenceanalysis.com/glossary.html

Science Magazine annotated gateway to glossaries has entries on bioinformatics at
http://www.sciencemag.org/feature/plus/sfg/education/glossaries.shtml#postgenomics and http://www.sciencemag.org/feature/plus/sfg/education/glossaries.shtml#medical

Oak Ridge Natl Lab. (ORNL) Genegateway's Glossary of Bioinformatic Terms at
http://www.ornl.gov/sci/techresources/Human_Genome/posters/chromosome/genejargon.shtml

---

[1]Compiled by Dr. **Ofer Markman**, Procognia (Israel) Ltd. 3 Habosem St., Ashdod 77601 Israel, http://www.procognia.com

Falcon Rosewel Park Cancer Institute cite the folowing bioinforatics glossaries at http://falcon.roswellpark.org/labweb/glossary.html; http://bioinformatics.utmem.edu/glossary.html; http://www.bscbioinformatics.com/Stu/Glo/glossary.html; http://www.brunel.ac.uk/depts/bl/project/biocomp/sequence/seqanal_guide/glossary.html

Some terms of relevance can be found in the bioinformatics rich glossary at the site of the **Buffalo Center of Excellence in Bioinformatics** http://www.bioinformatics.buffalo.edu/current_buffalo/glossary.html

Paracel Inc.'s bioinformatics glossary at http://www.paracel.com/c/bio-glossary.htm

Virginia Bioinformatics Institute (VBI) has a glossary at https://www.vbi.vt.edu/article/articleview/122

And finally, **The UK CCP11 project's** TBR Glossary gateway at http://www.hgmp.mrc.ac.uk/CCP11/directory/directory_glossaries.jsp?Rp=20

Biodata of **Joseph Seckbach**, senior editor of this book (with Eitan Rubin)

Professor Joseph Seckbach is the initiator and chief editor of the series *Cellular Origins, Life in Extreme Habitats and Astrobiology* (**COLE**) (Kluwer Academic Publishers, Dordrecht, The Netherlands) of which this book is the eighth volume. He is the author of several chapters in various books of this series. Dr. Seckbach has recently edited (with co-editors Julian Chela-Flores, Tobias Owen, and Francois Raulin) the volume entitled *Life in the Universe* (2004),

Dr. Seckbach earned his Ph.D. from the University of Chicago (1965) and spent his postdoctoral years in the Divison of Biology at Caltech. Then at the University of California at Los Angeles (UCLA) he headed a team searching for extraterrestrial life in Cytherean-like environments. He was appointed to the Science Faculty of the Hebrew University of Jerusalem, Israel, performed algal research, and taught biological courses. Dr. Seckbach spent his sabbatical periods in Tübingen (Germany), UCLA, and Harvard University and served at Louisiana State University (LSU) as the first occupant of the Chair for the Louisiana Sea Grant and Technology Transfer, at LSU (Baton Rouge, USA).

Among his publications are books, scientific articles concerning plant ferritin (phytoferritin), plant ultrastructure, cellular evolution, acidothermophilic algae, and life in extreme environments. He has also edited and translated several popular books. Dr. Seckbach is the co-author (with R. Ikan) of the *Chemistry Lexicon* (Dvir publication, Tel-Aviv, 1991, 1999) and edited other volumes, including the *Endocytobiology VII* (with E. Wagner et al., published by the University of Geneva, 1999); *Algae and Extreme Environments* (with J. Elster, W.F. Vincent and O. Lhotsky) published by Nova Hedwigia, 2001.

Dr. Seckbach's recent interest is in the field of enigmatic microorganisms and life in extreme environments as models for astrobiology.

E-mail: **seckbach@huji.ac.il**

Biodata of **Eitan Rubin** co-editor of this volume

Dr. Eitan Rubin is the Head of Bioinformatics at the Bauer Center for Genomics Research, Harvard University, Cambridge, USA. He has a Ph.D. in Biology from the Weizmann Institute of Science, Israel (1999) and a B.Sc. in Biology from Ben Gurion University in the Negev, Israel (1992). Dr. Rubin worked on deciphering the human transcriptom in Compugen Ltd., Tel Aviv, Israel, as a part of an interdisciplinary team of mathematicians, biologists, physicists and chemists.

Since 2001, Dr. Rubin focused on promoting the use of bioinformatics in life science research, first as the head of bioinformatics at the Weizmann Institute of Science, and later on at Harvard University. He helped form the Israeli Society for Bioinformatics and Computational Biology and The Israeli Center of Knowledge in Bioinformatics. Dr. Rubin is a member of the Scientific Advisory Board of the bioinformatics company Molecular Connections of a Bangalore, India, the news column editor for Briefings in Bioinformatics, and a member of the Education committee of the International Society for Computational Biology.

E-mail: **erubin@cgr.harvard.edu**

I. Opening

INTRODUCTION

Bioinformatics was not always as popular as it currently is. Sequence analysis, pattern recognition, and other "classical" bioinformatics activities used to be considered a part of "theoretical biology," a term used with scorn by many biologists. Yet today bioinformatics is very fashionable. Many researchers, students, post-docs, and professionals in the industry are shifting their interests to bioinformatics. Experimental biologists, computer scientists, physicists, chemists, and mathematicians abandon their existing fields of interests to seek a new career in bioinformatics. This translates into a flurry of education in the field: graduate and undergraduate degrees, training courses and even professional retraining programs for bioinformatics are sprouting in universities, colleges, and for-profit organizations. Looking at this "bioinformatics fever," we can't help but think about the Californian gold rush.

In the late 1840s, John Sutter and his employees discovered that the rivers and creeks of California were loaded with gold. The gold was "easily accessible to anyone with a few simple tools and a willingness to work hard" (Holliday, 2002). The gold was there, the ability to extract it was there, and all of a sudden the knowledge of its whereabouts was there. That's why tens of thousands of "49ers" left homes, businesses, and workplaces to migrate to California to become gold miners.

Gold Fever is defined as the surrender to the temptation of easy money, of immediate rewards that lay in gold prospecting. A similar fever swept the IT industry, with the inflation and collapse of the dot-com bubble. Many of the "01ers" in bioinformatics are IT prospectors in search of a new gold field. We suspect that they, and other newcomers, are in search of easy gold. They see the databases as rich gold fields, full of nuggets that can be extracted with no special tools (or at least with the tools they already have from their current profession). Just shovel some raw data into a certain algorithm or statistical method, wash a little, and you will get the gold. This dream is fueled by anecdotal success stories of this nature: the right person with the right tool hits the right data. Yet the history of the California gold fever makes us worry for them. In California, gold production shifted to mechanized mining by 1849. The days of the single prospector were glorious but short. Gold extraction quickly shifted to relying on dedicated equipment that required large capital investments, resoluteness, and professionals. Bioinformatics is undergoing the same process. (As a side note, it should also be mentioned that few prospectors if any got rich from digging gold; the only people who got rich from gold prospecting are those who supplied the prospectors or provided them with services. But we would like to return to the main point now.)

Today, very few Gold Nuggets still exist in bioinformatics. Most achievements come from carefully constructed data mines. This requires meticulous construction of basic methodologies as well as careful experimentation with different approaches for their application. To stretch the gold mine analogy a little further, much of the energy in bioinformatics is invested today in the engineering of the required machinery and processes and in the study of the nature of gold deposits.

Is bioinformatics going to disappoint us? Yes and no. There is a real, huge gold field out there. Only those who think it would take a shovel and a tin pan to find it will be disappointed. Those working on the careful construction of the equivalent of the mines will sustain and increase the value of their output. As long as they have the commitment and expertise required to improve the mining machinery or to apply it in life sciences, they will remain a valuable part of life science research.

J. Seckbach and E. Rubin (eds.), The New Avenues in Bioinformatics, 3–4.
© 2004 *Kluwer Academic Publishers. Printed in the Netherlands.*

In this book we try to present a realistic view of bioinformatics. We explore several of the more basic challenges in the field, which we find has already started settling into the "mine building" stages. Such are the works describing the state of the art in sequence repositories, protein classification, and transcript prediction. We touch upon areas where gold nuggets are still being found. These are the works we offer on protein structure prediction, population genetics, and plant post-genomic bioinformatics. Perhaps most exciting are examples of virgin areas where few have visited, and gold may still be running in the streams. Such are the works on glycomics (the study of the cell repertoire of glycoproteins), or the discussion of knowledge representation in bioinformatics. We touch upon the making of bioinformatics into gold by discussing its commercialization on the one hand (through the discussion of bioinformatics-related patenting issues), and the challenge of bioinformatics education on the other. Finally, we reflect a bit upon deeper questions about the nature of the field: where did biological information come from? How is biological information related to information theory?

Bioinformatics is still an exciting, vibrant field of research at the intersection of computer sciences, mathematics, and life science research. We suspect the gold rush days are nearly over, but we are convinced the golden days of bioinformatics have just begun. It is quickly settling in to become a structured science, and extracting insights from biological data is recognized as requiring special skills and expertise. This switch to "gold mines" only makes the field more exciting in our eyes: it will allow the full realization of its potential. Over time, the gold rush demise will see a blessed maturation of the field that will only benefit those who committed themselves to become professionals in the field.

The authors of this present book (the eighth volume in the series *Cellular Origins, Life in Extreme Habitats and Astrobiology*) are experts in this new field and come from Belgium, France, Hungary, Israel, Italy, Poland, The Netherlands and the USA.

References

Holliday, J.S., Lamar, H.R., and William Swain (2002) ***The World Rushed in: The California Gold Rush Experience***. University of Oklahoma Pr (Trd).

Cellular Origins, Life in Extreme Habitats and Astrobiology (COLE), Kluwer Academic Publishers, Dordrecht, The Netherlands. See: http://www.wkap.nl/prod/s/COLE.

April 30, 2004

Eitan Rubin

Joseph Seckbach

II. Overview & Origin of Information

Biodata of **Claude Gaudeau** and his co-authors, authors of ***"General Overview on Bioinformation: Genomic Modeling by Molecular Automata Grammars and Knowledge Based Systems."***

Claude Gaudeau is a Professor in the field of computer science and operational research at the Conservatoire National des Arts et Métiers in Tours, France and the director of the Bio-informatics and Bio-technology laboratory in Tours, France. After an engineer degree from the School of Transportation in 1956, he entered the French National Center of Scientific Research as a physician in the Center of Geophysical research where his principal interest was the relationship between geophysical parameters and human physiology. From 1961 to 1963, he was an assistant researcher in bio-statistics at the Biophysics Lab., Univ. of Minnesota, and also in the field of pattern recognition and learning techniques in the neurology section of the Electronic System Laboratory at the Massachusetts Institute of Technology. From 1965 to 1970, he was the head of a bio-informatics section oriented toward biomedical signal and image analysis and a scientific consultant at the computing center of Harvard Univ. where he developed new methods of computer analysis of vectocardiograms and prediction oh hemodynamic parameters for the Boston Children's hospital. He became a CNRS research engineer in 1971 and he obtained in 1975 a Ph.D. in human biology in the field of physiological regulation in cardiovascular and digestive systems. In 1981, he became a scientific consultant at the society of Bio-informatics and Biotechnology and its director since 1996, managing the design of a series of biomedical knowledge engineering and expert systems. Prof. Gaudeau is the leader of a topical team at the European Space Agency with regard to the space motion sickness, syndrome adaptation and the author of numerous publications in the biomedical and bio-informatics fields.

E-mail: **sbibi@netcourrier.com**

Philippe Bobola is a scientific consultant at the Bio-informatics and Biotechnology society in Tours, France. He obtained a Master degree and DEA in physics and chemistry at Paris VI Univ. in 1989 and 1990, for the study of molecular films in amphibious molecules, and a DESU, in the field of science, technology, problematic and world perspective in 1991. He obtained in 1996 a Ph.D. in Physics and Chemistry, for the application of asymmetrical hard spheres to the study of colloidal suspensions. His teaching experience include periods for Master degrees in sanitary science and environment engineering and chemistry at Paris VI and Cergy-Pontoise Univ. He is the author of several publications in the field of physico-chemistry and condensed materials.

Frederic Thevot is preparing a Ph.D. in the knowledge engineering field at the Bio-informatics and Biotechnology Laboratory in Tours, France. He acquired a double competence with a DEA in cellular biology—molecular and metabolic engineering obtained in 2001, combined with a DESS in computer science in 2002. His training periods include a study on bird metabolism in 200 at the national agronomic research center INRA, and at the national medical research institute INSERM in 1998 and 1999, in biochemistry and pharmaceutical biophysics laboratories, for the identification of gene mutations and chemical synthesis of therapeutic products. He is working on an expert system developed at the Bio-informatics Laboratory for space motion sickness and stress control and has contributed to a

J. Seckbach and E. Rubin (eds.), The New Avenues in Bioinformatics, 7–20.
© 2004 *Kluwer Academic Publishers. Printed in the Netherlands.*

study for the protection of human body against electromagnetic fields generated by mobile phone systems.

Yves Lucas is an associate professor at Bourges Institute of Technology of Orleans Univ., France where he is teaching image and signal processing, instrumentation and computer science at the Physics Measures department and at the Conservatoire National des Arts et Metiers. After a Master degree in the field of Discrete Mathematics in 1987, He obtained in 1993 at the National Institute of Applied Sciences in Lyon, France, a Ph.D. in the field of computer science and feedback control, related to the automatic learning of vision systems from CAD models and entered the artificial vision group of the Vision & Robotics Lab. in Bourges, France. His research interests include biomedical imaging and industrial applications of artificial vision systems. In particular, he has been working on the 3D reconstruction of human anatomical parts for diagnosis and therapeutic following, such as spinal deformities early detection, orthopedic sole design and vision assisted bedsore care. Concerning industrial vision aspects, he is interested in the automatic tuning of embedded vision systems, in particular for the detection of road collisions for cars, the dynamic volume measurement of packets, the automated inspection of products and the guidance of autonomous systems. He is also working on biomedical instrumentation systems related to experimentation on human physiological adaptation to extreme environments, especially high altitude and micro-gravity conditions, during several mountaineering expeditions and parabolic flights. He is the author of several publications in the field of image processing and machine vision systems.

KeyWords: automaton, auxiliary alphabet, auxiliary symbol, context- sensitive grammar, context-free grammar, formal system, grammar, knowledge-based system, logic of predicates, logic of propositions with global variables, logic of pure propositions, regular grammar, terminal alphabet, terminal symbol, unrestricted grammar

Claude Gaudeau

Philippe Bobola

Frederic Thevot

Yves Lucas

GENERAL OVERVIEW ON BIOINFORMATICS
Genomic Modeling by Molecular Automata Grammars and Knowledge Based Systems

CLAUDE GAUDEAU[1], PHILIPPE BOBOLA[2], THEVOT FREDERIC[2], LUCAS YVES[3] and MORIN MAGALI[2]
[1]Laboratoire de Bio-informatique et de Biotechnologie BIO ESPAS, Tours, France, [2]Société de Bio-informatique et de Biotechnologie, Tours, France, and [3]Laboratoire Vision et Robotique, IUT de Bourges, Université d'Orléans, France

1. Introduction

Bioinformatics is mainly concerned with the analysis of biological data and especially genetic sequences and protein structures. The two main objectives of bioinformatics are the identification of genes and the prediction of their function. So, the scope of bioinformatics covers completely functional genomics. Our research is guided by the "model-driven" approach.

Bioinformatics combines theoretical and practical methods allowing a comparative reflection between the foundations of computer science and biology and it contributes to a good approach to a better understanding of genetic mechanisms of the living to conceive the principles of new computer systems which are more adaptable, do better cooperate and are more tolerant to breakdowns and environment conditions by defining new kinds of **automata** and formal grammars which are at the root of the construction of machines and of the elaboration of programming languages.

For example, the analogy between Turing machine and the cellular ribosome allows a better understanding of the functioning of the latter. Professor Chauvet's works (Institute of Biology of Angers), concerning the cerebellum, are likely to lead to new concepts in robotics, to new types of formal neuronal networks (Chauvet 1987). To resume: biology inspires the science of computer program advances and reciprocally computer science, biological system modeling and understanding.

Genomic research aims at understanding the ways in which cells execute and control the great number of operations required for normal function and those in which cellular systems fail in disease. Biological systems function in a very similar way (Alberts 1996). Feedback and damping are routine in every case. Single gene perspectives are becoming limited for getting an insight into biological processes, whereas global, systemic, or network perspectives are becoming important to understand how genes and molecules collectively form a biological system, what will be a useful knowledge in educated intervention for correcting diseases. These require computational and formal methods to process massive amounts of data, understand general principles that govern the system and make predictions about system behavior (Shmulevich, Dougherty and Zhang 2002). To approach this problem, our laboratory (Laboratoire de bio-informatique et de biotechnologie) proposes a knowledge management system based on an expert system of simulation (Gaudeau 93). The genomic knowledge is formalized by a graphical representation derived from the actigram of the SADT method.

This approach models the genetic regulatory system and infers the model structure and parameters from real gene expression data. This expert system has two main objectives : first to understand the underlying gene regulatory mechanisms by inference from scientific, experimental and bibliographical data. This generally falls within the scope of computational learning theory (Anthony and Biggs, 1992) or system identification (Ljung 1999). The expert system Genesyx permits to infer different rules containing premises (inputs) and a conclusion (output), the relation of which we will calculate by methods of identification. Secondly, by using the inferred model, we make useful predictions as a result of the simulation. The inference must precede the analysis and simulation. This type of model-based analysis can give us a better knowledge about the physiology of an organism and disease progression, but also translate into accurate diagnosis, target identification, drug development, and treatment.

A fundamental question is: what levels of models should be chosen? A model class should be selected according to the data requirements and the objectives of the modeling and analysis. This involves classical engineering tradeoffs. For example, a "fine" model with many parameters will capture detailed "low-level" phenomena, but will require large amounts of data for the inference, for fear of the model being "over fitted " to the data, whereas a less complex "coarse" model with fewer parameters will capture "high-level" phenomena, but will require small amounts of data. Within a chosen model class, according to Occam's Razor principle, the model should never be made more complex than what is necessary to "explain the data". There are numerous approaches for modeling gene regulatory networks: it goes from linear models, Bayesian networks, neural networks, non linear ordinary differential equations, and stochastic models to Boolean models, logical networks, Petri nets, graph-based models, grammars, and process algebras. There have been several excellent survey papers on modelling and simulation of genetic regulatory networks (Smolen, Baxter, and Byrne 2000; Hasty, McMillen, Isaacs and Collins, 2001; De Jong 2002).

In the following, we present fundamentals of computing theory which are required before we can go into detail with concrete examples from biological systems. First, the general frames of formal systems are introduced. Then, 'Turing machines' and **automata** are described. Typical examples from the biological field will demonstrate how automata do precise model biochemical and genetic phenomena. Finally, the Genesyx **knowledge-based system** is presented. Based on SADT hierarchical description language and on the activation of rules, it is a powerful simulation tool for complex biological processes modeling and understanding.

2. Definition of a Formal System

To model a complex mechanism of reading, writing and "translation", we are going to rely on a very general concept of the formal logic: the concept of **formal system** which will be possibly used as a guideline for introducing the analysis methods of **knowledge-based systems**.

A Formal System (FS) is a mathematical object, composed of a set of data and operations, completely defined, through the rules of manipulation of its symbols. The considerations will be purely syntactical; at this level, we are not going to take semantic aspects into account. Practically, we can also consider a formal system as a modeling tool of a concrete biological reality. A formal system consists of four elements:

- A finite alphabet of symbols
- Rules of concatenation
- Axioms
- Rules of inference

2.1. THE ALPHABET

This alphabet will be constituted of a finite number of symbols and will be able to present itself in diverse ways: letters, forms, bi-dimensional or tri-dimensional objects. In a biological context:

- (C, G, A, . . .); $\updownarrow$, $+$, $\rightarrow$, $\neg$; ! [: diverse forms and symbols
- amino acids, macro-molecules (tri-dimensional objects).

We will distinguish the **terminal symbols** which appear in the actual language, and the **auxiliary symbols** (non terminal symbols) which will be used to build words of language, which allow the application of rules of transformation and which will represent in some cases the state in which the molecular system is. Then the notions of **terminal alphabet** {noted Vt or X} and **auxiliary alphabet** {noted Va} appear. Example:

- the terminal alphabet Vt will be constituted out the four basic nucleotides of ADN sequences: Vt = {C,G,A,T}.
- the Auxiliary alphabet will represent in some automata the molecular states Va = {w1, w2, w3}.

2.2. THE RULES OF CONCATENATION

These rules allow certain presentations of characters to form words (codon). In the genetic code, the letters from the alphabet are associated three by three (CGC, CTG), or by very long series of letters. The letters can associate themselves with each other linearly. If it concerns molecules, it deals with tri-dimensional objects which can connect themselves with each other in the space.

2.3. THE CONCEPT OF AXIOM

An axiom will be a starting datum from which we are going to build "valid data". This concept uses words. Example : "w" (auxiliary alphabet) or "wCGG" (auxiliary alphabet + terminal alphabet).

2.4. THE RULES OF INFERENCE

Among the rules of inference, we can find logically deduction, induction and the similarity rules of inference. This set of rules allows to deduce from a set of words another set of words, generally in the form: m1, m2, ..., mp → w1, w2, ..., wp

First example, the transcription from the ADN to the ARN transforms a set of words to another set of words:

ADN T T T A G C G A T G G => ARN A A A U C G C U A C C

2.5. NOTION OF LANGUAGE

Let us remember a few definitions:

- a monoid is a set provided with an associative T binary operation admitting a neuter element. Let Λ = neuter element and X∈A. Then, $XT\Lambda = \Lambda TX = X$
- a free monoid indicates that there is only a way to express an element X∈A.
- starting from an alphabet X = {C,G,A,T...}, we call a language on X (all parts of the free monoid X*), every set of words on X.

3. Molecular Grammars

The basic question is how a computer program can judge if a sentence is grammatically correct or not. Chomsky proposed the mathematical concept of **grammar** to formalize this question (Chomsky 1963). At first, there is a mechanism (rules) which produces sentences (sets of words for natural language or sequence of symbols for biological problems). We can now wonder if a given sentence can be produced by the set rules.

So, this kind of grammar consists of a list of symbols and rules of transformation. In the case of the analysis of sequences, we wonder if it comes from a given grammar, that is from given rules of production. Parsing refers to grammatical analysis, to the action of searching for a derivation of the sequence from the grammar. There is also an alignment between the sequence and the grammar. In practice, Chomsky has defined four types of grammars:

- **regular grammars**
- **context-free grammars**
- **context- sensitive grammars**
- **unrestricted grammars** ('Turing's machine')

For genetic modeling, we are principally interested in regular and context-free grammars (see table I)

A molecular **automaton** with finite states is a kind of machine with several internal states and possibilities of transition from a state to another one according to rules of production (see HMM, but more general).

TABLE 1. Examples of grammars

Grammar type	Regular	Context-free (algebraic)	Context-sensitive	Unrestricted
Grammar example	w → Aw	w→AwU+UwA+C wG+ GwC+□	w1→ACG+Aw2CG; w2C→Cw2; w2G→w3CGG Cw3→w3C;Aw3→AA w2;Aw3→AA	
Automata type	State finite automata	State and pile automata	Turing machine	Turing machine

3.1. EXAMPLE OF REGULAR MOLECULAR GRAMMAR

The gene FMR can contain an arbitrary number of under-sequences CGG which sometimes are substituted for the triplet AGG.

Which **automaton** or grammar could produce a typical sequence of FMR? A typical sequence is:

GCG CGG CGG ... CGG AGG CGG ... **CTG**

start end

The parsing **automaton** is the following: we start in the state S. We read the sequence one symbols by after the other from the left to the right. Each transition i → i + 1 is dictated by the presence of a symbol.

If the symbol read in the sequence doesn't correspond to a permitted transition, the sequence is rejected, and so can not be derived from the chosen grammar. If the **automaton** gets at the symbol Λ at the end of the reading, the sequence corresponds to the model (Figure 3).

3.2. EXAMPLE OF CONTEXT-FREE MOLECULAR GRAMMAR

To model the mRNA secondary structure or the means who it to retire within oneself, we can construct context-free grammars to describe the resultant topology of the folding up. We must identify the paired complementary bases (A-U, G-C) which we form the liaison. The grammar will generated palindromes such as AAGGAA or if we consider the sequence:

CAGUGCUUAGCGCUG

Equivalent formulation:

$$S \to Gw_1;\ w_1 \to Cw_2;\ w_2 \to Gw_3;\ w_3 \to Cw_4;\ w_4 \to Gw_5;\ w_5 \to Gw_6;\ w_6 \to C\,w_7 + Gw_4;$$
$$w_7 \to Cw_8 + Gw_5;\ w_8 \to Cw_n$$

The context-free grammar could be: w→AwU+UwA+CwG+GwC+Λ and applied at the middle of the sequence, the folding will be:

Then, we can define operations on automata (products, operations, complements) and more complex automata and grammars such as 2- and 3-dimensionnal ones.

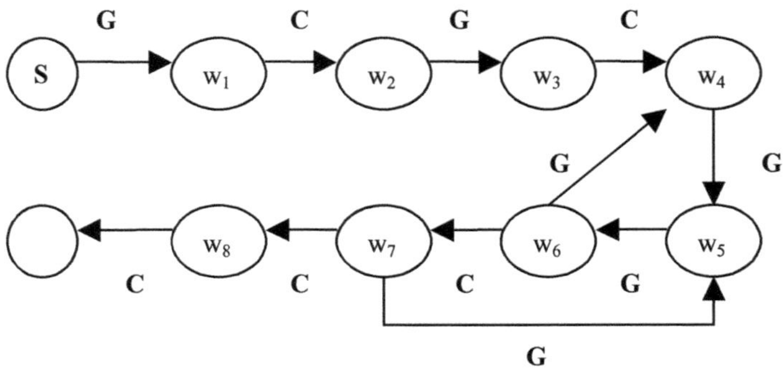

Figure 1. State automaton corresponding to the grammar producing a typical FMR sequence.

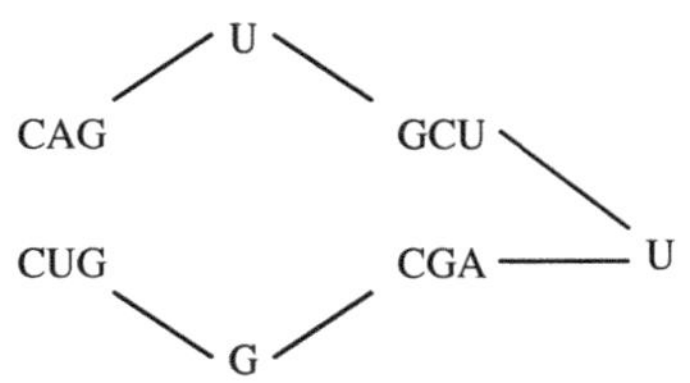

Figure 2. Result of the folding up.

3.3. EXAMPLE OF CONTEXT-SENSITIVE MOLECULAR GRAMMAR

Let X = {A, C, G} terminal alphabet and Va = {w1, w2, w3} auxiliary alphabet and the context-sensitive grammar defined supra in paragraph 4. The set of genetic words generated by this grammar is in the form: ACG, AACCGG, AAACCCGGG... and so on and the **automaton** associated will be able to recognize and interpret these sequences.

4. Knowledge Based Systems

One common point between **knowledge-based systems** and automata is that facts (resp. words) and rules are used to produce new facts (resp. words).

4.1. OBJECTIVES

One of the main purposes determined by experts in artificial intelligence consists in reproducing human ways of thinking by means of machines. Expert systems rank among the first developed software of artificial intelligence (Mariot 83). They are defined as computing programs elaborated with the aim of solving problems whose solution requires a logical reasoning. The expert system's purpose lies in simulating expert's way of thinking (an expert is a person who has a thorough knowledge in a special field and who is able to solve problems in an efficient manner). Specialist's knowledge represents a set of methods. Sometimes, it is difficult to make the connection between these different methods. In an expert system, the technique consists in gathering all the available knowledge in order to model them in the form of rules, which are as independent as possible. Then, it is necessary

TABLE 2. Matrix representation of Grammar

	S	w_1	w_2	w_3	w_4	w_5	w_6	w_7	w_8	Λ
S		G								
w_1			C							
w_2				G						
w_3					C					
w_4						G				
w_5							G			
w_6					G			C		
w_7						G			C	
w_8										C
Λ										

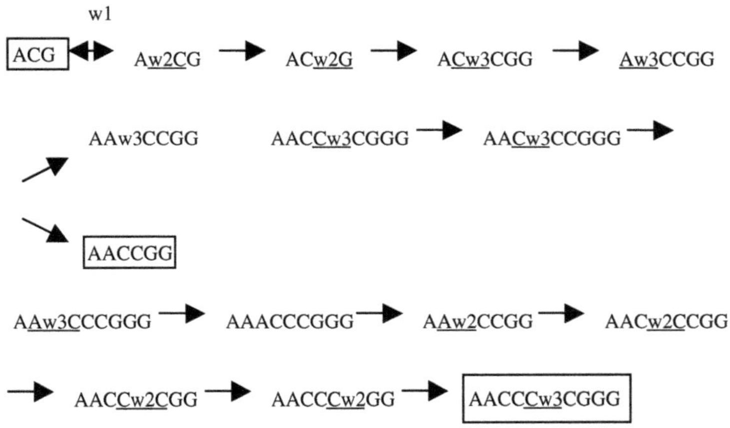

Figure 3. Derivation tree.

to try to define a working mechanism (an interference engine), which is able to manipulate this gathering of knowledge, in order to find a solution.

4.2. ORGANIZATION

An expert system is composed of three main components: a base of facts, a base of rules and an interference engine. These three elements are essential to the expert system functioning. They are derived from the elements of formal systems (Table 2).

4.2.1. The Base of Facts

The base of facts can be defined as being the whole collected data and known by the expert system. Therefore, it constitutes the knowledge which will help the system to make its deductions. Some examples are given by genetic sequences (CGG, GCU, . . .) , ribosomes, amino acid and phenylalanine . . . As we have seen in the previous example, facts can be represented in different ways:

The **logic of pure propositions:** the fact represents the existence of a datum or the state of a datum (for example: battery out of order). In this case, there is no available quantitative information about the datum.

TABLE 3. Organization of expert system functioning

Formal System	Knowledge Based System
RULES OF INFERENCE deduction, induction and the similarity rules of inference.	KNOWLEDGE BASE containing production rules and meta-rules.
AUTOMATA It uses the inference rules to produce valid words.	MOTIVE OF INFERENCE expressed in a language of programming. For example it uses the deductive mechanism of modus ponens.
AXIOMS	BASE OF ACTS, expressed in a natural language or natural pseudo.
A FINITE ALPHABET OF SYMBOLS	MANIPULATED VARIABLES concern usually the user specific alphabet, such as those utilized by a doctor in the medical class).
WORDS	DEDUCED FACTS (states and relations between the variables)

The **logic of propositions with global variables:** the fact is represented by means of three elements < actor, comparative, value > and we have at our disposal a quantitative information about the datum (example: surface = 9m2): the actor is a character string, the value is a numerical term and the comparative corresponds to one of the following signs: "+", ">", "<", "<>", "<=", >=" or "nl". This last comparative is the zero comparative. It is used, when it is necessary to indicate the absence of a comparative (in this case, we have to refer to the logic of pure propositions.) For example, the phrase "Antibody nl 0" only means the existence of antibodies. Our expert system uses this logic. The base of facts and the list of actors are shown in the first window of the program.

The **logic of predicates:** the fact is represented in the following way: "is father of (x, y)"; thus, it means that "x is the father of y". If we want the system to make a deduction, we have to choose different facts in the base of facts, in order to create an inference base. This base consequently represents hypotheses which will contribute to the deduction of other facts made by the system.

4.2.2. The Base of Rules: SADT Constituting Method

The base of rules contains the expert's know-how expressed in the form of production rules, which correspond approximately to the different elements of the expert's reasoning. The expert system is essentially composed of rules originating from scientific literature. The bibliographic analysis consists of formalizing scientific knowledge in the fields of cell biochemistry and genetic represented graphically by blocks containing input, output, and a function that acts upon the input. The integration of these rules forms knowledge base of the expert system. To model knowledge, the SADT method has been used because it can embrace a vast, complex problem, thus communicating results from analysis in clear, precise notation (Roth 93, Gaudeau 91).

This method is presented as a coherent group of methods and rules that make up modular, hierarchical processes for analysis and conception. Its is constructed on a group of closely connected concepts. The SADT method models a system by chaining together actions and concepts that are less complex than those at the starting point. Its application consists of beginning with the most general and abstract description of a system. If we consider this first concept as contained in a single module represented by a block, we can break it down, and so on. In this way, the conceptual and organizational stages of a project (e.g., a new medical therapy) can be clearly understood before its real, physical consequences are known. Its graphical representation provides the user with a base for methodical system construction that provides an image of reality. Controls are when rules are used. If there is an error in the result, the rule in a more precise manner. The principle of diagram modeling used in the SADT methods is used. Two types of diagrams can be used to model knowledge: the actigram and the datagram.

The actigrams model the observed activities and the datagrams model the observed data. To represent knowledge in the biology field, actigrams are more appropriate because they express directly the physiological functions. A SADT diagram is composed of a set of boxes (datagrams and actigrams) connected one to each others by arrows. A rule is presented in the following way:

IF <condition>

THEREFORE <conclusion>

In which <condition> is a fact or a gathering of premise facts of the rule and <conclusion> is a conclusion fact of the rule. Example: a rule extracted from Expert-Aids, a knowledge based system developed in our laboratory to simulate AIDS mechanisms: Rule Retrotranscription:

IF <Circulated matrix RNA> AND <Inverse Transcriptase>
THEREFORE <Sample Breath Retrotranscripted DNA>

Thus, when facts are propositions with global variables, the rule is represented in the following way:

IF <actor, comparative, value>
AND <actor, comparative, value>
Therefore <actor, comparative, value>

Actors of each premise permitting to obtain the conclusion are not necessarily different as it is illustrated by the following example:

IF TCD4 Lym $<200/mm^3$
Therefore HIV Stade = 3 (major opportunist infection)
IF TCD4 Lym $<50/mm^3$
Therefore HIV Viral Load >5000 copies/ml

These rules can be used for diagnostic and therapeutic action.

4.2.3. Inference Engine

As for formal systems the inference engine, whose role is to simulate an expert's thought process, detects whether a rule can be released. In the instance of Genesyx, the inference engine is 0+ order (logic motor with global variation) and is based on forward chaining. The inference engine will thus attempt to deduce new facts from facts and rules that is already knows or that have been introduced into the system by the user.

5. Discussion and Conclusion

Linguistic and combined approaches using formal grammars have been proposed to describe the genetic mechanisms in particularly the origin of life (Barbier and Brack, 1992). For example, it can act to count and to generate randomly combinatory structures: trees, graphs, words, etc. Also, there exists correspondences between these objects and formal languages. The definition of algebraic grammars called also context free grammars for these languages allows to deduce from one part formulae for the exact enumeration or asymptotic of structures, (deterministic or probabilistic) of efficient random generation.

The notion of algebraic grammar is fundamental for syntax analysis of texts: in this way, we can determine if a given set of characters respects the rules of a given grammar (compared with the function of compilers). In the same way, there exist research programs for certain types of RNA in sequences.

The work of D.B. Searls stretches the formalism of context free grammars to String Variable Grammars in a way as to take in to account the intra-molecular dependence presented by nucleic acids and proteins (Searls 1993). They introduce string variable grammar characters on the terminal and non-terminal term sides. Over the course of the analysis, the first encounter of a given variable leads to the recognition and "memory storage" of this variable (Searls 1995). When this variable is encountered again, there is a confrontation between the contents of this variable and the rest which has not yet been analyzed in the sequence. The SVG formalism authorizes the approached research of complex patterns (structures) in a long biological sequence and takes into account one single type of error: substitution.

Studies currently underway on the generalization of SVG's in so called morphic transformation grammars (MTG's). The MTG's use terminal and non-terminal symbols, string variables and operators: the morphic transformations (Sinoquet 1998). These operators permit to capture the inversions and repetitions of a sequence language, with eventual superimposition of character substitutions (morphisms). These grammars are adapted to the description of links between regions of a sequence and permit to take into account the three errors (substitution, insertion, and deletion). These grammars have been evaluated with success in the recognition of tRNA and of pseudo-nodes in the sequences of GenBank as well as in the structural study of certain regions of human histo-compatibility.

On the other hand, context free grammars associated with statistical tools have been applied in order to describe the code of sequences; the goal being to find the "hidden" significance of sequences in a statistical manner.

Tools treating the local and global structures of proteins, from speech recognition methods, have been developed. The human voice and proteins have similar hierarchies. In the case of speech, these are phonemes, words, phrases, paragraphs and significance. In the case of proteins, these are primary structures (Sagot and Myers, 1998), secondary structures, super secondary structures, tertiary structures and functions. Hidden Markov Models, (HMM's) currently used in speech recognition, are beginning to be applied to molecular biology.

The HMM's have been used to predict the structure of proteins, for multiple (Sakakibara and al. 1994; Asaï and al. 1996); sequence alignments, to classify protein sequences, to extract sequence patterns for proteins, to model the hidden sequences of DNA (we cannot see the hidden states which correspond to these structures). The HMM's represent the nucleic acid bases, the codons and the amino acids. The genetic words in the dictionary are described by the sequence of these HMM's and represent the exons, introns, protein patterns and the signals in the DNA sequences (Asaï and al. 1994). The statistics between these components is supported by grammar, which is a stochastic network of these genetic words. From the point of view of the theory of languages, the Markov models are regular stochastic grammars associated with probabilistic automata (Asaï and al. 1998).

The present expert system is based on rules generated from relations between the entries and the output who don't yet take not the states that the system happen. The **automaton** introduction in the rule requires to generate as much mathematical relations states. A fact enables the **automaton** enclosed in the system to change of state.

The modeling tools (grammars, automata, experts systems) rapidly reviewed enable to spread and test for example of the theoretic physic a theoretic bioinformatics in that we conceive molecular automata at time normal and pathologic, to conceive new software tool to analyze genetic sequence and to conceive news medicine of genetic engineering (writing, genetics codes, reading, etc . . .)

But there exist others bioinformatics tools such as two and three dimensional, fully automata which will allow new method for genome and protein interaction analysis. Some years ago, inspired from cybernetic works, Jacobs, Monod and Wolf succeeded in modeling the mechanism of molecular regulation and validated experimentally this approach. It appears that tools provided by bioinformatics should also help to model more finely biological process and guide the experimental identification of molecular automata.

6. Acknowledgements

We would like to thank Mr Jefferson Thomas, Mr Sarmadi Milade and Mr DY Sowannara for his technical assistance and the West Industrial Credit (C.I.O.) for its financial support.

7. References

Alberts, B., Bray, D., Lewis, J., Raff, M., Roberts, K. and Watson, J.D. (1996) *Biologie moléculaire de la cellule*. Médecine-Sciences, Flammarion, Paris.

Anthony, M. and Biggs, N. (1992) *Computational Learning Theory*. Cambridge, U.K. Cambridge Univ. Press.

Asaï, K., Yada, T. and Itou, K. (1996) *Finding Genes by Hidden Markov Models with a protein Pattern Dictionary*. Genome Informatics Workshop VII - GIW 96.

Asaï, K., Onizuka, K., Kenosha, M., Tanakam H. and Itou, K. (1994) *Modeling of Protein Structure by Hidden Markov Model*. International symposium on Fifth Generation Computer Systems - 5G Symp. 94.

Asaï K, Itou K, Yada T, (1998) Integrating Multiple Evidences by Hidden Markov Models.Genome Informatics, 347–350.

Barbier, B. and Brack, A. (1992) Conformation-controlled hydrolysis of polyribonucleotides by sequential basic polypeptides. J. Amer. Chem. Soc., **110** , 6880–6882.

Chauvet, C. (1987) *Traité de Physiologie et Théorique: Formalismes Niveaux Moléculaire et Cellulaire*. Masson, Paris.

Chomsky, N. (1963) *Formal Properties of Grammars*. Handbook of Mathematical Psychology, **2**, John & Sons Inc.

De Jong, H. (2002) Modelling and simulation of genetic regulatory systems: A literature review.J. Computer Biologic., **9**, n° 1, 69–103.

Gaudeau, C. and al., (1993) A modelisation expert system of chemotherapy and induced illness for the virus of the immuno-deficience acquired.Comm. XIV Cong. Living in Mathematical Biology, Paris.

Gaudeau, C., Rakotomalala, V., Gouthière, L., Ravaud, E. and Benoist, S. (1991), Fuzzy automata and grammars : manipulation of uncertain knowledge. Modelling complex data for creating information, Berlin.

Gaudeau, C., Rakotomalala, V. and Poton, J.P. (1991), Expert system of electromagnetic field action on biological tissues. Comm. 2nd International scientific Cong. on Microwaves in Medicine, Rome.

Hasty, J., Mc Millen, D., Isaacs, F. and Collins, J.J., (2001) Computational studies of gene regulatory networks: In numero molecular biology. Nature Reviews Genetics, **2,** 268–279.

Ljung, L. (1999) *System Identification: Theory for the User*. Upper Saddle River, NJ, Prentice-Hall.

Mariot, P., Regnier, F., Louvel, E. and Gondran, M. (1983), *Experts systems introduction*. Editions Eyrolles, Paris.

Roth, A. (1993) *SADT, A language for the communication*. I.G.L. Technology, Editions Eyrolles, Paris.

Sagot, M.F., Myers, E.W., (1998) give title of article In: S. Istrail S, P. Pevzner P. and M. Waterman M, (**eds.**) Identifying satellites in nucleic acid sequences. Proc. of the second annual international conference on computational molecular biology Recomb'98, ACM. Press, New York, USA, 234–242.

Sakakibara, Y., Brown, M., Hughey, R., Mian, I.S., Sjolander, K., Underwood, R.C., Haussler, D. (1994) Stochastic context-free grammars for tRNA modelling. Nucleic Acids Res., **22,** n°23, 5112–20.

Searls, D.B., (1993) *The computational linguistic of biological sequences*. Artificial intelligence and molecular biology, Editions AAAI / MIT Press.

Searls, D.B. (1995) String variable grammar: a logic grammar formalism for the biological language of DNA. Journal of logic programming, **24**, 73–102.

Shmulevich, I, Dougherty R.E., Zhang W. (2002) From Boolean to Probabilistic Boolean Networks as Models of Genetic Regulatory Networks in Proceedings of the IEEE, **90**, n°.11.

Sinoquet, C.(1998) *Grammaires à transformations morphiques, recherche de motif exacte ou approchée- adaptée aux séquences génétiques : le système GTM*. Ph.D. Thesis, University of Rennes 1, France.

Smolen, P., Baxter, D., and Byrne, J.(2000) Mathematical modelling of gene networks. Neuron, **26**, 567–580.

Biodata of **Juan G. Roederer**, author of ***"When and Where did Information First Appear in the Universe?"***

Juan G. Roederer is Professor of Physics Emeritus at the University of Alaska-Fairbanks. Italian born, raised in Austria and educated in Argentina, he received a doctorate in physical-mathematical sciences from the University of Buenos Aires in 1952. From 1956 to 1966 he was professor of physics at that university. In 1967 he emigrated to the United States where he became professor of physics at the University of Denver, Colorado. In 1977 he was appointed director of the world-renowned Geophysical Institute of the University of Alaska, a post held until 1986; during that time he also served four years as dean of the College of Environmental Sciences. Since 1987 he teaches and conducts research at the University of Alaska. From 1983 to 1897 he was chairman of the advisory committee on space and Earth sciences of the Los Alamos National Laboratory, and from 1986 to 1992 he served two United States presidents as chairman of the United States Arctic Research Commission. At present he is adviser to the director of the Abdus Salam International Center for Theoretical Physics in Trieste, Italy.

Roederer's research fields are space physics, psychoacoustics, informatics and science policy. He conducted pioneering research on solar cosmic rays, on the theory of Earth's radiation belts and on pitch perception, and is author of 250 articles in scientific journals and four university textbooks. He served as member and chairman of several US Academy of Sciences/National Research Council committees, and was president of the International Association of Geomagnetism and Aeronomy and of the ICSU Committee on Solar Terrestrial Physics.

Roederer is a member of the Academies of Science of Austria and Argentina, and the Third World Academy of Sciences, as well as a Fellow of the American Geophysical Union and the American Association for the Advancement of Science. He received the medal "100 Years in Geophysics" from the former Soviet Academy of Sciences, three awards from NASA for his collaboration in the "Galileo" space mission to Jupiter, and the Year 2000 Flinn Award of the AGU. He is also an accomplished organist.

E-mail: **jgr@gi.alaska.edu**

J. Seckbach and E. Rubin (eds.), The New Avenues in Bioinformatics, 21–37.
© 2004 *Kluwer Academic Publishers. Printed in the Netherlands.*

WHEN AND WHERE DID INFORMATION FIRST APPEAR IN THE UNIVERSE?

JUAN G. ROEDERER
Geophysical Institute,
University of Alaska-Fairbanks,
Fairbanks, AK 99775, USA

1. Introduction

Most scientists would assume that information has been playing a role right from the beginning—the Big Bang. As the Universe evolved, after the gradual condensation of atoms and molecules and the formation of planetary systems, "islands" of increasing complexity and organization appeared, containing discrete aggregates of condensed matter with well-defined boundaries and increasingly complex interactions with each other and their environment. Viewed this way, it indeed seems that the process of cosmic evolution itself is continuously generating information [Chaisson, 2001].

On second thought, however, aren't we talking here of information *for us the observers* or *thinkers*? Did information as such really play an active role in the fundamental physical processes that shaped the Universe? Was information and information-processing involved at all in the evolution of the Universe *before* living organisms started roaming around and interacting with it, and intelligent beings began studying it? When and where did information begin to play an active role, actually controlling processes in the Universe?

It is obvious that to address and answer these questions objectively we must first discuss in depth the concept of information and its meaning. We must find a definition of this ubiquitous and seemingly trivial concept that is truly objective and independent of human actions and human-generated devices. This is no trivial matter: for instance, one cannot tell by examining a complex organic molecule or a slice of brain tissue whether or not it possesses information (beyond that generated in our senses by its own appearance)—complexity and organization alone do not represent information [Davies, 1990]. Concomitantly, we must find answers to some very basic questions: Can the rigorous methods of physics adequately describe and explain that strange but most fundamental property of information, namely, that the mere *shape* or *pattern* of something—*not* its field, forces or energy—could trigger a dramatic yet quite specific change in a system, and do this over and over again even if the initial conditions are not all the same? Or is information irreducible to the laws of physics and chemistry? How are information and complexity related?

Traditional information theory does not help. It works mainly with communications and control systems and is not so much interested in an independent formal definition of information and its meaning as it is in a precise mathematical expression for the information content of a given message and its degradation during transmission, processing and storage. In general two classes of information are considered: (i) *statistical* or *semantic* information describing the outcome of expected alternatives for a process or event, and (ii) *algorithmic* information, defined as the minimum-bit statement that describes a given thing (e.g., see Zureck [1990]). Shannon's theory of information, related to the former, deals with an ensemble of possible events and analyzes the uncertainty of their occurrence; Shannon's expression $I_k = -\ln_2 p_k$ defines the information content I_k in bits of a message x_k that has a probability p_k to occur among N possible messages; the expectation value of the content of a single message in the ensemble is defined as $H = \sum p_k I_k = -\sum p_k \ln_2 p_k$ (also called the entropy of the source of information). The information content is a subjective entity in the sense that it requires prior knowledge of the probabilities p_k. In algorithmic information, on the other hand, the information content is expressed by the actual number of bits of the defining minimum-bit statement. This, in turn, requires prior knowledge of the given algorithm used in the description.

Shannon's measure is not apt to express the information content in many non-technical human situations, nor is it adequate to measure information in biochemical systems. Consider the case of a genome: each one of the innumerable combinations of nucleotides has an equal a priori probability of appearance in random chemical synthesis, and all sequences of the same length have the same Shannon information content. Yet the corresponding molecules would have drastically different functional significance; only a minute fraction would be biologically meaningful. In other words, what counts is *how information becomes operational*. For this purpose, the concept of *pragmatic information* was introduced, linking the pattern in a "sender" with the pattern-specific change triggered in a "recipient". This concept is objective and better suited for use in biology and bioinformatics, but it is more complicated to quantify (see Küppers [1990]).

It should be clear that information as a stand-alone concept has no absolute meaning: we can only speak of information when it has both a sender and a recipient between which some previous interconnection or "understanding" exists [Küppers, 1990]. I shall try to accomplish the task of finding a comprehensive and objective definition of information by choosing the process of *interaction* as the underlying basic, primordial concept. I shall identify two fundamentally different classes of interactions between the bodies that make up the universe as we know it, with the concepts of information and information processing appearing as the key discriminators between the two. As much as possible, I shall avoid appealing to "made things" such as computers, artificial communications systems, machines or robots.

2. Physical and Biological Interactions

It is our experience from daily life (and from precise observations in the laboratory) that the presence of one object may alter the state of other objects in some well-defined ways. We call this process an *interaction*—without attempting any formal definition (i.e., taking it as a "metaphysical primitive"). We just note that in an interaction a *correspondence* is established between certain properties of one object (its position, speed, form, etc.) and specific changes of the other. The interactions observed in the *natural* environment (i.e., leaving out all human-made artifacts) can be divided into two broad classes.

The first class comprises the physical interactions between two bodies, in which we observe that the presence of one modifies the properties of the other (its motion, structure, temperature) in a definite way that depends on the relative configuration of both components of the system (their relative positions, velocities, etc.). Primary physical interactions between elementary particles are two-way (true *inter*-actions) and reversible. The concept of *force* is introduced as the agent responsible for the change (acceleration or deformation) that occurs during the physical (non-thermal) interaction of two bodies isolated from the rest [Mach, 1893]. For fundamental interactions such as gravity and electromagnetism, the concept of a force *field* is introduced as a property of the space surrounding the interacting bodies. The end effect of a physical interaction will always depend on some initial conditions, such as the initial configuration (positions, velocities) of the interacting bodies. A most fundamental characteristic is the fact that during the interaction, there is a direct transfer of energy from one body to the other, or to and from the interaction mechanism itself. In other words, the changes that occur in the two interacting bodies are coupled energy-wise. I shall call the physical interactions between inanimate objects *force-field driven interactions*.

At the microscopic, subatomic level, everything can be reduced to four basic interactions between elementary particles. The environment with which humans and animals interact, however, pertains to the *macroscopic domain* in which objects consist of the order of 10^{20} or more molecules, and in which all physical quantities of relevance, such as those to which our senses respond, are averages over enormously large ensembles of particles. In the macroscopic domain, all physical interactions can be reduced to two elementary ones, namely gravitation and electromagnetism; the relevant objects are in general complex, consisting of many physically linked parts, and their interactions often involve complex chains of individual cause-and-effect mechanisms. The more complex and irreversible, the weaker the energy coupling will be between one end and the other of a chain.

As a first example, consider a mass point orbiting in the gravitational field of a massive central body. The motion is governed by the force of gravitational interaction, which is a function of position and masses—and of nothing else. The concept of information is totally alien to gravitational interactions—and indeed to *all* purely physical interactions between inanimate objects: they just happen and don't require intermediate operations of information processing. The configuration of a satellite orbit may vary greatly depending on the initial conditions (position and velocity), and, most importantly in our context, there is a direct energy coupling between the satellite and the gravitational potential of the central body.

Another example is the case of a low-energy electron interacting with a proton at rest. From the point of view of quantum electrodynamics, the interaction mechanism consists of the emission and absorption of virtual photons by the interacting particles. In a first-order Feynman diagram the electron (or the proton) emits a photon which is then absorbed by the proton (or electron); the energy and momentum balance at each node (emission/absorption process) accounts for a change in the motion of each particle. The end result, strictly reversible, is the sum total of all possible photon exchange processes; the total energy (and momentum) of the pair of particles is conserved. We are tempted to say (and often do so) that in this interaction, a photon (or the electromagnetic wave it represents) "carries information" from one charged particle to the other; however, this is purely subjective parlance: no information, information-processing and purpose are at work at either end of the emission/absorption process.

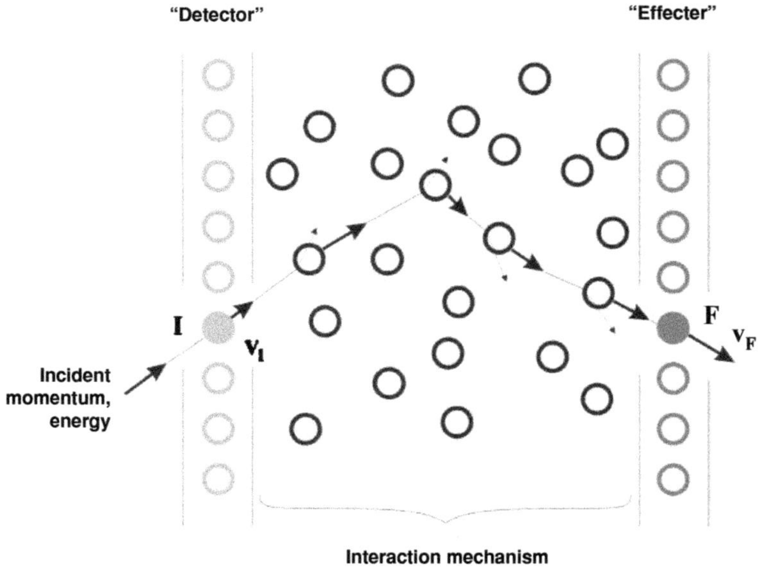

Figure 1. Example of a complex physical (force-based) interaction between a "detector" of incoming impulses and an "effecter". Initially, all balls are at rest. At each step there is transfer of energy, but the amount of energy transferred between detector and effecter could be very small if the interaction mechanism is complex.

As our last example, consider the situation shown in Figure 1. A set of billiard balls is at rest on a frictionless table. Ball I is given a velocity V_I, triggering a chain of collisions that finally imparts an impulse to ball F. We can view the intervening processes as the action of *one* interaction mechanism between I and F, consisting of many stages (cause-and-effect relationships); the interaction between I and F is therefore *complex* and *irreversible*. It is considered irreversible because, to occur in an exactly reverse way, it is not enough that F arrives with velocity $-V_F$: *many* other conditions must be fulfilled exactly (e.g., each collision partner must arrive at the right place and time with the right (reversed) velocity). There is energy coupling between balls I and F, but it will be weaker the more complex the interaction mechanism (the more intervening balls there are, even for purely elastic collisions). If some of the intermediate collisions are explosive (i.e., sources of energy), F may be completely decoupled energy-wise from I. I shall return to this example in section 5.

We now turn to the second class of interactions. Compare the above examples of force-field driven interactions with the example of a dog walking around an obstacle. The dog responds to the visual perception of the obstacle, a complex process that involves information-processing and decision-making at the "receiver's" end with a definite *purpose*. At the obstacle's (the sender's) side, we have scattering and/or reflection of incident light waves; no information and no purpose are involved—only physical processes are at work here. There is no energy coupling between sender and receiver; what counts is *not* the energy of the electromagnetic waves but the *pattern* of their spatial distribution (determined by the obstacle's shape). This is a prototype of interactions that involve *information extraction*: it is a fundamental element in the interaction of any organism with the environment. An equivalent example at the cellular or molecular level would be that of a cell that has sensed the lack of a given nutrient, and has responded by starting up the synthesis of enzymes necessary to utilize other nutrients. The signal (e.g., glucose concentration) has neither purpose nor information, but becomes information when detected by the metabolic machinery of the cell.

Another prototype example is that of an ant leaving a scent mark along its path. This is a reverse process of the preceding example, with *information deposition* in the environment for later use. The sender has a specific purpose and information is involved. At the other end of the interaction (the path) only physical (chemical) processes are at work. This is an example of deliberate environmental modification with a purpose (another ant can extract information from the scent signals). At the cellular or molecular level we can refer to the methylation process, which changes the cell's ability to transcribe certain portions of its DNA into RNA, without changing the information content of the DNA. This alters the program of the genes that they can express, thereby rendering these cells different from the others. Methylation is heritable, so this lets the change persist across multiple cell divisions.

We call the above class of interactions *information-based* or information-driven [Roederer, 2000]. Of course, the responsible mechanisms always consist of physical and/or chemical processes; the key aspect, however, is the *control* by information and information-processing operations. There is no direct energy coupling between the interacting bodies (obstacle-dog, ant-track), although energy must be supplied locally for the intervening processes. In the first example the electromagnetic waves (light) themselves do not drive the interaction—it is the *information* in the patterns of the wave trains, not their energy, which plays the controlling role; the energy needed for change must be provided locally (by the organism). Quite generally, in all natural information-based interaction mechanisms the information is the *trigger* of physical and chemical processes, but has no relationship with the energy or energy flows needed by the latter to unfold. Physical and chemical processes provide a medium for information, but do not represent the information *per se*. As we shall see in sections 5 and 6, the mechanisms responsible for this class of natural interactions must *evolve*; they do not arise spontaneously (in fact, Darwinian evolution itself embodies a gradual, species-specific information extraction from the environment). This is why natural information-driven interactions are all *biological* interactions.

3. Information and Life: Definitions

Let us now formalize our description of information-based interactions and related definitions, and point out again both the analogies and differences with the case of physical interactions between inanimate bodies. First of all, we note that information-based interactions occur only between bodies or, rather, between systems the complexity of which exceeds a certain, as yet undefined degree. We say that system A is in information-based interaction with system B if the configuration of A, or, more precisely, the presence of a certain spatial or temporal pattern in system A (called the sender or source) causes a specific alteration in the structure or the dynamics of system B (the recipient), whose final state depends *only* on whether that particular pattern was present in A. The interaction mechanism responsible for the intervening dynamic physical processes may be integral part of B, and/or a part of A, or separate from either. Furthermore: (a) both A and B must be *decoupled energy-wise* (meaning that the energy needed to effect the changes in system B must come from sources other than energy reservoirs or flows in A); (b) *no lasting changes* must occur as a result of this interaction in system A (which thus plays a catalytic role in the interaction process); and (c) the interaction process must be able to occur *repeatedly* in consistent manner (one-time events do not qualify). In other words, in an information-based interaction a specific one-to-one *correspondence* is established between a spatial or temporal feature or pattern in system A and a specific change triggered in system B; this correspondence depends only on the presence of the pattern in question, and will occur every time the sender and recipient are allowed to interact (in this basic discussion I will not deal with stochastic effects).

We should emphasize that to keep man-made artifacts and technological systems (and also clones and artificially bred organisms) out of the picture, both A and B must be *natural* bodies or systems, i.e., not deliberately manufactured or planned by an intelligent being. We further note that primary information-based interactions are *unidirectional* (i.e., they are really "actions", despite of which I shall continue to call them *inter*actions), going from the source or sender to the recipient. There is an irreversible cause-and-effect relationship in which the cause remains unchanged (the pattern in A) and the effect is represented by a specific change elsewhere (in B). While in basic physical interactions there is an energy flow between the interacting bodies, in information-based interactions any energy involved in the participating (but not controlling) physical processes in the interaction mechanism must be provided (or absorbed) by reservoirs *external* to the interaction process. Energy distribution and flow play a defining role in complex systems [Chaisson, 2001]; however, for information-based interactions, while necessary, they are only subservient, not determinant, of the process per se. In physical interactions the end effect always depends on the initial conditions of the combined system AB; in information-based interactions in general it does not—it is the presence of a pattern or sign that counts. Finally, information-based interactions are usually discontinuous, in the sense that if the original pattern at the source is modified in a continuous way, the response in the recipient may not vary at all in a continuous way.

Although we have been talking about information all the time, we must now provide a more formal definition: information is *the agent that mediates the above described correspondence*: it is what links the particular features of the pattern in the source system A with the specific changes caused in the structure of the recipient B. In other words, information represents and defines the uniqueness of this correspondence; as such, it is an irreducible entity. We say that "B has received information from A" in the interaction process. Note that in a natural system we cannot have "information alone", detached from any interaction process past, present or future: information is always there *for a purpose*—if there is no purpose, it isn't information. Given a complex system, structural order alone does not represent information—information appears only when structural order leads to specific change elsewhere in a consistent and reproducible manner, without involving any direct transfer or interchange of energy. Thus defined, we can speak of information only when it has both a sender and a recipient which exhibits specific changes when the information is delivered (the interaction occurs); this indeed is what is called pragmatic information [Küppers, 1990]. It is important to note that the pattern at the sender's site (system A) in itself does *not* represent information; indeed, the same pattern can elicit quite different responses with different interaction mechanisms or in different recipients (think of a Rorschach test!).

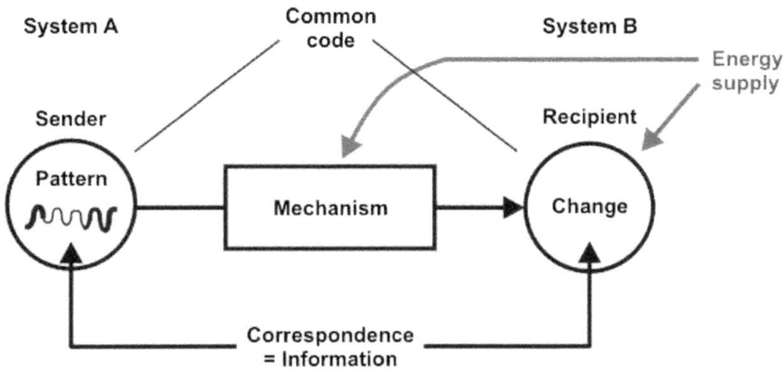

Figure 2. In an information-based interaction a *correspondence* is established between a pattern in the "sender" and a specific change (structural or dynamic change) in the "recipient". Information is the agent that represents this correspondence. The mechanism for a natural (not artificially made) information-based interaction must either emerge through evolution or be developed in a learning process, because it requires a common code (a sort of memory device) that could not appear by chance. There is no direct energy transfer between the sender and the recipient, although energy, to be supplied externally, is involved in all intervening processes.

Concerning the interaction mechanism per se, i.e., the physical and chemical processes that intervene between the sender's pattern and the corresponding change in the recipient, note that it does not enter in the above definition of information. What counts for the latter is the uniqueness of the correspondence—indeed, many different mechanisms could exist that establish the same correspondence. In the transmission from *A* to *B* information has to "ride" on something that is part of the interaction mechanism, but that something is not *the* information. As stated above, there can be information processing at the recipient's end (example of the dog in section 2), at the sender's end (example of the ant), or at both ends (two people communicating orally). In all cases an "accord" or common code must exist between sender and recipient so that an interaction can take place at all. As I shall explain later, the word "accord" used in this context is a simple anthropomorphic term designating a complex memory process that must operate within the interaction mechanism. While this latter mechanism is not mentioned explicitly in the definition of information, the accord or common code is integral part of the concept of information—and that of purpose. Information, information-processing and purpose are thus tied together in one package: the process of information-based interaction (Figure 2). We never should talk about information alone without referring to, or at least having present in our mind, the interaction process in which that information mediates a specific correspondence pattern→change. As an absolute, stand-alone concept, information cannot be defined.

Since in the natural world only biological systems can entertain information-based interactions, we can turn the picture around and offer the following definition [Roederer, 1978]: *a biological system is a natural (not-human-made) system exhibiting interactions that are controlled by information.* The proviso in parentheses is there to emphasize the exclusion of artifacts like computers and robots. More recently, Küppers [1990] stated (p. 66): "...systems of the degree of complexity of organisms, even at the molecular level, can arise and maintain themselves reproductively only by way of information-storing and information-producing mechanisms." By adopting the process of interaction as the primary algorithm, I am really going one step further and recognize information as the defining concept that *separates* life from any natural (i.e., not made) inanimate complex system.

4. Information and physics

The preceding section already answers the question asked in the title of this chapter: *information appears in the Universe only wherever and whenever life appears*. In the abiotic world there is no information, unless there is an interaction with a living organism. A rock lying on the lunar surface has been interacting physically during billions of years with the Moon's gravity, the soil on which it came to rest, solar radiation, meteorites, interplanetary dust, and solar wind and cosmic ray particles. But it only becomes a source of "information" when it is looked at or analyzed by a human being; information played no role in its entire past. In short, without "observers" or "users" such as life forms (or devices designed and built by intelligent beings), there is no information.

It is rather difficult to imagine a physical, prebiotic world without information. This happens because the mere act of imagination necessarily puts information into the picture. Yet all interactions in the inanimate world are governed by forces and force fields in which information as a controlling agent does not appear. To physicists in particular, the absence of information in a life-less world may seem quite odd. In statistical thermodynamics, for instance, doesn't information play a crucial role in how a system behaves? A loss of information (about the microscopic state of a given system) is accompanied by a gain in entropy and vice versa. And in a double-slit single electron diffraction experiment, doesn't the electron going through the open slit have information on whether or not the other slit is closed and "behave" accordingly?

Physics works with information obtained from the environment and sets up a mathematical framework that quantitatively describes the behavior of idealized *models* of reality. The extraction of information is done through a process called "measurement", which involves a measuring device and two types of interactions: a physical interaction with the system being measured at one end and an information-based interaction with the perceptional and cognitive apparatus of *a human being* at the other. In the case of an instrument which stores the information in some memory (data bank or graph) we still have, at that end, an artificial information-based interaction with the ultimate purpose of cognitive action. Based on the results of measurements, the human brain constructs simplified models of the systems under study and their interactions, and formulates physical laws. The spatial and temporal features of these models are in specific *correspondence* with some, but not all, features of the systems being modeled. In view of the definition given in the previous section, it is clear that information as such appears in physics because of the paradigms (correspondences) involved and because measurement and experimentation are processes imposed on nature with a pre-designed purpose—but not because information was present and controlled the physical world "before we looked".

Information, whenever it appears in physics, relates to the observer, experimenter or thinker, *not* to the system per se. Quite generally, the behavior of a physical system *under study* is affected by what we humans do to it or with it: it is us humans who prepare a system, set up its boundaries and dictate its initial conditions [Roederer, 1978; Bricmont, 1995]—even if only in thought experiments. As a matter of fact, what *we* consider the "initial conditions" for a *natural* system is only a result of its previous evolution, unless those initial conditions are deliberately set (or imagined) by a human being (the only cosmologically "true" natural initial condition is the state "right after" the Big Bang [e.g., after a Planck-time interval of 10^{-43} s]).

Our senses only respond to macrosystems, and we describe the physical interactions between macroscopic systems using mathematical relationships between state variables. On the other hand, we can infer from observations with special instruments that there also exists a micro-description of a thermodynamic system (unattainable in exact, non-statistical terms for practical reasons), in which the interactions between the constituent molecules are basically reversible, with information *per se* playing no role. It is only when we connect both descriptions that the concept of information creeps in as a participant. Here we can build a bridge between our focus on pragmatic information and the use of traditional information theory in statistical thermodynamics. Consider Figure 2 and view the mechanism as a measuring device or "detector" to determine whether or not the given pattern, to which the recipient is "tuned" via the accord, is present or not in the sender. In other words, take a "digital approach" to the information-based interaction scheme depicted in the figure. If we then apply this detector to an *ensemble* of senders each one of which may or may not carry the given pattern, it will be possible to assign an *information-content value* to that pattern in the ensemble. One can use Gibbs' paradox as an illustrative example (see Roederer [2000]).

Similar considerations apply to quantum mechanics: it is *us* who interfere with a quantum system by creating "unnatural" situations such as the application of a measuring device (a process that unavoidably changes the quantum system), or, for the probabilistic interpretation of the wave function, the need to consider (even if only in our minds) many identical systems prepared with identical set-ups of initial and boundary conditions.

5. The Transition from Physical to Information-Based Interactions

Let us return to the example of Figure 1. Suppose that there are many more balls initially at rest, and that instead of an initial impulse on just one incident ball at the detector level, there is a whole distribution, with a given, specific pattern such as $\{A, B, C, D\}$ shown in Figure 3. The result will be, at the other end, a whole distribution of outgoing balls $\{M, N, R\}$, which will also exhibit a specific distribution in space and velocity. A different incident pattern will lead to a different outgoing one, and we are certainly tempted to state that the outgoing distribution "carries information on the incident pattern"; that "information on the incident pattern has been transmitted and transformed into another pattern at the output"; that "there is a specific and unique correspondence between pattern and change elsewhere"; or, more generally, that "we have a complex interaction between a 'sender' and a 'receiver' that is guided by information" (on the distribution of incident impulses). However, note that we can say all these things legitimately only if we recognize a purpose for the interaction, namely that of obtaining the same specific correspondence between pattern and response every time we activate the system. For that, however, we must make sure that at the beginning the position of all balls (and the energy supply, if any) *is always the same*. This requires either human intervention, or, if this were some (weird) natural system, an additional mechanism that always *resets* the balls to the same initial positions. The existence and operation of such

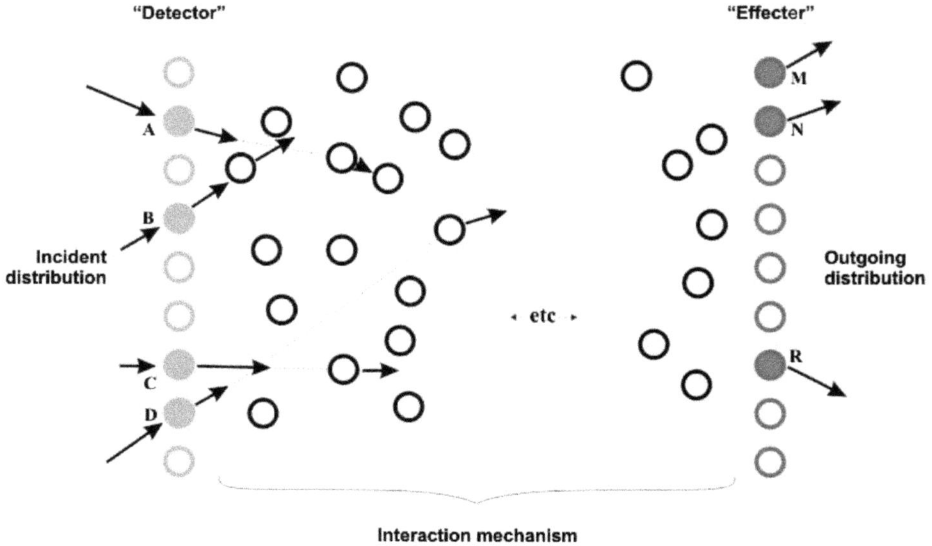

Figure 3. Similar to Figure 1. For each *given* pattern of initial impulses, there is *one* specific outcome (outflying balls). This will become an information-based interaction (between detector and effecter) *only* if in addition there is something in the interaction mechanism that resets all balls *to the same initial position* (i.e., a memory device) so that the process can be repeated many times. The energy relation between input and output becomes irrelevant. Indeed, input constellations that are totally equivalent energy-wise could have very different outgoing distributions. In any natural (not made) system with similar properties, such a mechanism could only emerge through evolution, or, in the neural system, also through a learning process.

an additional mechanism would represent the "accord" between sender and recipient, and would endow the interaction with a specific purpose. To function, it must "remember" the initial positions; in other words, what we have called an "accord" must involve both memory and some memory-reading device. Clearly, the whole mechanism still would be a purely "physical" one, i.e., there is no irreducibility here, as we had wondered in the Introduction. A question of irreducibility, however, does arise in relation to *how such a mechanism would actually be put together* by Nature! (See next section.)

We can drive the example of Figure 3 one step further. Suppose we have a clever arrangement of balls in the interaction mechanism such that whenever balls $\{A, B, C, D\}$ are struck, only ball R flies off, regardless of the initial velocities or energies; when another given initial group of balls is struck (i.e., for another initial spatial pattern) a different single ball responds at the effecter level. This would be an arrangement with a well-defined purpose: that of transforming certain complex input patterns from a given set, like $\{A, B, C, D\}$, into single, pattern-specific output responses every time they are presented. Again, we would have a genuine information-based interaction (between the detector or sender and the effecter or recipient), according to our definition in section 3. But such a mechanism, guaranteeing a consistent, reproducible pattern correspondence would never be found assembled by chance in nature—the "memory device" representing the accord between sender and recipient could only be assembled in a process of evolution, learning or set by human design (see next section). An optical or acoustical pattern recognition network in the brain, or an artificial pattern recognition device, works in a somewhat analogous way.

6. Biomolecular and Neural Information

There are two and *only two* distinct types of natural (not made) information systems: biomolecular and neural. Bacteria, viruses, cells and the multicellular flora are governed by information-based interactions of the biomolecular type; the responses of individual cells to physical-chemical conditions of the environment are ultimately controlled by molecular machines and cellular organelles manufactured and operated according to blueprints that evolved during the long-term past. In plants they are integrated throughout the organism with a chemical communications network. For faster and more complex information processing in multicellular organisms with locomotion, and to enable memory storage of current events, a nervous system evolved which, together with sensory and motor systems, couples the organism to the outside world in real time.

In any information-based interaction, the pattern in the sender can be any set of things, features or symbols that bear and maintain a certain relationship in space and/or time. We may represent this pattern in simplified form as a function of space and time $P_A = P_A(\mathbf{r}, t)$. P represents a physical variable that can be simple (e.g., a color, sound intensity or frequency, or the coordinates of a discrete set of objects) or extremely complex (the spatio-temporal distribution of electrical impulses in brain tissue). In particular, if the pattern consists of a spatial or temporal *sequence* of objects or events from a finite repertoire (e.g., the bases in the DNA molecule or the action potentials fired by a neuron), we usually call it a *code*. In general, spatial patterns do not require a constant supply of energy, but temporal patterns do. As already mentioned, a complex system A (the sender) could well exhibit many different patterns which, however, from the energetic point of view are all equivalent (*a priori* equally probable). In that case, changing a pattern would change the information-based interaction (i.e., the response of B) even if this would not imply any change of the initial thermodynamic state (energy and entropy) of the system.

Key to the evolution of life was the molecular synthesis of large and complex polymer-like macromolecules as *templates*, i.e., potential information carriers, whose effect on other molecules in their environment is to bind them through a catalytic process into conglomerates according to patterns represented by a code. Some of these polymers also served as templates for the production of others like themselves—those more efficient in this process would multiply. Replication and natural selection most likely already began in a pre-biotic chemical environment. Concerning these macromolecules, we can say that beyond a certain degree of complexity, *information as such* begins to play the decisive role in organizing the chemical environment.

Coding of information in molecular chains, and the translation of this information into a different kind of molecules, is, of course, the mainstay of molecular genetics. But there is no "transient" or "quick" storage of environmental information in molecular genetics: it is all the result of a slow process of *evolution* that shapes the informational content through chance mutations and elimination of the unfit. When I say that "information on the environment has been stored in the genetic structures of an organism" I am really describing a process of information extraction and storage "by default" (for lack of a better term). By this I mean that the information content in a molecule such as DNA is not derived from physical laws and *deliberately* stored, but that it emerges gradually in the course of a long Darwinian process of selective adaptation. It could be called an "unreflected learning process by trial and error" [Küppers, 1990]. Evolution is *not* goal-directed: there is no pre-existing information or purpose in adaptive evolution—it will happen whenever environmental circumstances are right. If the order in the nucleic acid chain were to be determined by physical processes alone, the molecule would not have the ability of storing information capable of evolution. It is through the evolutionary process that the necessary "accords" between senders and recipients are gradually built into the interaction mechanisms and information-handling components of the organism of a species [1].

In the nervous system, on the other hand, there are two basic ways in which information is represented or encoded [2]: (1) a dynamic, transient form given by the specific *spatio-temporal distribution of electrical impulses* in the neural network (analog postsynaptic potentials and standardized action potentials); (2) a static form represented by the spatial distribution of synapses (inter-neuron connections and their efficiencies) or *synaptic architecture* of the neural tissue. Type 1 is a dynamic pattern, changing on a time-scale of tens of milliseconds and involving millions of neurons, that requires a continuous supply of energy to be maintained; type 2 is static and in principle can subsist with no extra supply of energy. It is important to point out that the synaptic architecture of a neural network represents the interaction mechanism and the "accord" between different levels of dynamic neural representations (the "balls at rest" in Figure 3).

How is information actually represented in the brain? In this discussion we are mainly interested in the neural representation of higher-level cognitive information—maps that involve many processing stages in the brain and hundreds of millions of neurons. Concerning the dynamic mode of encoding neural information, there is now a convincing ensemble of data, obtained through both microelectrode probing of single neuron activity and macroscopic functional imaging that show that this encoding indeed occurs in the form of a *specific spatio-temporal distribution of neural impulses*. For instance, the mental representation of an object (visual, acoustic, olfactory or tactile) appears in certain areas of the cerebral cortex in the form of a specific distribution of electrical signals that is in one-to-one correspondence (albeit not at all a topological one) with the specific features sensed during the actual perception of this object. Concerning the principal information-processing and—storage operations in the nervous system, a summary is given in Roederer [2002].

With many levels or stages of neural information processing in the brain, many different inputs that belong to correlated environmental objects or events can lead to the formation of one single representation, image or map at one of the uppermost information-processing levels. In turn, because of a two-way coupling between most of the intervening levels, elicitation of that common higher level image can feed back to the lower stages and trigger maps that correspond to one or several individual components of the original input images. When that happens, we say that the system has *cognition* of some common properties that characterize the group of individual objects or events and define a specific category for them. Elicitation of the common image by any single input component represents the process of object *recognition* (for instance, elicitation of the common image corresponding to the

[1] This is why I said that in a natural (not made) equivalent of the system shown in Figure 3, the mechanism that resets the balls to their initial position would have to emerge through evolution.

[2] There is a third information system—the chemical neurotransmitters—which I will not discuss because its function is mainly that of information transmission and modulation of global synaptic function (a sort of neural "volume control").

category "apple" by either the smell of a bowl of apples, the sound of the word "apple", a picture of an apple orchard, or an apple pie).

In a neural system, the common code between environment and a sensory system's lower, prewired, stages has evolved during the slow course of evolution. In the associative memory process [e.g., Kohonen, 1988], on the other hand, the neural network hardware (interaction mechanism, Figure 2) changes during the learning process in real time, and so does the "accord" or common code. As a result, in highly concatenated networks, one and the same input pattern can trigger different images at the highest level, depending on circumstances such as the current state of information processing and previously stored information.

Concerning neural networks, the human brain is the most complex of all—as a matter of fact, it is the most complex and most organized system in the Universe as we know it. The brain evolved in a way quite different from the development of any other organ. Separate layers appeared, with distinct functions, *overgrowing* the older structures but not replacing them, thus preserving "older functions"—the hard-wired memories or *instincts* represented by a synaptic configuration "frozen in time" that the species has acquired during evolution. The outermost layer, or neocortex, executes all higher order cognitive operations, based on information acquired in real time during the life of the organism; its synaptic architecture changes as the animal interacts with the environment (synaptic "plasticity"). Subcortical structures such as the "limbic system" are phylogenetically old parts of the brain carrying information acquired during the evolution of the species. We thus have a system in which two basic "modes" of information coexist and cooperate. This "cortico-limbic cooperation" (we should say, more precisely, this "coherent, cooperative mode of interaction") leads to *consciousness* in higher mammals (see Roederer [2002] and references therein).

In the *human* brain, there is a third mode—not based on any "new" network hardware but representing a new information processing *stage* at which the neural activity of both lower processing modes, the instinctive and the cognitive ones, are combined and mapped together in tight coherence and synchrony into a "representation of representations", giving rise to the awareness of one's own brain function in the context of long-term memory, and to *self-consciousness*. This endowed the human brain with its most distinctive function: to recall stored information, manipulate it and re-store modified versions thereof without any concurrent sensory or somatic input—*the human thinking process*.

In summary, in an animal brain we have information from the long-term past of the species acquired during evolution, originally stored in the genes and expressed in prewired circuits, and information on the ontological past and the present acquired through the senses. Humans have the notion of future time and the ability of making long-term predictions—in informational terms, their brains also handle *information on the future* [Roederer, 2000]. It is here where we can make a link to our discussion in section 4 on how the human brain intervenes in the study of purely physical systems. If we consider, say, a dynamical problem with an equation of motion, we "plug in" a set of initial conditions to see what changes will occur in the motion. More than just changes of motion, we are really interested in the changes of *modes* of system behavior. In other words, we posit a particular outcome and it is the concept of the "final state" that makes us adjust things (in this case, the initial conditions) so that it is achieved. This is not much different from long-term planning such as designing a house: "the future determines the present" [Squires, 1990]. But note that it is very different from a beaver building a dam: the animal follows blueprints developed during the long process of evolution, merely adjusting the details of the construct to local circumstances.

7. Conclusions

This paper could have ended with section 3. At that stage, I posited that information is the defining concept that separates life from any natural (i.e., not made) inanimate system. In other words, *information begins when and where life begins*—which is the answer to the question in the title. Information and life go and evolve together. In a lifeless world, information plays no active, controlling role—if it is there, it is so because we, the observers of that inanimate world, are interacting with it with our senses or our instruments. But for living organisms, information is the very essence of their existence: to maintain a long-term state of unstable thermodynamic equilibrium with its surroundings, consistently increase its organization, reproduce and adapt to its ecological niche, an organism has to rely on information-based interactions. This latter class comprises biomolecular information processes controlling the metabolism, growth, multiplication and differentiation of cells, and, for animals, neural information processes controlling behavior and intelligence. The only way new information can appear in the Universe is through the process of biological evolution and, in the short term, through sensory acquisition and the manipulation of images by the nervous system. Machines planned and built by human beings (purposely ignored in this chapter dedicated to purely natural, not made, systems) also can create new information—but it still would be, albeit indirectly, a consequence of nervous system image manipulations.

Of course, to come to these conclusions it was necessary to examine and define the concept of information in a strictly objective and general way, detached from human artifacts and related algorithms and semantics, and not based on any mathematical formula. I have chosen the process of interaction as the primary concept and departing point of the discussion, which led us to consider pragmatic information as the most appropriate concept for bioinformatics. In our quest we came to realize that there is nothing mysterious about "information"—clearly, only physical processes are involved in information-driven interactions. What appears to be irreducible to presently known physical laws, though, is the way the mechanisms responsible for information-driven interactions actually arise in the natural world. We know it happens through Darwinian evolution, but it is not at all clear how this all started at the pre-biotic molecular level, i.e., how information and life emerged out of the proverbial "primordial soup". Specifically, the fundamental question of the origin of biological information is how under prebiotic conditions the nucleotide sequence of the protogene was selected out of innumerable, energetically equivalent, alternatives [Küppers, 1990].

There are many physicists who reject the idea that information plays no role in physical interactions. It is all a matter of defining and interpreting the concept of information in a strictly objective way—we must realize and get used to the idea that whenever we talk of information in a basic physical process, it refers to the information *we*, the observers or thinkers, have about it; the physical process itself doesn't use it. On the other hand, there are many biologists who believe that the difference between life and non-life in the natural world is mainly one of quantity: the vastly higher degree of complexity and organization exhibited by the former. However, in this chapter I have tried to convince the reader that at least one very fundamental difference does exist: it is called *information*.

8. References

Bricmont, J. (1995) Science of chaos or chaos in science?, Physicalia Mag. **17**, 159–208.

Chaisson, E. J. (2001) *Cosmic Evolution: the Rise of Complexity in Nature*, Harvard University Press, Cambridge Mass.

Davies, P. (1990) Physics and Life, in: J. Chela-Flores, T.Owen and F. Raulin (eds.), *The First Steps of Life in the Universe*, Kluwer Acad. publ., Dordrecht, The Netherlands, pp. 11–20.

Kohonen, T. (1988) *Self-Organization and Associative Memory*, Springer Verlag, Berlin.

Küppers B.-O. (1990) *Information and the Origin of Life*, The MIT Press, Cambridge Mass.

Mach, E. (1893) *Science of Mechanics* (Translation: The Open Court Publ. Co., La Salle, Illinois).

Roederer, J. G. (1978) On the relationship between human brain functions and the foundations of physics, Found. of Phys. **8**, 423–438.

Roederer, J. G. (2000) Information, life and brains, In: J. Chela-Flores, G. Lemarchand and J. Oró (eds.), *Astrobiology*, Kluwer Acad, Publ., Dordrecht, The Netherlands, pp. 179–194.

Roederer, J. G. (2003) On the concept of information and its role in nature, Entropy **5**, 3–33.

Squires, E. (1990) *Conscious Mind in the Physical World*, Adam Hilger, Bristol, New York.

Zurek, W. H. (ed.) (1990) *Complexity, Entropy and the Physics of Information*, Addison-Wesley Publ. Co., New York.

III. Biological Biodata, Evolutionary Data & Chirality

Biodata of **Manuela Pruess** author of ***"Biological Databases—Insights and Trends."***

Dr. Manuela Pruess obtained her Ph.D. in 1997 at the University of Bremen, Germany, for her work on genetic risk factors for chronic-obstructive lung diseases. She then went to the University of Magdeburg for a postdoctoral position in the Bioinformatics group of Prof. Hofestaedt at the Institute for Technical Information Systems, being involved in the development of an information system for the computer aided diagnosis of metabolic diseases. Then she had a postdoctoral position at the German Research Centre for Biotechnology (GBF) in Braunschweig in the Bioinformatics group of Dr. Wingender, where she was working on the extension of the TRANSFAC database to integrate pathologically relevant data. From there she moved to the company BIOBASE in Wolfenbuettel where she was responsible for the development of a database of pathologically mutated transcription factors. Since 2001 she is working at the European Bioinformatics Institute (EBI) in Cambridge, UK, in the Sequence Database Group of Dr. Apweiler as annotation ccordinator for different database projects.

E-mail: **mpr@ebi.ac.uk**

J. Seckbach and E. Rubin (eds.), The New Avenues in Bioinformatics, 41–55.
© 2004 *Kluwer Academic Publishers. Printed in the Netherlands.*

BIOLOGICAL DATABASES—INSIGHTS AND TRENDS

MANUELA PRUESS
EMBL Outstation, The European Bioinformatics Institute (EBI)
Wellcome Trust Genome Campus
Hinxton
Cambridge, CB10 1SD
United Kingdom

1. Introduction

Recent years have seen an explosive growth in biological data, which, instead of being published conventionally, is instead deposited in a database (for the number and size of completely sequenced genomes see Figure 1). Sequence data from mega-sequencing projects may not even be linked to a conventional publication. This trend and the need for computational analyses of the data have made databases essential tools for biological research. Biological databases must be managed and updated so that they contain minimal redundancy. They must also be easily accessible and searchable both by humans and by computer programs. But it is especially important that the data are reliable—otherwise tasks such as sequence similarity searches will fail. Furthermore, for a database to be useful it should be linked to other related sources of information.

Since the 1990s, not only has the number of known biological sequences increased rapidly, but the number of known mutations and polymorphisms has also grown. Furthermore, more and more information about gene expression, mass spectrometry, and protein-protein interactions is being produced and stored. This increase in the size and complexity of data—and thus in the size of databases—has created the need to integrate different types of data—providing the opportunity for 'databases' to evolve into 'knowledgebases'. Accordingly, a large number of biological databases have become accessible to the biological community. Universities, scientific institutes and commercial vendors offer access to all kinds of biological information systems via the Internet. These information systems differ with respect to their contents, reliability and ease of use. They also differ in size, from a few hundred megabytes to tens or even hundreds of gigabytes, and in the underlying technology that is used to maintain these databases.

In this chapter, the most important molecular biology databases available to researchers will be described. Only a fraction of the huge number of existing specialized databases will be mentioned, to provide a representative selection of the different fields of biology. The URLs under which they are accessible are shown in Tables 1 and 2. I will continue by addressing the specific challenges of the genomic and proteomic era, which encompass not only collecting and storing information, but also making use of it. Automatic annotation is

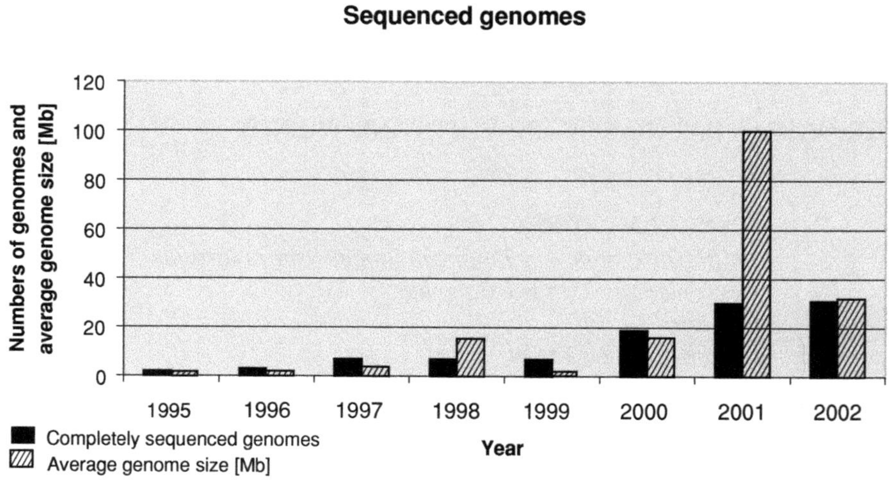

Figure 1. The growth in the number of completely sequenced genomes and average genome size in mega bases (Mb) over the years. The number of sequenced genomes per year is steadily increasing, while the number of average Mb per sequenced genome varies, depending on the completed genome projects. For example, the high number in 2001 is due to the deposition of the draft *Homo sapiens* genome (2910 Mb) in that year.

an important step towards improving the handling of huge amounts of raw data, and tools for computational analysis and data mining are needed to exploit the data. Finally, a section on prospects for the future how bioinformatics resources might develop in the future.

2. Molecular Biological Core Resources

The classical core resources for the molecular biologist are: bibliographic databases, which allow to search for literature under user-defined aspects and to access abstracts of articles; sequence databases, which store nucleotide and protein sequences; structure databases; and taxonomy databases. They provide the basic information needed for both the researcher at the bench and the bioinformatician.

2.1. BIBLIOGRAPHIC DATABASES

Services that abstract the scientific literature began to make their data available in machine-readable form in the early 1960s. However, one should be aware that none of the abstracting services covers the scientific literature comprehensively. The best known of these services is PubMed (formerly MEDLINE), which abstracts mainly the medical literature. PubMed is best accessed through the NCBI's ENTREZ browser, which also provides access to databases, online books, gene expression and microarray datasets and other information. EMBASE is a commercial product for the medical literature. It provides access to the Excerpta Medica database, which is about drug-related research literature; it covers all aspects of human medicine and related biomedical research. BIOSIS, the successor to Biological Abstracts, covers a broad range of biological subjects; the Zoological Record ("the world's oldest continuing database in the animal sciences") indexes the zoological literature. CAB

TABLE 1. URLs of the biological databases mentioned in the text in chapters 2 and 3.

Resources	Database (short name)	URL
Bibliographic Databases	AGRICOLA	http://www.nal.usda.gov/ag98/ag98.html
	BIOSIS	http://www.biosis.org/
	CAB International	http://www.cabi.org/
	EMBASE	http://www.bids.ac.uk/embase.html
	PubMed	http://www.ncbi.nlm.nih.gov/entrez/query.fcgi
Sequence Databases	Nucleotide Sequence DB	http://www.ddbj.nig.ac.jp/, http://www.ebi.ac.uk/embl/, http://www.ncbi.nlm.nih.gov
	PIR	http://pir.georgetown.edu
	Swiss-Prot	http://www.ebi.ac.uk/swissprot/, http://www.expasy.org/
	TrEMBL	http://www.ebi.ac.uk/trembl/
Structure Databases	CSD	http://www.ccdc.cam.ac.uk/prods/csd/csd.html
	MSD	http://www.ebi.ac.uk/msd/
	NDB	http://ndbserver.rutgers.edu/
	PDB	http://www.rcsb.org/pdb/
Taxonomy Databases	IOPI	http://plantnet.rbgsyd.gov.au/iopi/iopihome.html
	ITIS	http://www.itis.usda.gov/
	NCBI Taxonomy DB	http://www.ncbi.nlm.nih.gov/Taxonomy/
	NEWT	http://www.ebi.ac.uk/newt/index.html
	Species 2000	http://www.sp2000.org/
Genetic Databases—Human	Ensembl	http://www.ensembl.org/
	GDB	http://gdbwww.gdb.org/
	GeneCards	http://bioinformatics.weizmann.ac.il/cards/
	LocusLink	http://www.ncbi.nlm.nih.gov/LocusLink/index.html
	OMIM	http://www.ncbi.nlm.nih.gov/omim/
	RefSeq	http://www.ncbi.nlm.nih.gov/LocusLink/refseq.html
Genetic Databases—Model Organisms	FlyBase	http://flybase.bio.indiana.edu/
	MGI	http://www.informatics.jax.org/
	TAIR	http://www.arabidopsis.org/
	WormBase	http://www.wormbase.org/
	ZFIN	http://zdb.wehi.edu.au:8282//
Genetic Databases—Yeast	CYGD	http://mips.gsf.de/proj/yeast/CYGD/db/
	SGD	http://genome-www.stanford.edu/Saccharomyces/
	YPD	http://www.incyte.com/sequence/proteome/index.shtml
Genetic Databases—*E.coli*	CGSC	http://cgsc.biology.yale.edu/
	ECDC	http://www.uni-giessen.de/~gx1052/ECDC/ecdc.htm
	EcoCyc	http://ecocyc.org/ecocyc/ecocyc.html
Some Other Genetic Databases	ArkDB	http://www.thearkdb.org/
	Gramene	http://www.gramene.org/
	MaizeDB	http://www.agron.missouri.edu/
	Parasite Genome DBs	http://www.rna.ucla.edu/par/pfdb.html
	RGD	http://rgd.mcw.edu/
	RATMAP	http://ratmap.gen.gu.se/
	SubtiList	http://genolist.pasteur.fr/SubtiList/
Some Other Specialized Resources	BRENDA	http://www.brenda.uni-koeln.de/
	EMP	http://emp.mcs.anl.gov/
	ENZYME	http://www.expasy.ch/enzyme/
	EPD	http://www.epd.isb-sib.ch/
	IMGT	http://imgt.cnusc.fr:8104/
	KEGG	http://www.genome.ad.jp/kegg/
	LIGAND	http://www.genome.ad.jp/ligand/
	MENDEL	http://www.mendel.ac.uk/
	REBASE	http://www.psc.edu/general/software/packages/rebase/rebase.html
	TRANSFAC	http://transfac.gbf.de/TRANSFAC/

TABLE 2. URLs of the databases mentioned in the text in chapter 4. (For a comprehensive list of links to all sorts of biological databases see also http://www.expasy.ch/alinks.html.)

Resources	Database (short name)	URL
Proteomics Databases	ArrayExpress	http://www.ebi.ac.uk/microarray/Array Express/arrayexpress.html
	BIND	http://www.bind.ca/
	DIP	http//dip.doe-mbi.ucla.edu/
	MINT	http//cbm.bio.uniroma2.it/mint/
	SWISS-2DPAGE	http//ca.expasy.org/ch2d/
Protein Pattern Databases	Pfam	http//www.sanger.ac.uk/Pfam/
	PRINTS	http//www.bioinf.man.ac.uk/dbbrowser/PRINTS/
	ProDom	http//prodes.toulouse.inra.fr/prodom/doc/prodom.html
	Prosite	http//www.expasy.ch/prosite/
	SMART	http//smart.embl-heidelberg.de/
	TIGRFAMs	http//www.tigr.org/TIGRFAMs/
Proteome Analysis Databases and Tools	CluSTr	http//www.ebi.ac.uk/clustr/
	GO	http//geneontology.org/
	InterPro	http//www.ebi.ac.uk/interpro
	Proteome Analysis DB	http//www.ebi.ac.uk/proteome

International maintains abstract databases in the fields of agriculture and parasitic diseases. AGRICOLA is the most important bibliographic database for the agricultural field. With the exception of MEDLINE/PubMed, the bibliographical databases are available only through commercial database vendors.

2.2. SEQUENCE DATABASES

Sequence databases are comprehensive sources of information on nucleotide and protein sequences. The most important databases for DNA sequences are DDBJ, EMBL-Bank and GenBank, which form the International Nucleotide Sequence Database Collaboration. The most important protein sequences databases are the Swiss-Prot Protein Knowledgebase and the PIR Protein Sequence Database.

2.2.1. Nucleotide Sequence Databases

The Nucleotide Sequence Databases are data repositories that accept nucleic acid sequence data from the community and make it freely available. The databases strive for completeness, with the aim of recording every publicly known nucleic acid sequence. These data are heterogeneous; they vary with respect to the source of the material (e.g. genomic versus cDNA), the intended quality (e.g. finished versus single-pass sequences), the extent of sequence annotation and the intended completeness of the sequence relative to its biological target (e.g. complete versus partial coverage of a gene or a genome). The International Nucleotide Sequence Database Collaboration allows the participating databases—EMBL-Bank at the European Molecular Biology Laboratory in the UK, the DNA Data Bank of Japan (DDBJ) in Japan, and GenBank by the National Center for Biotechnology Information (NCBI) in the USA—to exchange data. Each of these collects, organizes and distributes nucleotide sequence data. EMBL, NCBI and DDBJ automatically update each other every 24 hours with the new sequences that they have collected or updated. This daily exchange is facilitated by shared rules, a unified taxonomy, and a common set of unique identifiers. The International Nucleotide Sequence Database Collaboration is a good example of the successful joining of forces. The three sites have developed independently, driven by different research bodies and with different funding. The collaboration allows them to share data instead of competing, thus making optimal use of the resources—although each site still has their own database format.

2.2.2. *Protein Sequence Databases*

The protein sequence databases are the most comprehensive source of information on proteins. It is necessary to distinguish between universal databases, which cover proteins from all species, and specialized data collections that store information about specific families or groups of proteins, or about the proteins of a specific organism. Two categories of universal protein sequence databases can be discerned: simple archives of sequence data, and annotated databases where additional information has been added to the sequence record. The second category is of particular interest for the researcher who wants to compare the sequences that s/he is working with, and their features, with related ones described by others.

The Swiss-Prot Protein Knowledgebase is an annotated protein sequence database that was established in 1986. It is maintained collaboratively by the Swiss Institute of Bioinformatics (SIB) and the European Bioinformatics Institute (EBI). It strives to provide a high level of annotation, a minimal level of redundancy and a high level of integration with other biomolecular databases. But because there is such a tremendous increase in sequence data, owing to technological advances (such as sequencing machines), the use of new biochemical methods (such as PCR technology) and the implementation of projects to sequence complete genomes, it has become impossible for Swiss-Prot to keep up with this information: careful manual curation is the rate-limiting step in the production of the database. To make new protein sequences available for the public as quickly as possible without relaxing the high editorial standards of Swiss-Prot, the EBI created a supplement to Swiss-Prot, known as TrEMBL (Translation of EMBL nucleotide sequence database), in 1996. TrEMBL is a database that consists of computer-annotated entries derived from the translation of all coding sequences (CDS) in EMBL-Bank, except for those already included in Swiss-Prot; it already contains nearly seven times as many entries as Swiss-Prot, and this discrepancy will increase in the future.

The PIR (Protein Information Resource) Protein Sequence Database, PSD, is the oldest of the protein sequence databases. It was established in 1984 by the National Biomedical Research Foundation (NBRF) as a successor of the original NBRF Protein Sequence Database, developed over a 20-year period by the late Margaret O. Dayhoff and published as the 'Atlas of Protein Sequence and Structure'. Since 1988 the database has been maintained by PIR-International, a collaboration between the NBRF (National Biomedical Research Foundation), the MIPS (Munich Information Center for Protein Sequences), and the JIPID (Japan International Protein Information Database).

2.3. STRUCTURE DATABASES

The number of known macromolecular structures is also increasing very rapidly, and these are made available through the Protein Data Bank (PDB). PDB is a repository for the processing and distribution of experimentally determined 3-D biological macromolecular structure data, including the structures of proteins, peptides, viruses, protein-nucleic acid complexes, nucleic acids, and carbohydrates. The PDB was established over 20 years ago at the Brookhaven National Laboratory, where it was also maintained until 1998. Since 1998, the PDB has been maintained by the Research Collaboratory for Structural Bioinformatics (RCSB). The Macromolecular Structure Database (MSD) is the European project for the collection, management and distribution of data about macromolecular structures. One of its objectives is the development of procedures to aid the deposition of structures in the PDB, but it also develops tools to aid the retrieval and analysis of PDB data. These include tools for secondary structure matching, finding small molecules that bind particular structures, and finding multimeric structures.

The Nucleic Acid Database (NDB) is a database for structural information about nucleic acid molecules. The NDB maintains the macromolecular Crystallographic Information File (mmCIF) web site, which is the IUCr-approved data representation for macromolecular structures. Databases for monomer units and ligands have been created by a specific tool, à la mode (A Ligand And Monomer Object Data Environment for building models). The NDB project is maintained at Rutgers University, New Jersey. The Cambridge Structural Database (CSD), maintained by the Cambridge Crystallographic Data Centre, is a database of structures of 'small' molecules that are of interest to biologists concerned with protein-ligand interactions. It contains crystal structure information for organic and metal organic compounds; all of these crystal structures have been analyzed using X-ray or neutron diffraction techniques.

2.4. TAXONOMY DATABASES

Databases about taxonomy, the formal nomenclature and description of organisms, don't seem to be very trendy in the "post genomic era". But nevertheless they are of great importance: They provide valuable information necessary to understand which genome is under investigation, and where in the tree of life it is positioned. Taxonomy helps to identify, classify and compare organisms, and to build up meaningful systematics. However, taxonomic databases are rather controversial: different groups regularly question the soundness of the taxonomic classifications done by other groups. Various efforts are going on to create taxonomy resources with different focuses.

The two projects Species 2000 and the Integrated Taxonomic Information System (ITIS) have the objective to provide taxonomic information on plants, animals, fungi and microbes. Both programmes joined forces in 2001 to create the 'Catalogue of Life'. The International Organization for Plant Information (IOPI) manages a series of cooperative international projects that aim to create and link databases of plant taxonomic information. The most widely used taxonomic database is the one maintained by the NCBI. This hierarchical taxonomy is used by the Nucleotide Sequence Databases, and is curated by an informal group of experts. NEWT is a taxonomy viewer maintained by the Swiss-Prot group. It represents the NCBI's taxonomic data for species with protein sequences stored in Swiss-Prot, but it names species using the slightly different Swiss-Prot nomenclature.

3. Specialized Databases

There are a huge number of specialized biological databases, covering a variety of different fields. Genetic databases contain data about the genomes of certain organisms (some of them being "official" projects, closely linked to genome projects). Other specialized databases collect and display information about certain topics, which can be specific classes of sequences or specific functions.

3.1. GENETIC DATABASES

For organisms of major interest to geneticists, there is a long history of conventionally published catalogues of genes or mutations. In the past few years, most of these have been made available in an electronic form. Several new organism-specific databases are also being developed as more and more organisms arouse the interest of sequencing projects.

3.1.1. Human Databases

There are two major databases for human genes, but one of them doesn't seem to have a bright future. McKusick's Mendelian Inheritance in Man (MIM) is a catalogue of human genes and genetic disorders; it is available online (OMIM) from the NCBI and is well maintained. The Genome Database (GDB) was established at Johns Hopkins University in Baltimore in 1990 and transferred in 1998 to the Hospital for Sick Children (HSC) in Toronto, Canada. It was intended to become the main international repository for genome mapping information, but has not been well maintained more than a year (Bonetta, 2001). Until the end of 2002, GDB was looking for an institution interested in assuming responsibility for its maintenance and curation—but it seems as if no institution has been intersted so far (February 2003). There will be no further curation of this resource; the HSC is continuing to make GDB available as a static database for at least 90 days in 2003, possibly with reduced performance. The GeneCards Encyclopedia at the Weizmann Institute, Rehovot, Israel, automatically integrates a subset of the information stored in 37 major data sources that deal with human genes and their products, with a major focus on medically relevant information.

Some other interesting resources are available that do not exclusively deal with human data: Ensembl is a joint project between the EBI and the Wellcome Trust Sanger Institute. It produces and maintains automatic annotation of eukaryotic genomes. Human data are available, as well as data for mouse, zebrafish and the malaria mosquito (*Anopheles*); rat and the puffer fish (fugu) are in preparation. LocusLink from NCBI organizes information around genes to generate a central hub for accessing gene-specific information for fruit fly, human, mouse, rat and zebrafish, and NCBI's RefSeq (Reference Sequence project) provides reference sequence standards for genomes, transcripts and proteins (human, mouse and rat mRNA).

3.1.2. Model Organism Databases

Two of the best-curated genetic databases are FlyBase, the database of genetic and molecular data for the fruit fly *Drosophila melanogaster*, a joint project of the Berkeley Drosophila Genome Project, and the Mouse Genome Informatics resource (MGI) at the Jackson

Laboratory in Bar Harbor. The Berkeley Drosophila Genome Project includes data from the Drosophila Genome Projects and data curated from the literature, and MGI provides integrated access to data on the genetics, genomics, and biology of the laboratory mouse.

WormBase, a database for the nematode *Caenorhabditis elegans*, is maintained by an international consortium of biologists and computer scientists based at Cold Spring Harbor Laboratory, California Institute of Technology, Washington University at St Louis, and the Wellcome Trust Sanger Institute (UK). The Arabidopsis Information Resource (TAIR) at the National Center for Genome Resources (NCGR) provides genomic and literature data about *Arabidopsis thaliana*, and ZFIN is the database for another important model organism, the zebrafish Danio rerio, and is maintained at the University of Oregon. The *Saccharomyces* Genome Database (SGD) is an important resource for information on the yeast genome and its products. The MIPS yeast database (CYGD) is another major yeast database; both of these offer genome comparisons between different kinds of yeast. The detailed curated YPD, Proteome's yeast protein database, now belongs to the company Incyte and is only commercially available.

There are several databases for *Escherichia coli* available. CGSC, the *E.coli* Genetic Stock Center (Yale), maintains a database of *E.coli* genetic information, including genotypes and reference information for the strains in the CGSC collection. The *E.coli* Database collection (ECDC) at the University of Giessen, Germany, maintains curated gene-based sequence records for *E.coli*, and EcoCyc, the "Encyclopedia of *E.coli* Genes and Metabolism", is a database of *E.coli* genes and metabolic pathways.

3.1.3. Other Genetic Databases

There are many other databases for genetically important organisms. MaizeDB, for example, is the database for genetic data on maize; Gramene is a comparative mapping resource for grains. There are also genome databases available for several animals of economic importance, such as pig, cow, sheep, chicken and salmon, provided by ArkDB at Roslin Institute, Edinburgh. RATMAP is a database of genes that have been physically mapped to chromosomes in the laboratory rat, and is mainly dedicated to rat gene nomenclature. The Rat Genome Database (RGD) curates and integrates rat genetic and genomic data and provides access to this data to support research using the rat as a genetic model for the study of human disease. Parasite genome databases are summarized at some special sites, and include databases for *Plasmodium falciparum, Toxoplasma gondii, Leishmania major* and others, and for the *Bacillus subtilis* genome, for example, there is the SubtiList resource.

3.2. SPECIALIZED SEQUENCE AND PROTEIN DATABASES

Many specialized sequence and protein databases are available, too; some are quite small whereas others are wider in scope and larger in size. The ENZYME nomenclature database, for example, is an annotated extension of the Enzyme Commission's publication. There are also databases of enzyme properties—BRENDA is a collection of enzyme functional data that intends to give a representative overview on the characteristics and variability of each enzyme, LIGAND is a database of chemical compounds and reactions in biological pathways, and the Database of Enzymes and Metabolic Pathways (EMP) is encoding the contents of original publications on the topics of enzymology and metabolism and is providing pictorial representations of metabolic pathways. LIGAND is linked to the metabolic pathways in KEGG, a database dealing with the information pathways that consist of interacting molecules or genes.

Some of the specialized sequence databases deal with particular classes of sequences, for example IMGT, the ImMunoGeneTics database, contains immunoglobulin sequences. MENDEL is a plant-wide database for plant genes. Others are focusing on particular features, such as TRANSFAC for transcription factors and transcription factor binding sites, EPD (Eukaryotic Promoter Database) for promoters, and REBASE for restriction enzymes and restriction enzyme sites.

4. The Challenge of the Proteomic Era

After the often-cited genomic era, we have already entered the proteomic era. This doesn't mean a complete change in the type of data available; genome research is continuing at a rapid pace. But it does mean that new types of information about the protein complement of a particular organism or cell type are being produced in parellel with genomic information. This new era also created a further challenge—the more raw data there are, the more there is to analyze, and the more information can be extracted out of the data.

Proteomics aims to uncover all proteins in the biochemical or biological contexts of all organisms, and their structure, function and expression. This has been made possible—or at least more possible than it was some years ago—by the technological advances in protein science. Proteomics plays important roles in modeling the systems of a living cell, discovering the causes of diseases, and therefore in target and drug discovery. So, in addition to the information described above, new types of data now being produced that have to be stored, made available and searchable and linked in a logical way in data collections. For such large-scale efforts, standards have to be set. The Proteomics Standards Initiative (PSI) for example, which was founded in 2002 and reflects the activities of HUPO (the Human Proteome Organization), aims to define community standards for data representation in proteomics to facilitate data comparison, exchange and verification. As a first step, the PSI will develop standards for two key areas of proteomics: mass spectrometry and protein-protein interaction data (Kaiser, 2002).

4.1. PROTEOMICS DATABASES

Public repositories are needed to store information about 2D-PAGE, microarray and mass spectrometry data, preferably analogous to the international nucleotide sequence database collaboration. Some have already been in existence for quite a while. SWISS-2DPAGE stores experimental data from human, mouse, *Arabidopsis thaliana*, *Dictyostelium discoideum*, *Escherichia coli* and *Saccharomyces cerevisiae*. ArrayExpress is a public repository of microarray based gene expression data, which is aimed at storing well annotated data in accordance with specific, so-called "MIAME" recommendations, a set of guidelines that aim to outline the minimum information required to unambiguously interpret microarray data and to subsequently allow independent verification of this data at a later stage if required. Several well-established databases for protein-protein interaction data are in place, including the Biomolecular Interaction Network Database (BIND), the Database of Interacting Proteins (DIP), and MINT, a Molecular INTeractions database. In contrast, widely accepted public repositories for mass spectrometry data do not yet exist; experimental results are usually compared against a database of theoretical spectra calculated from protein databases.

4.2. DATABASE INTEGRATION AND TOOLS FOR PROTEOME ANALYSIS

Rapidly growing amounts of data that lacks experimental determination of the biological function enhances the need for computational analyses of the data. Therefore, tools for computational analysis and data mining are needed. Also, special 'meta-databases', which integrate the data from various sources in order to represent a certain topic as comprehensively as possible, are of increasing importance. Such meta-databases should save the user long searches in different resources, and should extract the most important data.

The number of proteome analysis tools and databases is increasing and most of them are providing high quality resources of computational analysis and annotation. InterPro, which was created in 1999, is a very successful integrated documentation resource for protein families, domains and functional sites. It integrates data from six major pattern databases (Pfam, PRINTS, ProDom, Prosite, SMART and TIGRFAMs), which use different pattern recognition methods. By combining the individual strengths of the member databases, InterPro provides a powerful tool for protein classification. The Proteome Analysis database was set up in 2000 to provide comprehensive statistical and comparative analysis of the predicted proteomes of fully sequenced organisms. It integrates protein sequences with annotation from Swiss-Prot and TrEMBL, protein classification from InterPro, 'clustering' of proteins into groups of related ones from the CluSTr database, and biological functions of proteins according to GO Slim, a slimmed version of the three Gene Ontology (GO) ontologies—a set of controlled vocabularies that are used by many biological databases to describe the functions of gene products in a consistent way. One more example of the recent joining of forces to provide a comprehensive resource is the IntAct project, created in 2002 by a European consortium with the aim of defining a standard for the representation and annotation of protein-protein interaction data, providing a public repository, populating the repository with experimental data, and providing it with modular analysis tools. IntAct will use the PSI standards.

The user of proteome analysis tools should always bear in mind that *in silico* tools and databases, like wet lab technologies, have their own pitfalls. The material in databases and the output of tools is trustworthy only to a certain point. It is important to re-emphasize that most sequence data today comes from large-scale sequencing efforts and lacks experimental functional characterization. Many sequencing centers still annotate the predicted coding sequences only on the basis of automated high-level sequence similarity searches against protein sequence databases.

4.3. AUTOMATION OF ANNOTATION

The main value of many databases, especially the protein sequence databases, is the careful manual annotation of entries. In the Swiss-Prot protein sequence database, for example, curators provide summaries of the literature describing the functions of a protein, post-translational modifications, domains and sites, protein-protein interactions, pathway information, diseases associated with deficiencies in the protein, and sequence conflicts, variants, etc. This annotation work is a time-consuming task: on average, the number of new protein sequences per month in the TrEMBL database in 2002 was about 15,000, whereas the number of newly annotated sequence entries in Swiss-Prot was about 2,000. So annotation clearly is the bottleneck for curated databases. To speed up the process of annotation, efforts are made to enhance automated ways of annotation. This is not a substitute for manual annotation, but it can help to bring "raw" entries to a higher annotation standard.

Rule-based systems for automated annotation attribute manually curated information to groups of entries in a database. These can then be deduced for new entries that fulfil the criteria of this group. In an approach where a standard data-mining algorithm was applied to gain knowledge about the keyword annotation in Swiss-Prot, the coverage rate of the keyword annotation could be increased to 60% by tolerating an error rate of 5%, and to 33% with an error rate of 1.5% (Kretschmann *et al.*, 2001). Other recent approaches for automatic keyword annotation using symbolic machine learning techniques (Bazzan *et al.*, 2002) and for automated annotation of functional properties based on an adaptive algorithm (Leontovich *et al.*, 2002) also have been successful. The GeneQuiz system (Andrade *et al.*, 1999) is an automatic system for preliminary functional annotation of protein sequences, useful for the analysis of sequences from complete genomes. Of course, methods of automated annotation are not without risks: for multifunctional proteins there is the danger of loss of information and outright errors; there is no coverage of position-specific annotation such as active sites; the annotation is not constantly updated and is therefore quickly outdated; there might be inconsistencies in the data; and, as the best hit in pairwise sequence similarity searches is often a hypothetical protein, a poorly annotated protein or one with a different function, the propagation of incorrect annotation is widespread. These caveats should not discourage users of bioinformatics tools and databases from making the most of these important resources, but they should bear in mind the potential pitfalls, check all data carefully, and not blindly rely on the data. As a matter of principle, all databases appreciate suggestions for improvements and the reporting of errors, which allows the data custodians to improve their resources.

5. Future Prospects

Both 'Genomics' and 'Proteomics' are hot topics at the moment, and so are databases in this field. The amount of available data is growing rapidly: Genomic information is generally freely available, and Proteomics is generating even more data—for the human alone, we have to deal with roughly 3 billion nucleotides, 35 thousand protein-encoding genes, up to 10 variants per protein, and over 250 tissues in which proteins can be expressed. The strategical problem is to make biological use of these billions of data points.

The demand for a broad definition of Proteomics, for quality controls and standards for experimental methods, and for standardized analysis software and annotation has recently been identified and addressed by a group of experts in the field (Kenyon *et al.*, 2002), who also pointed out the importance of database infrastructures that allow the efficient and biologically intuitive storage and retrieval of data, and also the communication and interaction of different databases. So Proteomics knowledgebases will certainly be a main focus within the field of molecular biological database development and exploitation in the near future. Another important issue will be public access to private data, and how the trend for (mainly private) database operators to evolve into drug discovery companies will affect data access.

5.1. PUBLIC FUNDING OF DATABASES

Despite the acknowledged need to store molecular biological data in public repositories, the public funding of databases isn't something that can be taken for granted. Some of the most widely used databases have undergone funding crises in the past, including Swiss-Prot in 1996 and BRENDA in 2001. In these cases, the solution has been both to restart public funding and to bring the database to market. In those cases where databases have become commercial products, non-commercial users can generally access the full data free of charge, whereas commercial users require a license. Another strategy, followed by the TRANSFAC database for example, is to make only part of the database (or an older version of the database) freely available to non-commercial users, and to license the full version (the fee is generally much lower for non-commercial users). However, when the well-known and widely used yeast database YPD was commercialized in 2002, a substantial access fee was charged for all users. Many researchers protested at this sudden restriction in access to an important ans unique data set—albeit to no purpose (Abbott, 2002). GDB is an example for a database project that failed for several reasons. It was not adequately maintained, and efforts to commercialize it ran into legal problems (Bonetta, 2001).

It is especially difficult for taxonomy databases to get funding because this discipline still has to struggle against its relatively poor image among biologists. But taxonomy is now shaking off its old image and, in collaboration with the journal Nature, taxonomists are encouraging the creation of a high-profile, publicly accessible, centralized repository of taxonomic nomenclature (Nature News, 2002). To avoid funding problems and fruitless competition in a small field, databases can join forces. A recent example is the conjugation of Swiss-Prot/TrEMBL and PIR to create the United Protein Databases, or UniProt, a publicly funded project (Butler, 2002).

5.2. TRENDS IN THE DATABASE BUSINESS

The market is growing for all sorts of '—omic' platform technologies, and databases are used to store the resulting data. Some years ago, a lot of companies started up with the aim of collecting molecular biological data, setting up databases and selling subscriptions, mainly to the pharmaceutical industry. But this concept has become more and more problematic as public organizations have increasingly provided free access to very comprehensive databases and tools. Furthermore, the number of companies that can afford the often high fees is limited, and the costs of new developments are considerable. For this reasons, database companies are now having to find ways of making the transition from databases to profitable businesses. The current trend for companies is to make use of their data themselves: a lot of them are making the transition into drug discovery companies. To achieve this, they are reducing the emphasis on databases as their main source of income and instead investing in chemistry and biology assets. But the way from proteins to drugs isn't easy, advances in technologies are costly, and this change in the focus of a company generally is combined with the redundancy of staff. Nevertheless, for some companies this will surely be the way to overcome the current crisis in the database business. A first step to make biological sense of sequences is to develop smart data-mining systems to find patterns in large datasets, and then to combine informatics and statistics in order to derive commercial applications. Another general trend is the integration of databases. In the public domain, this is mainly done by consortiums. This helps to overcome the data fragmentation over many databases with differing structures and locations, allowing easier data access and thus the concurrent exploitation of different resources.

6. References

Abbott, A. (2002) Biologists angered by database access fee. *Nature* **418**, 357.

Andrade, M. A., Brown, N. P., Leroy, C., Hoersch, S., de Daruvar, A., Reich, C., Franchini, A., Tamames, J., Valencia, A., Ouzounis, C., Sander, C. (1999) Automated genome sequence analysis and annotation. *Bioinformatics* **15**, 391–412.

Bazzan, A. L. C., Engel, P. M., Schroeder, L. F., Da Silva, S. C. (2002) Automated annotation of keywords for proteins related to mycoplasmataceae using machine learning techniques. *Bioinformatics* **18** Suppl 2, S35–S43.

Bonetta, L. (2001) Sackings leave gene database floundering. *Nature* **414**, 384.

Butler, D. (2002) NIH pledges cash for global protein database. *Nature* **419**, 101.

Kaiser, J. (2002) Public-private group maps out initiatives. *Science* **296**, 827.

Kenyon, G. L., DeMarini, D. M., Fuchs, E., Galas, D. J., Kirsch, J. F., Leyh, T. S., Moos, W. H., Petsko, G. A., Ringe, D., Rubin, G. M., Sheahan, L. C. (2002) Defining the mandate of proteomics in the post-genomics era: Workshop report. *Mol. Cell. Proteomics* **1**, 763–680.

Kretschmann, E., Fleischmann, W., Apweiler, R. (2001) Automatic rule generation for protein annotation with the C4.5 data mining algorithm applied on SWISS-PROT. *Bioinformatics* **17**, 920–926.

Leontovich, A. M., Brodsky, L. I., Drachev, V. A., Nikolaev, V. K. (2002) Adaptive algorithm of automated annotation. *Bioinformatics* **18**, 838–844.

Nature News (2002) Genomics and taxonomy for all. *Nature* **417**, 573.

Biodata of **Isaiah T. Arkin**, the co-author (with Hadas Leonov) of the chapter entitled "***Bioinformatic modeling of transmembrane α-helical bundles based on evolutionary data***".

Dr. Isaiah Arkin is a Professor for Biological Chemistry at the Institute of Life Sciences, Hebrew University of Jerusalem. He obtained his Ph.D. in 1996 from Yale University, and soon thereafter joined the staff of the Biochemistry Department at Cambridge University (UK) where he was a lecturer. In 2000 he left Cambridge and joined the Hebrew University, where his major research focus is structural biology of membrane proteins, combining both computational and experimental approaches. Computational tools include molecular dynamics simulations and bioinformatics analysis of membrane protein structure. The main experimental tool used in his lab is Fourier transform infrared spectroscopy, where he developed a new approach based on site-specific dichroism analysis.

E-mail: **arkin@cc.huji.ac.il.**

Biodata of **Hadas Leonov**, the author (with Isaiah T. Arkin) of the chapter entitled "***Bioinformatic modeling of transmembrane α-helical bundles based on evolutionary data***".

Hadas Leonov is an M.A. student for Structural and Molecular Biology at the Hebrew University of Jerusalem. She obtained her B.A. in Computer Science in 2002, from the School of Computer Science and engineering at the Hebrew University of Jerusalem. Her research interests involve computational Biology and Bioinformatic approaches to study membrane proteins.

E-mail: **hleonov@cs.huji.ac.il**

Isaiah Arkin

Hadas Leonov

J. Seckbach and E. Rubin (eds.), The New Avenues in Bioinformatics, 57–68.
© 2004 *Kluwer Academic Publishers. Printed in the Netherlands.*

BIOINFORMATIC MODELING OF TRANSMEMBRANE α-HELICAL BUNDLES BASED ON EVOLUTIONARY DATA

ISAIAH T. ARKIN and HADAS LEONOV
Department of Biological Chemistry, The Alexander Silberman Institute of Life Sciences. The Hebrew University, Givat-Ram, Jerusalem, 91904, Israel

1. Introduction

Membrane proteins comprise an important family of proteins due to the following two reasons: (i) their biomedical importance and (ii) their genomic abundance. The paramount importance of membrane proteins in biomedicine stems from the fact that they serve as targets for the majority of drugs in medical use. The reason being is that in order to target membrane proteins, drugs do not need to traverse a lipid bilayer. Furthermore, membrane proteins often function as the starting point of many cellular signal transduction cascades. As such, terminating the entire cascade is best achieved by blocking the initial signal at the membrane. In terms of genomic abundance, it has recently been estimated using hydropathy algorithms, that putative membrane proteins comprise 20–30% of the proteomes of all organisms sequenced so far (Stevens and Arkin, 2000). This diversity is remarkable considering the relatively small volume that membranes comprise out of the entire volume of the cell. By analogy it would seem that membranes are the corral reefs and rain forests of the cellular world.

Structural studies have so far shown that membrane proteins fold into one of only two topologies: β-barrels or α-helical bundles. Since α-helical membrane proteins are far more abundant, as well as pharmaceutically more important, the following discussion will be restricted to this family.

Predicting membrane protein structure is of significant importance because despite of the pharmaceutical importance that they posses, out of nearly 20,000 protein structures solved using crystallographic or NMR methods, only a few dozens are membrane proteins.

Knowledge based homology methods that rely on structural information are difficult to implement for membrane proteins, simply due to the lack of solved structures. On the other hand, other modeling methods are relatively easy to implement compared to water soluble proteins, due to the overall simplicity of membrane proteins, in particular those formed from α-helical bundles. Furthermore, assignment of the different helices in an α-helical bundle (the more abundant and pharmaceutically important family) is relatively straightforward (Engelman *et al.*, 1986). Thus, we can conclude that while the structures of α-helical membrane proteins are the most difficult to determine experimentally, fortunately they are the easiest to predict computationally.

Two different approaches to model membrane proteins based on the overall simplicity of the protein are at hand. In structures containing few helices (so far up to four) or homo-oligomers it is possible to exhaustively search the entire conformational space of the structure. Such a search normally produces several candidate structures that due to limitations in the force fields are normally indistinguishable. Selection of the prime candidate is then achieved by a set of external restraints. In more complex instances, in which exhaustive searching can not be undertaken, modeling procedures are employed based on a set of initial constraints thereby limited the number of possible conformations. Thus in simple cases the restraints are used prior to modeling in an objective manner, while in complex systems restraints are used to limit the extent of the initial modeling efforts.

In this review we will focus on modeling simple transmembrane helical bundles, describing a method applicable to all such systems. The reader is refered to the following examples of modeling studies of more complex systems based on a low resolution EM structure of the protein (Stoilova-McPhie *et al.*, 2002, Nunn *et al.*, 2001), or homology data (Baldwin, 1993, Radresa *et al.*, 2002, Muller, 2000, Capener *et al.*, 2000), or both (Unger *et al.*, 1997, Stahlberg *et al.*, 2000, Radresa *et al.*, 2002).

1.1. GLOBAL MOLECULAR DYNAMICS SEARCHING

Due to their structural simplicity, transmembrane α-helical bundles represent a simple topology that can be described by a relatively small number of parameters: (i) helix tilt, the angle between each helix axis and the normal to the bilayer, (ii) rotational position, the angle between the shortest vector from the helical axis to the middle of the C=O bond and the plane that contains both the helical axis and the normal to the bilayer and (iii) translational shift (register) (see figure 1). Thus, for any hetero-oligomer $3 \times n$ parameters are needed to describe the overall structure, while for any symmetrical homo-oligomer only 2 parameters are generally sufficient to describe the structure: helix tilt (β) and a rotational pitch angle (ϕ).

The reduced number of degrees of freedom, allows an exhaustive search of each of the above parameters computationally in a procedure coined Global Molecular Dynamics Search (GMDS) (Adams *et al.*, 1995).

GMDS is used to find a structure of a helical bundle by exhaustively searching the conformational space. The search starts with an arbitrary crossing angle (β_i) and an arbitrary register (r_i) for each of the helices. Then, Multiple bundles of helices are constructed, each differing from the other by the rotation (ϕ_i) of the helices about their axes. Note that ϕ_i will be equal for all helices if the protein is a homo-oligomer. These initial bundles are then used as starting positions for molecular dynamics and energy minimization protocols. The output structures from these simulations are compared with a parameter called root mean square deviation (RMSD), calculated as follows:

$$\sqrt{\frac{1}{n}\sum_i (a_i - b_i)^2} \tag{1}$$

Where i sums over all backbone atoms, n is the number of backbone atoms, a_i is a backbone atom from one output structure and b_i is its corresponding backbone atom from a second output structure. The RMSD computation is performed for each pair of structures, only after

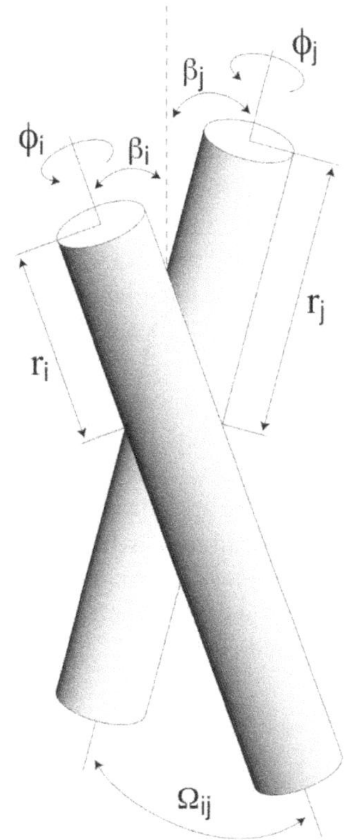

Figure 1. In a bundle with n transmembrane α-helices (a dimer with helices i and j in this case), $3n$ parameters can be used to describe the general structure, assuming rigid helices: (i) the inclination of the helices with respect to the bundle axis, β_i, related to the commonly used crossing angle Ω, (ii) the rotational angle about the helix director, ϕ_i, which defines which side of helix i is facing towards the bundle core and (iii) the helix register, r_i, which defines the relative vertical position of the helix.

superimposing the backbone atoms of both structures. (Otherwise even if the structures are identical, the RMSD may be very large, since they are simply located at different locations in the 3D space). RMSD comparisons yield local clusters of structures where each pair of structures in a cluster is spatially similar (i.e. The RMSD between them is smaller than a certain threshold). The clusters indicate the presence of a possible oligomeric form. At this point, it is still hard to determine which of the clusters found is the real cluster of the protein in question. The interface that is used for these simulations is CHI (Adams *et al.*, 1995), which uses the CNS program (Brunger *et al.*, 1998).

2. Model Selection

As stated above model selection based on energetic criteria are difficult due to the inaccuracies in the force fields. Nevertheless, several attempts have been made to implement energy values as one of the criteria used in the selection (Brunger *et al.*, 1998). The researcher is thus faced with several competing models, amongst which, hopefully the correct model exists. Initially, mutagenesis studies were employed as a model selection scheme, as was the case with the dimerizing human glycophorin A (Lemmon *et al.*, 1992, Treutlein *et al.*, 1992) or pentamerizing human phospholamban (Arkin *et al.*, 1994). More recently, orientational data derived from site-specific infrared dichroism measurements have been used as refinement energy terms employed during the simulation process (Arkin *et al.*, 1994). Using this methodology it is possible to increase convergence towards the correct model and solve the backbone structure of a number of transmembrane helical bundles (Kukol and Arkin, 1999, Kukol and Arkin, 2000, Torres *et al.*, 2000, Torres *et al.*, 2002b, Torres *et al.*, 2002c). However, the method requires isotopic labeling at specific sites and is therefore not readily applicable towards many systems (Torres *et al.*, 2000, Torres *et al.*, 2001b, Torres Arkin, 2002).

The most readily available source of information on membrane protein (or on any other protein for that matter) is evolutionary data. Membrane proteins are particular suited for this sort of analysis (Stevens and Arkin, 2001), since the biggest obstacle in sequence comparison between two proteins is the appropriate handling of gap (i.e. insertions). Since gaps occur rarely, if ever, in transmembrane α-helices (or any other secondary structure component) sequence analysis of membrane protein is capable of yielding powerful and discriminative information. Below we describe a recent method that we have developed to make use of such data in an unbiased strategy.

2.1. SILENT SUBSTITUTION ANALYSIS

This method allows the use of evolutionary conservation data made available by genomic sequencing efforts (Briggs *et al.*, 2001). Multiple sequence alignments can determine which residues in the protein have been conserved throughout evolution and which have not. This method focuses on silent mutations, i.e., residues which are not conserved during evolution, or are non-disruptive in mutagenesis experiments. This is based on the premise that such mutations, by definition will not affect the native structure, but may destabilize non-native structures. Multiple possible structures are generated using a global searching molecular dynamics protocol. Independent global searches are carried out using variants which carry silent mutations, or which correspond to different homologous sequences. If enough variants are used, all non-native structures identified will have been destabilized by at least one of the silent mutations. Consequently, only one structure, corresponding to the native structure, will have persisted in all searches. (Figure 2).

3. Modeling Examples

This method has been successfully tested on glycophorin A (Briggs *et al.*, 2001), a protein of known structure (MacKenzie *et al.*, 1997) upon which extensive mutagenesis analyses have been conducted (Lemmon *et al.*, 1992) and evolutionary conservation data is available. For the simulations, the human glycophorin A transmembrane domain consisted of residues 73–91: ITLIIFGVMAGVIGTILLI. Trials were carried out starting from both left and right 25° crossing angles, with the helices rotated 360° about their helical axes in 10° increments. Four trials were carried out from each starting configuration using different initial random velocities, making a total of 288 trials. Clusters of output structures were identified,

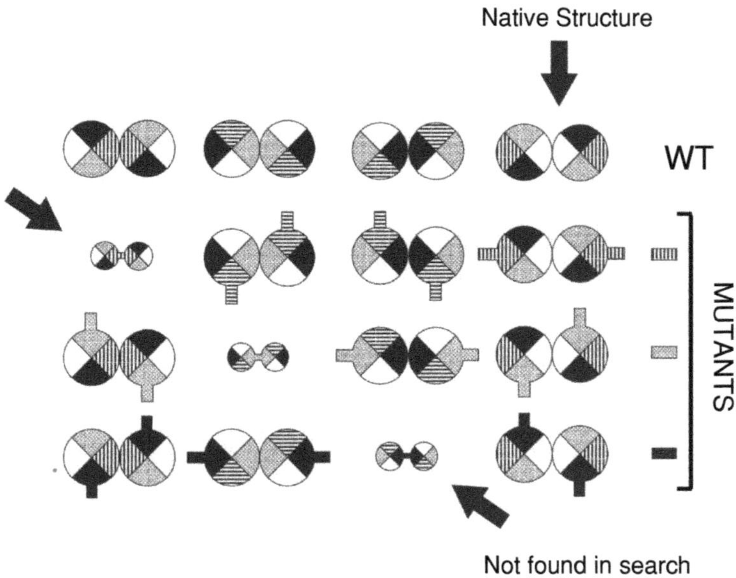

Figure 2. Schematic representation of a transmembrane helical homo-dimer showing the result of a global search protocol that identifies various dimeric structures. The top row is representative of the results obtained for a wild type protein, and the rows below correspond to mutant proteins. The location of the mutation is indicated by a knob, and each helix is dissected into four quadrants to illustrate the different sides of the helix. Note that only one structure, which is therefore taken as the native structure, persists in all variants, i.e., mutants and wild-type (Briggs *et al.*, 2001).

containing 10 or more structures within 1Å RMSD from any other structure within the cluster. Consequently some clusters overlap, and output structures may be members of more than one cluster. The output structures in a cluster were averaged and subjected to a further simulated annealing protocol (as used in the initial search). The cluster average was subsequently taken as the representative of the cluster.

The results of the global search molecular dynamics protocol for human glycophorin A are shown in Figure 3. A total of 6 clusters were identified (six circles). The representative structures of the clusters have energies between −32 and −52 kcal/mol. Based on energy alone it is not possible to reliably determine which cluster or group of clusters is representative of the native structure. Even if the different averaged structures exhibited substantially different energies, it would still be difficult to select the native structure, both because calculations of energy depend on the accuracy and applicability of the force-fields used, and because the simulations are undertaken in vacuo (due to CPU time limitations). We note that repeating the global search molecular dynamics protocol yields the same results.

The global search molecular dynamics protocols for different dimerizing mutants of glycophorin A (Lemmon *et al.*, 1992), produce similar results. The mutants selected were those which exhibited significant levels of dimerization in both the TOXCAT assays (Russ and Engelman, 1999) and SDS-PAGE assays (Lemmon *et al.*, 1992), and which are therefore likely to adopt the same native structure as wild type glycophorin A.

It is the comparison of results between all of these global searches that is most revealing. At only one position, $\Omega = -20°$, $\phi = -80°$ (Figure 3, see rectangles) are clusters found

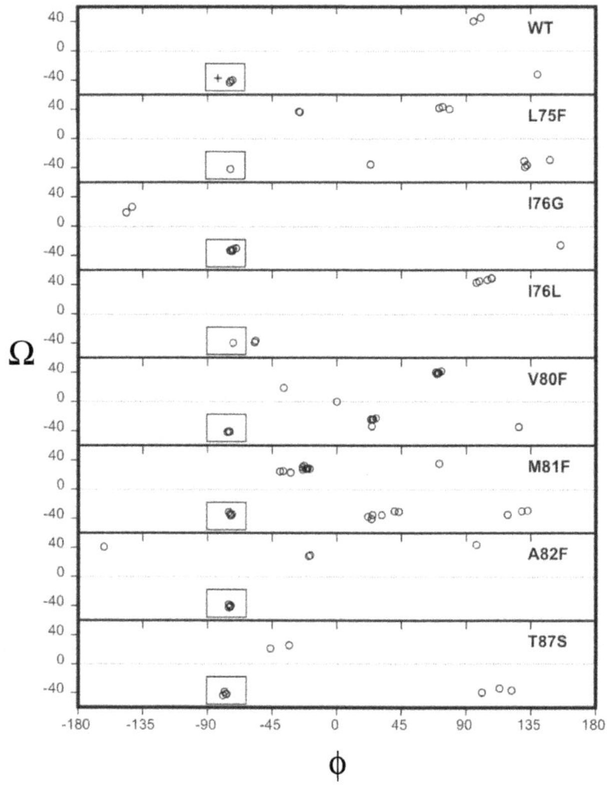

Figure 3. Results of the global search molecular dynamics protocol for the wild type and several mutants of human glycophorin A (L75F, I76L, I76G, V80F, M81F, A82F, and T87) all shown to be non-disruptive towards dimerization (Lemmon *et al.*, 1992). The clusters obtained are indicated in terms of their rotational and inter-helix crossing angle. The rectangle indicates the position at which clusters are identified in all searches.

in all of the eight proteins (wild-type and seven mutants). At this position, a set of clusters can be found that contains representatives from simulations in all variants (a "complete set"). The members of the set differ from all others by less that 0.7Å Cα RMSD. This set incorporates almost all the clusters found within the rectangles. No such set can be defined at any other position even after increasing the allowed RMSD to 3Å Cα RMSD. Essentially, the same final structure is identified in all seven dimerizing mutants and in the wild type protein. The members of this set that were obtained during the wild-type search were averaged and subjected to a simulated annealing protocol. The resulting structure differed from the published structure of glycophorin A (MacKenzie *et al.*, 1997) by an Cα RMSD of <1.0Å

The same approach was then taken using close evolutionary homologues carrying non-disruptive mutations. Searching the OWL data base (Bleasby *et al.*, 1994) for homologues of human glycophorin A identifies a number of sequences. Use was restricted to close homologues (Figure 4) in order to ensure that all sequences modeled represent species with the same structure and function.

```
ITLIIFGVMAGVIGTILSI   gibbon
ITLIIFGVMAGIIGTILFI   gorilla
ITVIILGVMAGIIGIILLL   horse
ITLIIFGVMAGVIGTILLI   human
MILIILGVMAGIIGTILLI   mouse
ITLIVFGVMAGVIGTILLI   orangutan
ITGIIFAVMAGLLLIIFLI   pig
IALIIFGVMAGVIGTILFI   rhesus monkey
```

Figure 4. Sequence alignment of close homologues taken from the OWL data base (Bleasby *et al.*, 1994) of human glycophorin A focussing on the transmembrane domain. Shaded positions are identical in all homologues and represent a consensus.

The results of global searching molecular dynamics protocols for different glycophorin A homologues (Figure 5), again point to a single structure that persists in all instances. This structure is the same as that identified using the mutagenesis data (the rectangles in figures 3 and are positioned at the same coordinates). Again, a complete set can be found where the RMSD between any pair within this set is less than 0.7Å Cα RMSD. No other such set can be defined within 2Å Cα RMSD. Thus, both procedures, using mutagenesis data or evolutionary conservation, point to the same structure that is identical to that obtained from NMR (MacKenzie *et al.*, 1997), which emphasizes the validity and wide applicability of the approach.

4. Conclusions

The results obtained making use of either mutagenesis or evolutionary conservation data, demonstrate that this method can successfully identify the structure of a transmembrane domain while making no assumptions as to where the mutated residues are located. More importantly it can be directly applied to the modeling of polytopic membrane proteins. We note that the concept is applicable to any protein structure prediction in which the selection between several competing models is required.

There are a number of possible extensions of this method. For example, silent mutations could be generated by mutagenizing bacteria that were engineered to require a functional copy of the target protein for survival; virus-encoded membrane proteins could be analyzed using sequences from different virus serotypes; or the consensus structure of families of transmembrane domains could be investigated using sequences from more distant homologues. Low resolution structures available from electron crystallographic studies can be used to position the helices laterally, followed by sampling of different rotational conformations to generate multiple competing models.

Despite the apparent ease in which it is possible to simulate membrane proteins using molecular dynamics, there is one issue that has can potentially raise difficulty: the presence or lack of a lipid bilayer. In the simulations of membrane proteins using molecular dynamics in CHI (Adams *et al.*, 1995) no lipids or solvent molecules are employed, because of the prohibitive computational cost. However it is possible to argue that the most import stabilizing force in any oligomeric bundle will be the interaction between the helices themselves (Torres *et al.*, 2001a). Thus, there is some justification in the simulation procedure

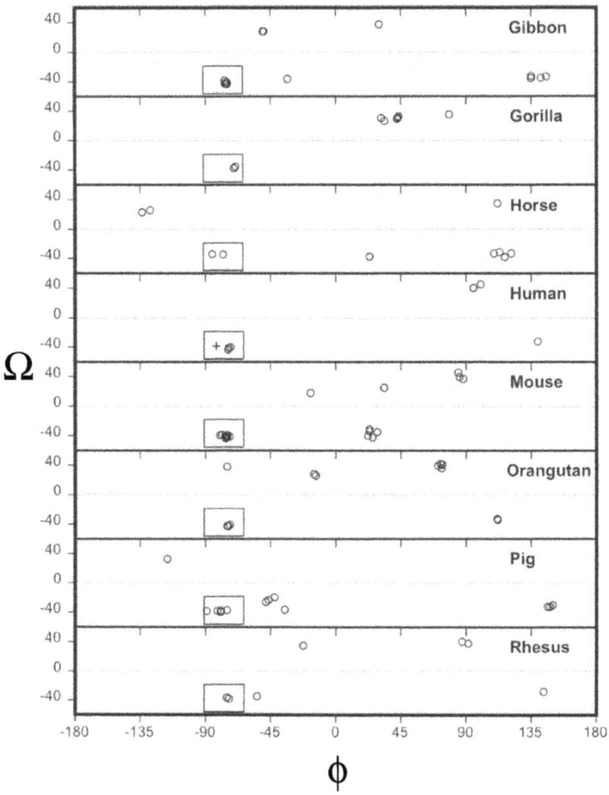

Figure 5. Results of the global search molecular dynamics protocol for close homologues of glycophorin A. The clusters obtained are indicated in terms of their rotational and inter-helix crossing angle. The rectangle indicates the position at which clusters are identified in all searches.

we have described, although the lack of a lipid environment should always be borne in mind.

The wealth of readily available evolutionary conservation data provides potential for the widespread application of this method, not only to membrane proteins using global searching molecular dynamics, but in concert with any technique that generates multiple candidate protein structures.

5. Acknowledgment

This research was supported in part by a grant from the Israel Science Foundation (784/01) to ITA.

6. References

Adams, P. D., Arkin, I. T., Engelman, D. M., and Brunger, A. T. (1995) Computational searching and mutagenesis suggest a structure for the pentameric transmembrane domain of phospholamban. *Nat Struct Biol*, **2(2)**, 154–62.

Arkin, I. T., Adams, P. D., MacKenzie, K. R., Lemmon, M. A., Brunger, A. T., and Engelman, D. M. (1994) Structural organization of the pentameric transmembrane α-helices of phospholamban, a cardiac ion channel. *EMBO J*, **13(20)**, 4757–64.

Baldwin, J. M. (1993) The probable arrangement of the helices in G protein-coupled receptors. *EMBO J*, **12(4)**, 1693–703.

Bleasby, A. J., Akrigg, D., and Attwood, T. K. (1994) OWL—a non-redundant composite protein sequence database. *Nucleic Acids Res*, **22(17)**, 3574–7.

Briggs, J. A., Torres, J., and Arkin, I. T. (2001) A new method to model membrane protein structure based on silent amino acid substitutions. *Proteins*, **44(3)**, 370–5.

Brunger, A. T., Adams, P. D., Clore, G. M., DeLano, W. L., Gros, P., Grosse-Kunstleve, R. W., Jiang, J. S., Kuszewski, J., Nilges, M., Pannu, N. S., Read, R. J., Rice, L. M., Simonson, T., and Warren, G. L. (1998) Crystallography & NMR system: A new software suite for macromolecular structure determination. *Acta Crystallogr D Biol Crystallogr*, **54(Pt 5)**, 905–21.

Capener, C. E., Shrivastava, I. H., Ranatunga, K. M., Forrest, L. R., Smith, G. R., and Sansom, M. S. (2000) Homology modeling and molecular dynamics simulation studies of an inward rectifier potassium channel. *Biophys J*, **78(6)**, 2929–42.

Engelman, D. M., Steitz, T. A., and Goldman, A. (1986) Identifying nonpolar transbilayer helices in amino acid sequences of membrane proteins. *Annu Rev Biophys Biophys Chem*, **15**, 321–53.

Kukol, A., Adams, P. D., Rice, L. M., Brunger, A. T., and Arkin, T. I. (1999) Experimentally based orientational refinement of membrane protein models: A structure for the Influenza A M2 H+ channel. *J Mol Biol*, **286(3)**, 951–62.

Kukol, A. and Arkin, I. T. (1999) vpu transmembrane peptide structure obtained by site-specific fourier transform infrared dichroism and global molecular dynamics searching. *Biophys J*, **77(3)**, 1594–601.

Kukol, A. and Arkin, I. T. (2000) Structure of the influenza C virus CM2 protein transmembrane domain obtained by site-specific infrared dichroism and global molecular dynamics searching. *J Biol Chem*, **275(6)**, 4225–9.

Lemmon, M. A., Flanagan, J. M., Treutlein, H. R., Zhang, J., and Engelman, D. M. (1992) Sequence specificity in the dimerization of transmembrane alpha-helices. *Biochemistry*, **31(51)**, 12719–25.

MacKenzie, K. R., Prestegard, J. H., and Engelman, D. M. (1997) A transmembrane helix dimer: structure and implications. *Science*, **276(5309)**, 131–3.

Muller, G. (2000) Towards 3D structures of G protein-coupled receptors: a multidisciplinary approach. *Curr Med Chem*, **7(9)**, 861–88.

Nunn, R. S., Macke, T. J., Olson, A. J., and Yeager, M. (2001) Transmembrane alpha-helices in the gap junction membrane channel: systematic search of packing models based on the pair potential function. *Microsc Res Tech*, **52(3)**, 344–51.

Radresa, O., Ogata, K., Wodak, S., Ruysschaert, J. M., and Goormaghtigh, E. (2002) Modeling the three-dimensional structure of H+−ATPase of Neurospora crassa. *Eur J Biochem*, **269(21)**, 5246–58.

Russ, W. P. and Engelman, D. M. (1999) TOXCAT: a measure of transmembrane helix association in a biological membrane. *Proc Natl Acad Sci USA*, **96(3)**, 863–8.

Stahlberg, H., Braun, T., de Groot, B., Philippsen, A., Borgnia, M. J., Agre, P., Kuhlbrandt, W., and Engel, A. (2000) The 6.9-A structure of GlpF: a basis for homology modeling of the glycerol channel from Escherichia coli. *J Struct Biol*, **132(2)**, 133–41.

Stevens, T. J. and Arkin, I. T. (2000) Do more complex organisms have a greater proportion of membrane proteins in their genomes? *Proteins*, **39(4)**, 417–20.

Stevens, T. J. and Arkin, I. T. (2001) Substitution rates in α-helical transmembrane proteins. *Protein Sci*, **10(12)**, 2507–17.

Stoilova-McPhie, S., Villoutreix, B. O., Mertens, K., Kemball-Cook, G., and Holzenburg, A. (2002) 3-Dimensional structure of membrane-bound coagulation factor VIII: modeling of the factor VIII heterodimer within a 3-dimensional density map derived by electron crystallography. *Blood*, **99(4)**, 1215–23.

Torres, J., Adams, P. D., and Arkin, I. T. (2000) Use of a new label, $^{13}C{=}^{18}O$, in the determination of a structural model of phospholamban in a lipid bilayer. Spatial restraints resolve the ambiguity arising from interpretations of mutagenesis data. *J Mol Biol*, **300(4)**, 677–85.

Torres, J. and Arkin, I. T. (2002) C-deuterated alanine: a new label to study membrane protein structure using site-specific infrared dichroism. *Biophys J*, **82(2)**, 1068–75.

Torres, J., Briggs, J. A., and Arkin, I. T. (2002a) Contribution of energy values to the analysis of global searching molecular dynamics simulations of transmembrane helical bundles. *Biophys J*, **82(6)**, 3063–71.

Torres, J., Briggs, J. A., and Arkin, I. T. (2002b) Multiple site-specific infrared dichroism of CD3-zeta, a transmembrane helix bundle. *J Mol Biol*, **316(2)**, 365–74.

Torres, J., Briggs, J. A., and Arkin, I. T. (2002c) Convergence of experimental, computational and evolutionary approaches predicts the presence of a tetrameric form for CD3-zeta. *J Mol Biol*, **316(2)**, 375–84.

Torres, J., Kukol, A., and Arkin, I. T. (2000) Use of a single glycine residue to determine the tilt and orientation of a transmembrane helix. A new structural label for infrared spectroscopy. *Biophys J*, **79(6)**, 3139–43.

Torres, J., Kukol, A., and Arkin, I. T. (2001a) Mapping the energy surface of transmembrane helix-helix interactions. *Biophys J*, **81(5)**, 2681–92.

Torres, J., Kukol, A., Goodman, J. M., and Arkin, I. T. (2001b) Site-specific examination of secondary structure and orientation determination in membrane proteins: the peptidic (13)C = (18)O group as a novel infrared probe. *Biopolymers*, **59(6)**, 396–401.

Treutlein, H. R., Lemmon, M. A., Engelman, D. M., and Brunger, A. T. (1992) The glycophorin A transmembrane domain dimer: sequence-specific propensity for a right-handed supercoil of helices. *Biochemistry*, **31(51)**, 12726–32.

Unger, V. M., Hargrave, P. A., Baldwin, J. M., and Schertler, G. F. (1997) Arrangement of rhodopsin transmembrane alpha-helices. *Nature*, **389(6647)**, 203–6.

Biodata of **Ralph Lorenz** author of ***"Zipf, Zipping and Melting Points: Entropy and DNA."***

Dr Ralph Lorenz is a Senior Research Associate in the Lunar and Planetary Lab at the University of Arizona, USA. He has a B. Eng. in Aerospace Systems Engineering from the University of Southampton, UK (1990) and a Ph.D. in Physics (1994) from the University of Kent, UK. Dr. Lorenz worked for the European Space Agency at ESTEC, Noordwijk, The Netherlands on the payload of the Huygens spacecraft destined for Saturn's moon Titan. His interests include astrobiology and planetary climate, through which he began exploring non-equilibrium thermodynamics, and in particular the connections between heat and information, and the conjecture that complex systems may seek a state of Maximum Entropy Production. He is the author of "Lifting Titan's Veil", a popular book on Saturn's largest moon [Published in 200x).

E-mail: **rlorenz@lpl.arizona.edu**

J. Seckbach and E. Rubin (eds.), The New Avenues in Bioinformatics, 69–80.
© 2004 *Kluwer Academic Publishers. Printed in the Netherlands.*

ZIPF, ZIPPING AND MELTING POINTS: ENTROPY AND DNA

R. D. LORENZ
Lunar and Planetary Laboratory, University of Arizona, Tucson, AZ 85721, USA

1. Introduction

This paper discusses some basic statistical properties of the genetic language, linking a number of concepts in information theory and thermodynamics. These properties are discussed for three reasons. First, bioinformatics has progressed to the point where such basic properties may no longer receive the attention they perhaps deserve—many recent texts (e.g. Claverie and Notredame, 2003) do not include the word 'entropy' in their indices, but focus on pattern-matching. Second, speaking from the perspective of an astrobiologist, these statistical properties are likely to be the first tools to be applied to an alien 'message', whether that message is merely a collection of potentially self-organized molecules, a genome of an organism discovered on Mars (whether of terrestrial origin or not), or transmitted radio messages. Finally, they are in themselves interesting as general properties of language and dynamical systems, being essentially independent of the physical or chemical substrate or channel in which the information is encoded.

2. Frequency Analysis

The first property to be discussed is the relative frequency of the different symbols in a message. Typical written english has a characteristic distribution of letter frequencies. The most frequent letter is 'e' with a probability of 12.7%, then 't' at 9.1% closely followed by 'a' and 'o' at 8.2 and 7.5%. At the tail of the league are 'q' and 'z' at 0.1%. (It may be noted that the scores for each letter in the game 'Scrabble'—wherein players form interlocking words by laying letters on a grid—relate to the frequency, with a,o,s,t etc. with score 1, and q,z scoring 10.) Other languages have different letter frequency distributions—Hawaiian, for example, uses fewer letters, with vowels and h,k,l appearing much more frequently than in english.

These letter frequencies offer the key to breaking simple substitution ciphers. (A cipher, strictly speaking, substitutes tokens for letters, whereas a code substitutes whole words.) By studying the frequencies of the letters or symbols in the coded message, and knowing the relative frequencies of the letters in the language (taking into account not only the language itself, but also the specialized terms that might be used in, for example a Naval cipher), one can make an educated guess of what the substitutions are. The redundancy does the rest—like a crossword puzzle, once a large enough part of the cipher is broken, the other letters fall into place.

This technique of codebreaking is attributed to the Arab Al-Kindi in the ninth century, and became significant in Renaissance Europe. Unfortunately for the Kingdom of Spain, Spanish cryptographers were the last to realise how this technique rendered simple substitution ciphers vulnerable (e.g. Singh, 1999). French cryptographers read Spanish communications with such ease that King Philip II petitioned the Vatican to intervene, believing the French to be in league with the devil!

The assembly instructions for living things are written in a language of four characters. Each character is a purine or pyrimidine base: Guanine, Cytosine, Adenine and Thymine, G, C A and T for short. Long strings of these characters are held on a backbone of ribose sugar—the assembly named Deoxyribose Nucleic Acid, or DNA.

Before the technology to permit sequencing was developed, relative frequency information was available from bulk chemical analysis. This information—Chargaff's Rules—proved essential in resolving the genetic code and the structure of DNA. The fact that the A and T appeared with a virtually equal amount, but significantly larger than G & C suggested a pairing, which Crick and Watson suggested might be the mechanism for DNA replication. If nature were more efficient, with all bases occurring with equal probability, the A-T and G-C pairing clue would not have been obvious, and the structure of DNA—a double helix with two ribose backbones, linked across the middle like a twisted ladder by hydrogen bonds between pairs of bases—might have not been discovered so early. It may be noted that while the proportions in human, wheat or yeast DNA are roughly 30% A & T, 20% G & C, for the bacterium E. coli the proportions are much more nearly equal—about 24% A & T and 26% G & C.

In a gene (i.e. a DNA sequence that encodes information for the construction of proteins in a living organism), the bases are grouped in sets of three, called codons. In a set of three characters, each with 4 possibilities, there are 4^3 or 64 possibilities. These 64 instructions translate into 61 results, expressed as 61 different amino acids to be inserted into the protein for which the gene is coding. The three remaining possibilities are used to terminate the sequence (just as a clear email message will have spaces as well as letters). These codons are the more useful basic unit of genetic information.

Just as the individual bases have certain probabilities of appearing, so too do codons. Before discussing probabilities further, we must first introduce more formal measures of information content.

3. Measures of Information—Signal Entropy and Kolmogorov Complexity

The formalization of information content was accomplished by the engineer Claude Shannon, in an attempt to understand the information-carrying capacity of a noisy telephone line (Shannon, 1948; Shannon and Weaver, 1949). His breakthrough was to consider the message (or, for that matter, every symbol in a message) as simply the correct choice between all possible messages. The rate at which the encoded message can be transmitted depends only on the transmission link itself, whether the message is voice telephony, telegraph messages or even television signals (while the principles are applicable generally, they are most easily understood if the message is encoded digitally as a string of binary digits, i.e. successive choices of 1 or 0.)

One aspect of the message was the predictability of the symbols, or redundancy. Shannon called the complement to this property, a measure of the efficiency of the message, the 'entropy' of the message, recognizing a similarity between his information measure and the entropy defined by Boltzmann. The entropy of a message has the formal definition $S = -\Sigma(p_i \log_2 p_i)$ where p_i is the probability of the symbol i.

It can be seen, considering as an example a coin-toss, or choice between symbols 0 and 1, that the entropy is maximized if the probability of the two symbols appearing (which must sum to 1) are equal. The heart of the 'maximum entropy' technique of resolving unconstrained problems is that, in the absence of other information, the solution with maximum entropy as defined above is the correct one. If we do not know that a coin is biased, we should assume the maximum entropy possibility, that it is fair. 'MaxEnt' techniques for deconvolution of chromatograms, astronomical images etc. are simply formalizations of this approach.

Since, as noted earlier, the probabilities of letters in human written languages are not equal, it follows that language is not an efficiently-coded scheme of conveying information. Since a choice among 26 letters requires about 4.5 bits ($2\hat{\ }4.5 = 27$), the entropy of a string of a random (or efficiently-coded) letters would be about 4.5 bits per symbol, but the entropy of written english is only about 2 bits per symbol. The entropy of some other languages (notably Hawaiian) is rather lower. The redundancy present in the distribution of letters makes possible not only code-breaking as mentioned earlier, and data compression which will be discussed in a later section, but also—as Shannon noted—crossword puzzles.

Other entropies can be defined than the simple unconditional probabilities of each letter; the probabilities of 2- or 3-letter combinations, or conditional probabilities (e.g. Ewens and Grant, 2001) such as the probability a letter appearing after a specified one (e.g. since U after Q in english is almost inevitable, the conditional entropy of letters after Q is very low, and the information content is small—one could delete every u following q in a text and its readability would be scarcely impaired.)

Similarly, the information content of DNA depends both on the information measure applied, and on the sequence itself. In particular, the entropy (as a reciprocal measure of compressibility of the message) of long sequences falls compared to short sequences. If genetic information were encoded in the most efficient way possible, there would be two bits per base, regardless of sequence length, since each base encodes one of 4 possibilities. In reality, studies of sequences yield entropies of around 1.9 bits per base. This indicates some slight order or redundancy, although only a little more than might be assumed from the slightly nonuniform frequencies of the four bases (which indicate around 1.95 bits per base.) However, some studies (Loewenstern and Yianilos, 1999) indicate that long sequences have entropies closer to 1.6 bits per base pair.

What this means is that there is long-range redundancy in the sequence—not repeated words or sentences so much as pages or chapters. In this regard, the genetic code looks much like a computer program written by a scientist. A scientist building a program will often try one thing, see that it sort-of works, but then wants to try something a little better. But in case that doesnt work out, he or she will often 'comment out' the old algorithm, instructing the computer to skip over the old instructions, rather than deleting them altogether. This approach leads to longer, messier programs, but by removing the commenting the old instructions can be reinstated if the new version doesnt offer any improvement. Another tendency is to 'cut and paste' chunks of code from one part of the program into another.

Professional programmers are perhaps more likely to have tidier code, deleting extraneous instructions rather than commenting them out (since professionals keep meticulous documentation, it is easy to recover the code from an archived version should it be needed again.) Another good programming habit is to modularize—if some sequence of instructions is to be used on a number of occasions, a good programmer will put them into a subroutine, a self-contained program that can be called from the main program. Only the call to the subroutine need be repeated in the main program, not the whole content.

Replacing instructions in this way is exactly like compressing a message, and increasing its entropy. Good programmers produce high-entropy code. Programmers may sometimes seek the most compact way of writing a program with a specified goal, purely for the challenge.

Another measure of information content is the Kolmogorov complexity, essentially the size of the smallest Turing machine (or crudely, the shortest computer program) that can reproduce the message (e.g. Li and Vitany, 1997). While fundamental, and related to the other measures described later, this specific description of information content is perhaps not very useful in bioinformatics.

Some codes constructed for human use seek higher entropies than would appear from simple coding, examples include morse code (with shorter symbol strings such as dot-dash for frequently-used letters such as 'e') and the Videoplus codes used to program video tape recorders, with shorter codes for more popular programmes. But, just as living things more complex than a wristwatch evolved without a watchmaker, the pursuit of efficiency in natural systems such as language also tends to introduce some compression. Examples abound, e.g. abbreviations such as 'e.g.', and jargon. The mobile telecommunications revolution and text messaging [where, unlike in a typewritten letter, there is a need for speed, and unlike verbal communication, there is a specific cost per letter, rather than per syllable] has similarly prompted the emergence of compact idioms such as 'C U L8ER' (see you later.) We might therefore expect more evolved biological messages to be less redundant, and have higher entropies. One can note, however, that the application of such compression, while reducing the information storage and transmission requirements, needs a more sophisticated processor—only mature language-users can act this way. Languages without a certain degree of redundancy would be impossible to learn.

A formal redundancy can be defined for words ('n-tuples') of a length n made from m symbols. The 'n entropy' of a text is $H(n) = -\Sigma(f_i \log_2 f_i)$ where the sum is over $i = 0 \ldots m^n$ combinations of symbols and the redundancy $R(n) = 1 - H(n)/kn$ with $k = \log_2 4 = 2$.

Mantegna *et al.* (1994) found that noncoding sequences from Yeast and C. Elegans had higher redundancy than coding sequences from those species. For C. Elegans, the noncoding redundancy increased from 4 to 6% as n varied from 1 to 6, while coding sequences had only 1–3% redundancy over the same range. Noncoding Yeast DNA was similarly more redundant (3.5% vs 1.5%) than coding DNA, although with less sensitivity to 'word length' n.

The implications are not altogether clear, although would be consistent with evolutionary pressure to compress coding DNA, with no penalty for accumulating redundancy (perhaps in a sense a history of unapproved drafts and transcription errors) in noncoding regions. Long repeats may be interesting to study in this regard.

4. Zipf's Law

In the last 20 years, an underlying rationale for a widespread statistical property in nature has been discovered. The rationale is termed 'Self-Organized Criticality'. Real-world effects as diverse as earthquakes, stock prices and wildfires—as well as simple mathematical models of them—termed cellular automata since simple rules are applied to a matrix of cells. An emergent property of systems appears to be that disturbances follow a $1/f$ kind of distribution, of the form $F = aR^k$ where F is the frequency of occurrence, a is a constant, R is the rank and k an exponent we will refer to as the Zipf exponent.

One of the earliest discussions of $1/f$ statistics is by Zipf (1949). This book notes that the frequency of occurrence of a word relates inversely to its rank—in other words, the hundredth most popular word occurs ten times less often than the tenth most popular word, the thousandth, one hundred times less often, and so on. It seems an innocent enough idea, but has some very interesting implications. For example, one neat corollary of Zipf's Law is that in a sequence of text which strictly observes the $1/f$ function, the total vocabulary (i.e. the number of different words) equals the number of occurrences of the most frequent word.

Zipf explores an impressive array of statistics in a similar way. One striking set of data is that of city size. In most nations, a logarithmic plot of city size against rank has the familiar -1 slope. However, some empires—and Zipf argues that this is somehow a characteristic of instability and imminent failure/revolution—do not conform to this pattern. In particular, some empires grow such that the capital city jumps far above the $1/N$ curve—there are no real second or third cities comparable with the capital, only a set of diminutive satellite cities.

More pertinent to the present discussion, however, Zipf also notes that there are significant deviations from the $1/f$ pattern in language. While the vocabularies of most english speakers do follow a $1/f$ curve, significant deviations occur in schizophrenics or those with other related disorders, where one or a number of words or phrases is repeated out of all proportion to normal discourse.

Mandelbrot (1961), and others, have made the pertinent observation that a completely random sequence of letters (and considering a space as a letter) will contain a vocabulary of words that have an inverse rank-frequency distribution (i.e. a Zipf exponent of 1.) However, this is not the case if fixed-length n-tuples are used, such that the n-tuples may include a space at any point.

Mantegna *et al.* (1994) performed a Zipf analysis of DNA sequences using n-tuples of bases. A purely random sequence of symbols would have a rank-frequency distribution that is flat, i.e. a Zipf exponent of zero. English texts analyzed this way have a Zipf exponent of about 0.57, rather than the more classical value around unity.

Mantegna *et al.* found that coding DNA sequences had consistently lower Zipf exponents (around 0.2) than noncoding regions (0.3 to 0.56). In other words, noncoding regions resemble human language with some symbol sequences repeated more often than others, consistent with the entropy measures described above. Coding sequences are 'stripped down', with a more efficient structure. Notably, the lowest exponents of all (0.158) were found in bacteriophages, arguably minimalist living structures. Interestingly, they also computed a Zipf exponent for a piece of computer code (specifically a 9MB Unix binary executable), and found an exponent of 0.77—consistent with rather sloppy, redundant coding!

A note of caution is in order, however. Martindale and Konopka (1996) have pointed out that in fact a Yule distribution $F = aR^k b^R$ (which, when $b = 1$ reduces to Zipf's law) is in fact a better fit to oligonucleotide sequences than Zipf's law. This is to be expected in the sense that an additional parameter should improve the fit, but they found that there was no correlation of the quality of fit with whether a sequence is coding or noncoding. While their arguments about the Yule fit are correct, it should also be pointed out that their Zipf 'fits' are better for noncoding regions (introns) than for exons, as Mantegna *et al.* originally noted, and thus Zipf analysis may still have utility.

5. Data Compression—Entropy Again

The 'zip' program used on many computers to compress files such that they use less disk space uses a simple but powerful algorithm whereby strings that are repeated in a text are instead encoded with a number that points to them. The algorithm, termed LZW after its inventors Ziv-Lempel-Welch (Ziv and Lempel, 1978; Welch, 1984) is widely used in a variety of data compression utilities (e.g. Unix compress or gzip) and image file formats (compressed TIFF) but for convenience we will refer hereafter to LZW compression by its most popular incarnation (on DOS and Windows systems) as 'zipping'.

The ratio of the length of a zipped file divided by the length of the raw file is not exactly equal to the entropy (defined in the Shannon sense) of the raw file, but is close to it. The longer the raw file, the better this approximation holds, and without writing a dedicated program to calculate a file's entropy, the zipping technique is an excellent compromise between convenience and accuracy.

Clearly, it is possible to recognize such strings of characters in text written in a language like english—individual words are the most obvious examples. Frequently-occurring words like 'the' would be encoded with the shortest pointer.

Now, if an algorithm attempts to compress some string of unknown text, by using a library of tokens generated from a base text, then it will be more effective if the target text contains many of the same tokens (i.e the same words) as the base text. If the target and base text are written in the same language, for example, it is likely that the target text can be compressed effectively and will have an entropy similar to that of the base text.

This effect was realised or discovered by Benedetto *et al.* (2002) who investigated the zipped file lengths of test texts appended to standard texts. In the standard text (for which they found an adequate length was 32–64 kilobytes, say 10,000 words) the zipper would 'learn' the vocabulary of words and phrases associated with it, and the test file could be from 1–15 kilobytes (a few hundred to a few thousand words) without affecting the results.

They defined a relative entropy as follows—beginning with two long files A and B, with a short file b (from the same source as B). File b was appended to the two files A and B and the zipped file length D_{Ab} and D_{Bb} measured. The relative entropy is simply $(D_{Ab} - D_{Bb})/|b|$, where $|b|$ is the length of the raw file b.

Not only did they find that a low relative entropy was a good indicator that b and A were in the same language (and even with as small a sample b of 20 characters, the algorithm picks out the correct one of ten possible languages) but the relative entropy indicated the correct author of texts in the same language in 93% of cases.

So, this one quantity, the relative entropy (really a normalized incremental entropy), is a powerful measure of sameness, or origin. It is likely that this mutual entropy technique will find application in genetics.

6. Melting Points—Entropy Once More

Just as entropy had an original thermodynamical definition, before information theory even existed, it is worth pointing out another thermodynamic connection with bioinformatics.

Our DNA differs by only 1.6% from that of chimpanzees Diamond (1991). This amount was known before the 1990s technologies of gene sequencing became developed—it turns out that there is a simple, if delicate, way of estimating the difference in structure between two sets of DNA. A pure sample of DNA from a single organism will have, like any other pure material, a well-defined and sharp melting point. However, a mix of two kinds of DNA will have a broader melting point, lower than either of the pure samples themselves. (In fact, the laboratory technique used—where DNA molecules are hybridized with radioisotope labelled tracer DNA to determine the number of single strands eluted from the hybrid duplexes at different temperatures—is rather more elaborate than simple melting, but of course follows similar kinetics, see e.g. Sibley and Ahlquist, 1984.)

A powerful technique in investigating melting behaviour is called Differential Scanning Calorimetry (usually abbreviated to DSC—e.g. Höhne *et al.*, 1996) in which two small pans that could be heated at a controlled rate, while monitoring the heat required to heat the sample pan and an empty control at that rate. The difference between the sample pan and control allows the isolation of the heat needed to warm the sample, and any exothermic or endothermic reactions or phase transitions that occur in it. A common application is to investigate pharmaceuticals. For example, many drugs will degrade over time, decomposing under the action of sunlight or air, and it is important to understand how fast this happens, because not only is the effectiveness of the original drug reduced as the dose per pill decreases, but sometimes the degradation products can be harmful. Rather than making a laborious chemical analysis at intervals to assay the pure drug content, a more convenient diagnostic technique is to check the melting behaviour in a DSC. A smudging and lowering of the melting point is an indicator that the sample is no longer pure—that some of the drug has turned into something else.

As an astrobiological side-note, although typical laboratory DSC's are desk-sized instruments and require careful operation and sample handling, a robust, miniaturized instrument was sent to the planet Mars on the Mars Polar Lander (MPL) mission, which was sadly lost on arrival at Mars in 1999. It carried a 6kg soil analyzer called TEGA (Thermal and Evolved Gas Analyzer) to measure the amount of ice in the soil, and detect the presence of water-bearing minerals and carbonates in the soil by DSC and evolved gas analysis.

We can see an analogy between this melting-point technique and the zipping technique. If we add one chemical sample to a known one, we lower the melting point if the unknown is the different from the known one. Similarly, the entropy of the combination of an unknown text with a known reference increases if the unknown is different from the reference.

7. Zipping Back....

Let us now return to the Benedetto paper about zipping. This paper didnt stop at demonstrating the success of a relative entropy measure using zipped file length at identifying authors. It also explored the relative entropy of the same piece of text (in fact, the Universal Declaration of Human Rights), written in different languages. The entropy measures for different pairs of languages could be compared, and (using the same techniques used by geneticists) a 'family tree' of languages built up.

Remarkably, the tree constructed (automatically) in this way agrees almost exactly with the tree painstakingly constructed by linguists from historical and other evidence. There are clear branching families of languages, like Germanic, Romance, Celtic and Slavic. Oddballs like Basque (which unlike virtually all the others is not an Indo-European language) stand out in the tree, unconnected to the others.

We may expect that mutual entropies of this sort may be useful in identifying the heritage of unknown sequences and constructing phylogenetic trees.

An alternative application of genomic entropy measures is the segmentation of sequences into coding and noncoding regions by an algorithm developed by Bernaola-Galvan *et al.* (2000). Here a sequence is split into two parts, and a relative frequency vector for each of the parts is calculated. The vector F_{lj} is of 12 quantities, namely the relative number of occurence of each of the 4 bases ($l = A, T, C, G$) in each of 3 positions ($j = 0, 1$ or 2 where $j = i \bmod 3$ and i is the position of the base in the sequence). From the relative frequency vectors F_1 and F_2 for the two sequences of length n_1 and n_2, a parameter named the Jensen-Shannon divergence is constructed, defined as

$$C(F_1, F_2) = 2\log_2[NH(F) - n_1H(F_1) - n_2H(F_2)]$$

where $N = n_1 + n_2$, $F = (n_1/N)F_1 + (n_2/N)F_2$ and H(F) is the Shannon entropy given by $H(F) = -\Sigma(f_{lj} \log_2 f_{lj})$. $C(F_1, F_2)$ is relatively insensitive to the length of the sequences being compared.

Experiments with genes of E. coli show that this parameter C reaches a maximum at the boundary between coding regions, or between a coding region and a noncoding region. An unknown sequence can be broken down into discrete coding and noncoding regions by recursively splitting the sequence at positions where C is maximized, the splitting being repeated until the statistical significance of the maximum value of C is less than some specified threshold.

8. Information Density and Energy Costs

The presence of fans on modern computers attests to the thermal dissipation associated with information processing. Dissipated powers of 60W are not uncommon for GHz processors: the increase in heat dissipation compared with early microcomputers is due to the higher throughput—the energy cost in terms of J/bit has decreased due to technological improvements in chip design, but this is offset by the orders of magnitude increased processing rate.

Although reversible computing operations in principle require no energy, reading and writing (and erasing) data from storage in real physical systems requires the expenditure of energy. The implied processing performance above of $\sim 10^{-8}$ J/bit or $\sim 10^{12}$ kT is far below the theoretical limit of kT ln 2, or about 4×10^{-21} J/bit at room temperature. We can see then that digital computers are pitifully inefficient. The firing of a neuron (with some indeterminate associated number of bits) requires about 10^{11}kT. The thermodynamic limit is only remotely approached by molecular machines such as that associated with DNA replication. There the energy cost is between 20 and 100 kT per base pair—just an order of magnitude or so above the thermodynamic limit.

As a thought experiment (Morisson, 1964) to put these numbers in perspective, consider a message written as a series of black and white patches, like a square bar code. Conceptually, this could comprise a monomolecular layer of black material on a white background; if the patches (bits) were 1mm across, they would require $\sim 10^{14}$ atoms, or $\sim 10^{-10}$ moles. The energy required for their deposition or removal would relate to a typical latent heat of vaporization, or reaction (e.g. with oxygen). Taking this as $\sim 10^{4}$ J/mole, this message takes $\sim 10^{-6}$ J/bit. Taking a cell size of ~ 1 micron, comparable with the state of the art of microelectronics yields 10^{-12} J/bit, a few orders of magnitude better than that actually realized. Only as the molecular scale is approached, say with cell sizes of 1 Ångstrom (10^{-10}m), does the energy associated with the colored patch fall to ~10–20 J/bit or a few times kT. Equivalently, the thermal entropy associated with the physical or chemical read-write process becomes comparable with the information entropy only for microscopic systems.

9. Conclusion

We have introduced and reviewed some statistical properties of symbol sequences, both in human languages and in DNA. Various entropic measures can be defined, with interesting parallels in thermophysics, and these can be used to quantify sequence similarities and heritage, and to identify coding from non-coding regions. Another, less formal, property is Zipf's law, where recurring sequences have an inverse rank-frequency distribution, or one related to it (a Yule distribution). Coding regions appear to contain less redundancy (i.e. they have higher entropies) than noncoding regions. We have also noted that information storage at the molecular level, as implemented in DNA, closely approaches the thermodynamic limits on information density.

10. References

Benedetto, D., Caglioti, E. and Loreto, V. (2002) Language Trees and Zipping, Physical Review Letters **88** 048702-1 to 048702-4.

Bernaola-Galvan, P. Grosse I., Carpena, P., Oliver, J., Roman-Roldan R. and H. E. Stanley (2000) Finding Borders between Coding and Noncoding DNA Regions by an Entropic Segmentation Method, Physical Review Letters **85**, 1342–1345.

Boynton, W. V., Bailey, S. H., Hamara, D. K., Williams, M. S., Bode, R. C., Fitzgibbon, M. R., Ko, W., Ward, M. G., Sridhar, K. R., Blanchard, J. A., Lorenz, R. D., May, R. D., Paige, D. A., Pathare, A. V., Kring, D. A., Leshin, L. A., Ming, D. W., Zent, A. P., Golden, D. C., Kerry, K. E., Lauer, H. V. Jr. and Quinn, R. C. (2001), Thermal and Evolved Gas Analyzer: Part of the Mars Volatile and Climate Surveyor integrated payload Journal of Geophysical Research, **106**, 17, 683–17, 698.

Claverie, J.-M. and Notredame, C. (2003) *Bioinformatics for Dummies*, Wiley.

Diamond, J. (1992) *The Third Chimpanzee*, Harpercollins, New York.

Ewens, W. J. and Grant, G. R. (2001), *Statistical Methods in Bioinformatics*, Springer, New York.

Höhne, G., Hemminger, W. and Flammersheim, H.-J. (1996) *Differential Scanning Calorimetry*, Springer, Heidelberg, 1996.
Li, M. and Vitanyi, P. (1997) An Introduction to Kolmogorov Complexity and its Applications, Springer, New York.
Loewenstern, D. and Yianilos, P. (1999) Significantly Lower Entropy Estimates for Natural DNA Sequences, Journal of Computational Biology, 6, 125–133.
Mandelbrot, B. (1961). On the theory of word frequencies and on related Markovian models of discourse. In *Structure of Language and its Mathematical Aspects*, vol. XII, *Proc. Symposia in Applied Mathematics*, 190–219. American Mathematical Society, Providence, RI.
Mantegna, R. N., Buldyrev S. V., Goldberger, A. L., Havlin, S., Peng, C.-K., Simons, M. and Stanley, H. E. (1994) Linguistic Features of Noncoding DNA Sequences, Physical Review Letters, **73**, 3169–3172.
Martindale, C. and A. Konopra (1996) Oligonucleotide Frequencies in DNA follow a Yule Distribution, Computers and Chemistry, **20**, 35–38.
Morrison, P. (1964) A Thermodynamic Characterization of Self-Reproduction, Reviews of Modern Physics, **36**, 517–524.
Shannon, C. (1948) A mathematical theory of communication,' Bell System Technical Journal, **27**, 379–423 and 623–656.
Shannon, C. and Weaver, W. (1949) *The Mathematical Theory of Communication*, University of Illinois Press, Urbana.
Sibley, C. G. and Ahlquist, J. E. (1984) The Phylogeny of the Hominoid Primates, as Indicated by DNA-DNA Hybridization, Journal of Molecular Evolution, **20**, 2–15.
Singh, S. (1999) *The Code Book*, Fourth Estate, London.
Welch, T. A. (1984) A Technique for High-Performance Data Compression, Computer, **17**, 8–19.
Zipf, G. K. (1949) *Human Nature and the Principle of Least Effort*, Addison-Wesley, New York.
Ziv, J. and Lempel, A. (1978) Compression of Individual Sequences via Variable-Rate Coding, IEEE Transactions on Information Theory, **24**, 530–536.

Biodata of **Lajos Bencze** coauthor (with author Gyula Palyi, Claudia Zucchi and I. Caglioti) of the chapter ***"Biological Chirality: A Tool of Information, In Vivo and In Vitro."***

Professor Lajos Bencze is the head of the Müller (Computational Chemistry) Laboratory at the University of Veszprém (Hungry). He studied engineering, in 1964 completed his PhD and joined the department of Organic Chemistry of the University of Veszprém. In 1978 he went to the Queen's University of Belfast (UK) for one year as a visiting professor. He received the Doctor of the Chemical Sciences award (Hungarian Academy of Sciences) for his research on the activation mechanism of non-carbenoid metathesis catalysts. He has founded the five-year university major course of Chemical Informatics (Computation Chemistry) in 2001 at the University of Veszprém. His research interests are homogeneous catalysis, organometallic chemistry, design and modelling of nanomechanical devices.

Biodata of **Luciano Caglioti**, coauthor (with author Gyula Palyi, C. Zucchi and L. Bencze) of the chapter ***"Biological Chirality: A Tool of Information, In Vivo and In Vitro."***

Professor Luciano Caglioti obtained his "diploma" in Chemistry in 1956. From, 57 he spent 3 years at the ETH in Zurich and then he worked at the Institute of Chemistry at Polytechnic of Milan, directed by Prof. A. Quilico. In 1965 he became full professor in Chemistry of Natural Substances at the University of Camerino and in 1968 he moved to University of Bologna and later (in 1971) to Rome ("La Sapienza"). He is author of more than 120 scientific publications, well-known in popularising sciences with several publications in this field (he cooperates with the main Italian newspapers, reviews and encyclopaedias) 8 industrial patent, 3 didactic books and editor of 3 books. His research activity is linked to two different fields: the chemistry of the natural substances and new reactions in organic chemistry. He is in charge by the Italian Government regarding fine chemicals, pharmaceutical chemistry, energy, environment, cultural goods, etc. From 1997 he is member of the Scientific Secretary of the President of the National Research Council (CNR) Member of Scientific Academies of New York, honorary member of the Academy of Sciences of Hungary and Member of the "Accademia dei XL", he has received gold medals of: Public Instruction Ministry (for Science, Culture and Art), Società Chimica Italiana (1984) Divisione di Chimica Industriale della Società Italiana (1988), Premio speciale Federchimica "Per un futuro intelligente" (1989) and prizes of: Council Presidency (for Culture), Glaxo-CEE (for scientific dissemination, 1980), Fregene (for scientific dissemination, 1983) Laurea Honoris Causa from Polytechnic of Budapest (1987) and from the University of Urbino (1988). Designated by the President of Italy "Grande Ufficiale al merito della Repubblica" (1999).

J. Seckbach and E. Rubin (eds.), The New Avenues in Bioinformatics, 81–96.
© 2004 *Kluwer Academic Publishers. Printed in the Netherlands.*

Biodata of **Gyula Palyi** author (with coauthors: C. Zucchi, L. Bencze and I. Caglioti) of ***"Biological Chirality: A Tool of Information, In Vivo and In Vitro."***

Professor **Gyula Pályi** is since 1987 a full professor of Inorganic Chemistry at the University of Modena and Reggio Emilia (Italy). He obtained his PhD at the Technical University of Budapest, then PhD at the University of Pisa (Italy), then DSc from the Hungarian Academy of Sciences. He started his carrier in (organic chemical) industry, became full professor of Organic Chemistry at the University of Veszprém (Hungary) in 1980. His research activity started with detergent chemistry, continued with organometallic catalysis and cluster chemistry. Beyond the latter two, actually he is involved also in problems of intramolecular transfer of chirality and (chemical aspects of) origins of life. Author of about 180 publications, Editor of 3 books. Prof. Palyi is a member of Scientific Academies of Bologna (1978), Catania (1982), New York (1994), Italian National (Modena, 2000), European (Paris, 2001).

E-mail: **palyi@unimo.it**

Biodata of **Claudia Zucchi**, coauthor (with author Gyula Palyi, L. Bencze and I. Caglioti) of the chapter "***Biological Chirality: A Tool of Information, In Vivo and In Vitro***."

Dr. Claudia Zucchi works since 2001 at the Department of Chemistry of the University of Modena and Reggio Emilia. She obtained her PhD at the University of Modena in 1993 and wined a postdoctoral fellowship (from the Ministry of University and Research, MURST) at the same University from 1994 to 1996. Her research activity started with organometallic catalysis and cluster chemistry. Dr. Zucchi is involved also in problems of intramolecular transfer of chirality and chemical aspects of origins of life. She was the secretary of the (first) Interdisciplinary Symposium on Biological Homochirality (Serramazzoni, MO, Italy in September 6–12, 1998). Member of the Organizing Committee of the Symposium on Biological Chirality (Szeged, Hungary, August 27–31, 2000); secretary of the Workshop on Life (Modena, Italy, September 3–8, 2000); secretary of the 3rd Interdisciplinary Symposium on Biological Chirality (Modena, Italy, April 30–May 4, 2003). Dr. Zucchi is a member of the *International Advisory Board* for Symposia on Biological Chirality since 1998. She is an author of 52 publications and editor of 3 books.

E-mail: **zucchi@unimore.it**

Luciano Caglioti

Gyula Pályi

Claudia Zucchi

BIOLOGICAL CHIRALITY
A Tool of Information, In Vivo and In Vitro

G. PÁLYI[1], C. ZUCCHI[1], L. BENCZE[2] and L. CAGLIOTI[3]
[1]Department of Chemistry, University of Modena and Reggio Emilia, Via Campi, 183, I-41100 Modena, Italy; [2]Müller Laboratory, Institute of Organic Chemistry, University of Veszprem, Egyetem u. 6, H-8200 Veszprem, Hungary; [3]Department of Chemistry and Technology of Biologically Active Compounds, University "La Sapienza" of Rome, P.le A. Moro, 5, I-00185 Roma, Italy

1. Preliminary Statement

The deterministic view of Natural Sciences requires causality chains behind every phenomenon. In other words, this means that the Universe is "saturated" with information. This information and its expressions as physical, chemical or biological phenomena are hierarchically distributed over various levels of size, time and energy. The present paper deals with molecular-level systems bearing importance for biology, that is: for living organisms.

Another limitation of our analysis is the stereochemistry (the 3-dimensional, 3D, shape of our subjects): only chiral (and prochiral) molecular systems will be discussed from viewpoints of structure, reactivity and kinetics.

2. Chirality

Chirality, as a concept of (molecular) geometry, means that the object (molecule) in question can not be brought in superposition with its specular image (Janoschek, 1991; Mezey, 1991). Chirality as a molecular feature, was recognized and studied by Pasteur (1848a; 1848b) in the first half of the XIXth century. Pasteur used the word *dissymmetry* (Pasteur, 1922), the term *chirality* was coined later by Lord Kelvin (1904) (from the Greek "hand"). A broader, mathematical approach views chirality as a feature of functions showing specific behaviour against some transformations (Feynman *et al.*, 1964).

More recently, the holographic theory of density functions (Mezey, 1999a) enabled tracing molecular (geometric) chirality down to very small (almost infinitely small) sections of atomic/molecular (orbital) electron density functions (Mezey, 1999b; 1999c) This approach was proved to be particularly useful in theoretical analysis of biomolecules (Mezey, 2002).

There are two fundamental types of material (molecular) manifestation of chirality (Janoschek, 1991; Mezey, 1991; Dodziuk, 1995): (a) *configurational* and (b) *conformational*. The former is derived from the relative spatial distribution of atoms and/or groups

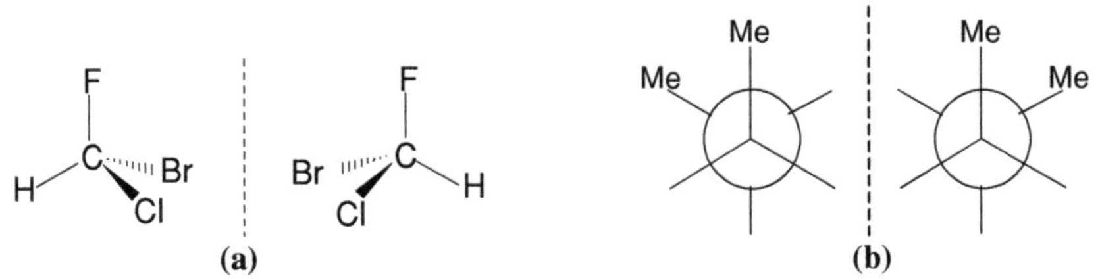

Figure 1. Examples of (a) configurational and (b) conformational chirality.

linked directly to each other by chemical bonds. The latter is derived from the geometric form (shape) of the molecule itself (Figure 1). A very "practical" consequence of this distribution is, that the geometry of configurationally chiral molecules is generally more stable (its change needs the transformation of regular chemical bonds), while the conformational chirality is (usually) derived from intramolecular rotations around single bonds and consequently these chiral "rotamers" could be transformed (interconverted) much more easily. The activation energies of changes in the configuration are usually in the order of 10^2 kJ/mol, while changes in conformations are normally 1–2 orders of magnitude easier. In spite of this energy difference (sometimes just therefore), chiral conformations are of fundamental importance in biochemistry (Voet and Voet, 1990; Elliot and Elliot, 1997).

3. Chirality, Information and Entropy

Thermodynamic *entropy* (Kondepudi and Prigogine, 1998), deduced from particle statistics, was proposed by Boltzmann around the end of the XIXth century. A half century later Claude E. Shannon developed a mathematical analysis of signal transmission, essentially on the basis of probability theory (Shannon and Weaver, 1949). Shannon obtained a function of a probability variable, which has a very similar mathematical form to that of Boltzmann's formula for entropy. Shannon followed the advice of J. von Neumann and he started to call this function **informational entropy**. The similarity between these two functions lies much deeper than the analogy of the statistical mathematical formalism: both functions are expressing distribution of order/disorder, Boltzmann's entropy in physical (material) systems, while Shannon's entropy in symbolic systems describing characteristics of the former. As order increases, both entropies decrease.

Chirality enters in the picture just at this point. Chiral (molecular) objects, beyond a certain spatial distribution of their parts (atoms, groups) are characterized also by a specific *direction* of this distribution (called often "helicity"). The directional (vectorial) distribution represents a more ordered situation. The corresponding energy difference can be calculated (Kondepudi and Prigogine, 1998), as well as the thermodynamic "cost" of information can be computed too (Shannon, 1951; Zurek, 1989).

Chirality (especially its configurational variant) has a particular advantage from the point of view of information. This can be demonstrated as follows. The information content (**h**) of a signal pair depends on their relative probabilities (**p**), according to relation (1) (Shannon and Weaver, 1949):

$$\mathbf{h}(\mathbf{p}) = \mathbf{p}\log_2 1/\mathbf{p} + (1-\mathbf{p})\log_2 1/(1-\mathbf{p}) \tag{1}$$

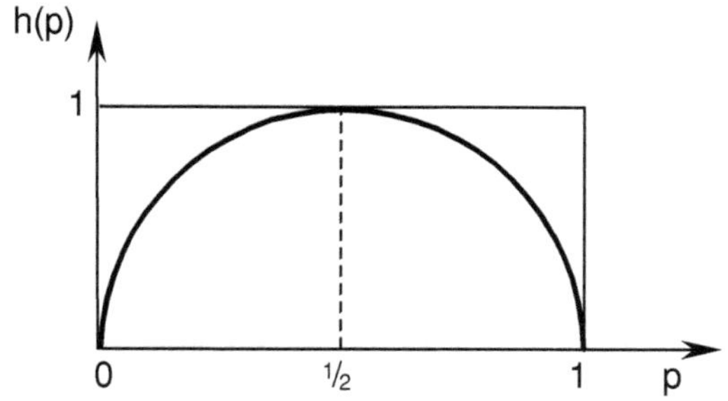

Figure 2. Information content of a signal pair (**h(p)**) in terms of their relative probabilities (**p**), according to equation (1).

This function is shown in Figure 2. It can be seen easily that the *maximum of the information content* (**h(p)** = 1) is reached at **p** = 1/2, where the *probabilities of both signals are equal*. In "molecular language" this corresponds exactly to an *enantiomeric pair*, at least according to the best of the presently available measurement possibilities. (Asymmetry of weak nuclear forces causes an energy difference of the order of $10^{-14} - 10^{-17}$. This effect could be detected at heavy atomic nuclei (Bi, Pb, Tl, Cs) (Crowe *et al.*, 1980; Fortson and Wilets, 1980; Emmons *et al.* 1983) and crystallization of racemic Co and Ir complexes (Szabo-Nagy and Keszthelyi, 1999a; 1999b)). It should be pointed out that the above argumentation does not mean that an enantiomer pair (racemate) would be more informable than a pair of molecules from the same enantiomer.

4. Chirality, Information and Knowledge

Information theory evolved following the (practical) requirements of signal transduction (communication). Consequently it became a theory of signals (codes) *without* any respect to the **meaning** of the message (Shannon, 1948; Shannon and Weaver, 1949).

Nevertheless, in living systems (as in all kinds of "machines" (von Neumann, 1966)), information serves in an ordered form, called also **knowledge** (Kuhn, 1972; 1976, 1988). Kuhn (1988) defines *knowledge* as follows: A specified system originates and evolves under continuous influence of a complex operational environment [and] by continuously testing [the] environmental properties, accumulates *knowledge*.

Knowledge is measured as the total number of bits to be discarded by throwing away carriers of information, until the evolutionary stage under consideration is reached. This parameter, however can be quantified (and therefore calculated) much more difficulty, than "naked" information. On the other hand, the critical threshold indicating the transition from non-living to living could be defined just in terms of knowledge accumulation (Kuhn, 1988). The fundamental difference between these concepts is shown schematically in Figure 3, by comparison of the elementary steps of information processing and of an evolving system.

Information processing (computer) can be reduced to (a) switching, (b) storage and (c) reset phases (Landauer, 1971; 1987), symbolized by single- and double-well potentials, corresponding to bits. The practical realization of such systems use double-well potentials of the same depth (Fig. 3/a). In the switching phase of an evolving system, on the other hand,

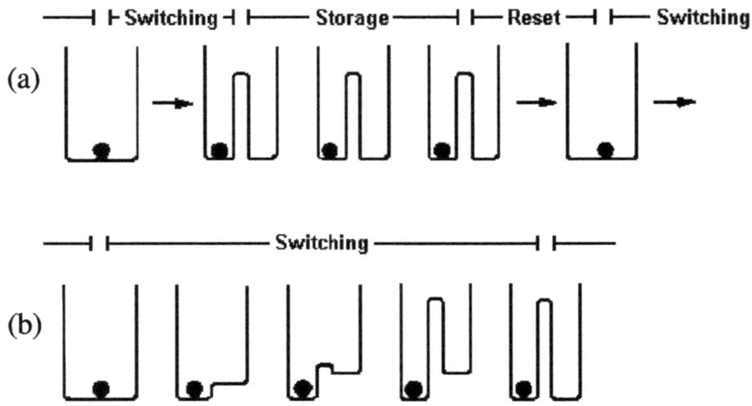

Figure 3. Schematic representation of the potential wells (a) utilised in a modulated potential computation and (b) in an evolving (template polymer) system. In case (b) the switching phase is that phase in which the (entering) monomer, under the determining field of the template strand gets fixed by its "complementery" monomer in the "new" strand; left well: "fit" monomer, right well: "unfit" monomer. (According to Kuhn (1988) and Landauer (1971; 1987))

the potential wells are not equal and their relative depths change during the switching phase (Fig. 3/b). In biological (and probably also prebiotic) systems this feature develops (first) as metabolic and related reaction network mechanisms (e.g. regulation of expression, inhibition and degradation of proteins or RNAs, etc.), as well as (later) in template polymerization of so-called information-carrier macromolecules.

Chirality (configurational) of monomeric units of the information-carrier macromolecules provides two very important advantages:

(i) Chirality facilitates the *molecular recognition* of the monomeric units in template polymerization (von Kiedrowski, 1986; Orgel, 1992; Ponnamperruma and Chela-Flores, 1993; Maynard Smith and Szathmary, 1995; Bolli *et al.*, 1997) (or even also in any other kind of polymerization (Green *et al.*, 1999)). These features are valid for monomers with (enantiopure) configurationally chiral "side chains". The facilitation means, that the activation energy of the "successful" approach to the "right" place of the monomeric unit (or of the growing chain) becomes lower, consequently the corresponding reaction rate increases (exponentially) and this results in *increased precision* of the template polymerization.

(ii) Chirality in terms of Fig. 3/b, increases the energy separation of the evolving potential wells and, consequently, it leads to easier selection. This aspect underlines the importance of the presence of configurationally chiral centers in biological macromolecules. The stability of these centers leads to more distinct potential wells.

Research efforts dealing with the origins of life are focused on several topics (Pályi *et al.*, 2002), however, only a very few of these are of such importance, which could be compared to the question of the mechanism(s) and of the driving force(s) of the selection of information-carrier macromolecules. A statistical-combinatorial analysis (Yockey 1992;

2000) yields such enormous numbers of possibilities, which *ab ovo* exclude a full-scale prebiotic combinatorial chemistry (Dyson, 1985, Kauffmann, 1993, Bolli *et al.*, 1997, Segre and Lancet, 1999). Configurational chirality, through the mechanisms described above, halves the statistical possibilities through chiral selection in polymerization (Yockey, 1992, see pp. 254–255). In other words it can be said, that configurational chirality **doubles the chances** of polymers built from chiral monomers. In present-day living organisms this resulted in the so-called homochirality of biological macromolecules (prominently nucleic acids, peptides and proteins) (von Kiedrowski, 1986; Orgel, 1992; Ponnamperruma and Chela-Flores, 1993; Maynard Smith and Szathmary, 1995; Bolli *et al.*, 1997; Pályi *et al.*, 1999). These macromolecules are not only homochiral, but all of these are of the same sense not only in a certain living organism, but *the whole living nature "uses" the same sense of chirality* (D-sugars, L-amino acids, etc.) (Voet and Voet, 1990; Keszthelyi, 1995; Cline, 1996; Elliot and Elliot, 1997; Pályi *et al.*, 1999; 2001). The uniform biological chirality in all living organisms is one of the most powerful arguments for a common origin of all living beings found presently on Earth (LUCA (Last Universal Common Ancestor): Woese, 1987; 1998; Woese *et al.*, 1990). This common "genealogy", on the other hand, provides a strong (indirect) proof for the supposition that biological homochirality must have been originated in a very early stage of terrestrial life, very probably even in the so-called prebiotic (Bengtson, 1994) period. Evidence is accumulating, that biological homochirality is not only an archaic accompanying factor in life processes, but one of the fundamental chemical requirements of life at all (Keszthelyi, 1995; Gilat, 1996). This approach can be considered, from the viewpoint of information theory, as a statement, that *evolution of biological homochirality was one of the most important steps in the evolution of biological information.* Even more, the evolution of biological homochirality can be identified as a **very early step in the evolution of biological knowledge** (Kuhn, 1988; Maynard Smith and Szathmary, 1995) that is: useful hereditary information.

5. The Chirality Code

Technical manifestation of information needs a system of symbols, a **code**, which enables its recording or transmission. The **bi**nary digi**t** (bit) code system, used in computer science and technology is, by no means a universal one. Molecular systems, for example, use a quite different code language. (Code, and information carried by this code, are only valid if there is an interpreter of this information [The Authors acknowledge a Referee pointing out this important distinction.] At molecular systems reactivity as well as intra- and intermolecular self-organization phenomena can be regarded as "interpreters" of the "molecular code" and the information carried by this code.) This code is composed of the following elements:

(i) *Chemical composition.* The number and kinds (including isotopes) of atoms in a molecule can be regarded as primary carriers of information. Biochemistry in this respect displays an important difference from abiotic chemistry: isotope effects (and consequently also the isotope ratios in biogenic molecules (Holland and Schidlowski, 1982; Schidlowski *et al.*, 1992) display in biochemical processes a very characteristic pattern, frequently used as a proof of (suspected) biogenic origin of fossils, which are morphologically no more typical (Schidlowski, 2002).

The isotopic signature of biogenic materials is regarded as one of the most ancient features of life processes (Some leading references: Schidlowski, 1987; 1998; Mojzsis *et al.*, 1996). It is interesting to compare the characteristics of biochemical isotope effects with the observed enantioselectivity of such reactions, both require a very fine tuning of reactivity. Both effects were successfully modeled in the same reaction: enantioselective polymerization of chiral monomers was achieved, where the monomers were of configurational chirality, containing an organic group with hydrogen isotopes (H and D) in different orientations (Green *et al.*, 1996). This model shows a high level of enantiomeric induction, due exclusively to the H/D substitution stereochemistry.

(ii) *Chemical structure*. The sequence and relative position of chemical bonds, the hybridization states of atoms involved in these bonds as well as delocalization of orbitals and intramolecular charge transfer effects represent another type of molecular information codes. The energy range of these codes is very broad, the primary chemical bonds are relatively strong, ranging between cca. 100 to 1000 kJ/mol, while delocalization and charge transfer effects are generally much weaker (from a few to cca. 150 kJ/mol). These structural codes, at the first sight, do not differ significantly in biogenic and abiotic chemistry. A more through analysis, however, shows, that the inter- and intramolecular charge transfer effects have a prominent role in biochemistry. The study of this aspect of biological chemistry led in the last few decades to the development of the so-called *supramolecular chemistry* (Wolf *et al.*, 1937; Lehn, 1988; 1990; 1995). These structural codes are, obviously, deeply interrelated with effects of chirality, but for the present analysis, we shall discuss these aspects separately in the following points.

(iii) *Two dimensional (2D) configurational chirality*. If three different atoms (or groups) are linked to a fourth one in *one plane* the resulting object (molecule) is achiral in the three dimensional (3D) space. If, however, this planar ensemble is approaching to an other object (molecule, ion, crystal, etc.) with one of its flat sides, the two faces of this planar molecule become different and consequently it is becoming chiral. Such interaction between 2D+2D, or 2D+3D objects, or even between more-less planar faces of two 3D objects, combined with chirality is an extremely powerful tool aiding **molecular recognition** (Rebek, 1994; 2000; 2002; Mecozzi and Rebek, 1998), which is driving in a highly selective manner several biochemical processes (adsorption, complexation, chemical transformation, signal transduction, etc.). The key element of coding the information (even knowledge) about selectivity of these 2D+2D, 2D+3D or (facial) 3D+3D processes is the 2D chirality of the interacting objects. (It should be added that in stereochemistry, this 2D chirality is termed generally as prochirality and the two faces of the 2D chiral object as prochiral or enantio-faces.)

(iv) *Three dimensional (3D) configurational chirality*. Different atoms or groups, linked to a central one in (approximately) tetrahedral, trigonal bipyramidal, octahedral, etc. geometric distributions result, that the mirror image of these molecules cannot be brought in superposition with the "original" one. These atoms (or groups) ordered according to arbitrary rules, indicate a certain spatial *direction* of the distribution. This additional parameter provides a new kind of informational code in addition to the 3D structure of the molecule. As a consequence of the specular relation, the directional vectors of the distribution point towards opposite directions. The two opposite directions are called also as helicity , while the two mirror images form enantiomers. This kind of mirror-twin enantiomerism is conceptually very close to the (0,1) binary digital symbolism of information (with not yet utilised possibilities in molecular computation . . .). As we pointed out earlier, biochemical evolution caused that the biochemistry of terrestrial living organisms is using exclusively only one of these "statistically" equally possible isomers (Keszthelyi, 1995; Cline, 1996; Pályi *et al.*, 1999; 2001). *This phenomenon is one of the most rigorously and most precisely followed informational rules in molecular biochemistry*. Interestingly the biological homochirality does not exclude the biosynthesis of the "opposite" enantiomers, which is fairly common at amino acids (Nagata, 1999; Kreil, 1999). The products of these "reversed" syntheses serve generally defence purposes (cell membranes, poisons, antibiotica, etc.). Other sources of the "opposed" isomers are spontaneous or chemically induced (e.g. metal ions) racemisation processes. These degenerative reactions represent a serious danger for the biochemistry of the normal organisms, cause serious health problems, as for example it is suspected at Alzheimer's disease (Majer *et al.*, 1999). Theoretically a "specular life", based on the opposite enantiomers of amino acids, carbohydrates, etc., could be imagined, but no trace of such organisms have been found on Earth yet. It should be pointed out, that such organisms would be highly poisonous for the existing living beings.

(v) *Conformational chirality*. The majority of covalent molecules contain several single chemical bonds. Rotation of groups of atoms around these bonds is generally a fairly smooth process: the various positions are separated only by low energy barriers, as it was pointed out earlier in this paper. Some of the positions are corresponding to more-less profound potential energy minima, resulting thus more-less stable conformations (rotamers). The interconversion of these conformations is (generally) a relatively easy process in the temperature range of living systems (usually from +5 to +40 °C, in extreme cases from −5 to +110 °C). Some conformations might be more stable, allowing isolation and structural characterization of the isomers. The reasons for stability include group-to-group repulsion (of the corresponding electron densities), charge transfer effects, as partial chemical bonds and H-bonding. The last one is of extraordinary importance in biochemistry. Several of these conformations display asymmetric geometry, corresponding to the requisites of chirality, yielding a considerable number of less stable isomers. The two kinds of chirality could be paralleled with the two kinds of information: discrete and continuous (Shannon and Weaver, 1949). The former is related to configurational, while the latter to conformational molecular chirality. Due to the relatively low energy barriers, the formation, function, population and interconversion of these conformational isomers allow a very fine "tuning" controlled by structural and charge distribution effects. These effects are coding the catalytic, chemo-, regio- and enantioselectivity effects in a very sophisticated manner in life chemistry (Voet and Voet, 1990;

Elliot and Elliot, 1997). The conformationally controlled evolution and function (usually catalysis) of active biomolecules is carefully concerted by particular timing mechanisms (BioEssays, 2000), which will be discussed later in this paper. Screw-like spatial distributions are called *helicity*. These are sometimes classified as a separate kind of chirality (Rowan and Nolte, 1998), others use this concept in a general sense as described earlier in this paper. Screw-like conformations are of particular importance in biochemistry, especially for DNA, RNA and proteins (Voet and Voet, 1990; Elliot and Elliot, 1997).

(vi) *Reaction chirality*. The two most characteristic features of biochemical processes are: (a) cyclic reactions and (b) irreversibility (Gánti, 1984; 1989; 1997; Kondepudi and Prigogine, 1998). Both features code implicitly for a time coordinate, which is one of the principal factors controlling the function of any living organism. The time irreversibility (non-reversal character) is a special kind of generalized chirality feature and it is essential for any living organism (Keszthelyi, 1995; BioEssays, 2000; Pályi *et al.*, 2001). These timing mechanisms are governing all chemical events in a (healthy) living organism by a *highly concerted program*, based on a well-established **time code**, in the form of relative reaction rates, cyclic reaction frequencies, and cycle connectivities (Gánti, 1984; 1989; 1997), marking out the direction of the "arrow of time" (Kondepudi and Prigogine, 1998), in an asymmetric manner (only one direction), similar to the selection of only one enantiomer by processes leading to homochirality. This aspect of biochemistry is very close to the operation principle of all machines, analysed masterfully in the automata-analogue description of living organisms by von Neumann (1966) and Gánti (1984; 1989; 1997). The study of so-called *molecular machines*, i.e. molecules imitating the constitution and function of mechanical and other devices, provides thus a powerful tool in the exploration of life processes and in the understanding of the most basic principles of life (Gánti, 1984; 1989; 1997; Lehn, 1988; 1990; 1995; Bencze *et al.*, 2002b). Some recent results concerning chiral molecular machines will be reviewed in the next section of the present paper.

6. Chiral Molecular Devices

The analogy between operation principles of machines and movements in some molecular systems attracted many research efforts in the last two decades (Lehn, 1988; 1990; 1995). The conceptual basis of this analogy is a kind of negative entropy flux: energy, obtained from a diffuse source, gets forced (by parts of the machine or molecule) to take an ordered (directed) path, usually (but not necessarily) performing work. The directionality of the resulting energy flux renders these devices conceptually (and often also geometrically) *chiral*. The machine/automata description of living organisms (von Neumann, 1966; Gánti, 1984; 1989; 1997) focused the attention of theoretical biology and of molecular sciences upon this aspect of life phenomena. The study of molecular devices became a kind of model research of living organisms (besides technological goals of miniaturization ambitions). It is of particular interest for the present review, that the parts of any machine embody a

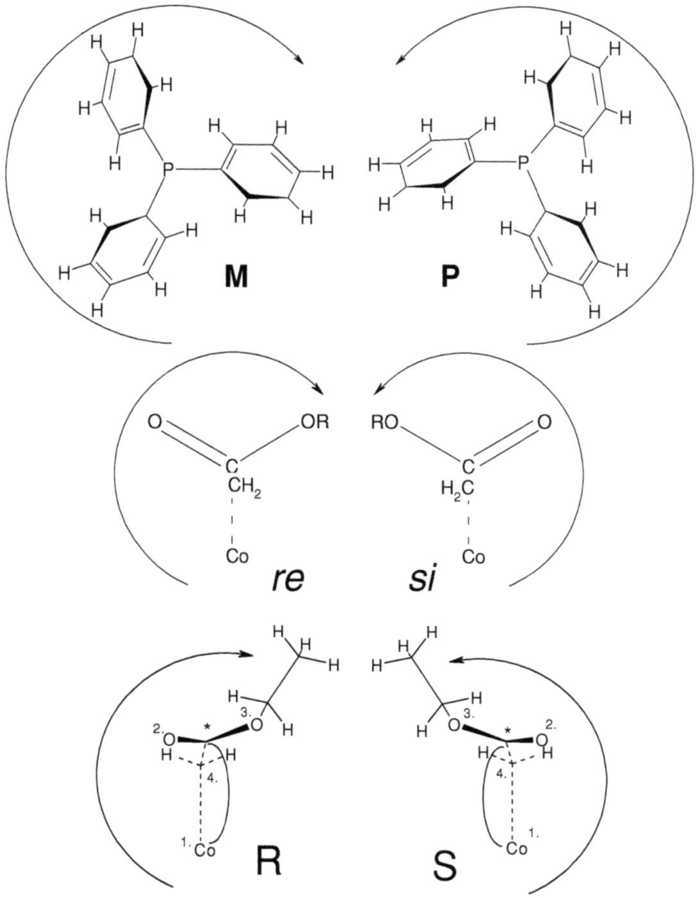

Scheme 1. Configurational (R, S), conformational (P, M) and 2D (re, si) chirality in alkylcobalt tricarbonyl triphenylphosphine compounds.

kind of *informational code*, while the cooperation of these parts corresponds to a piece of *knowledge*, both coded in the particular code language of the geometric shape and relative position of these parts.

Some of these molecular machines show a striking analogy with the first really complicated mechanisms ever built by mankind: the mechanical clockworks. The accidental discovery of this analogy at some organocobalt complexes (Pályi *et al.*, 1992/93; 1993; 1996) (Scheme 1) prompted our groups to a systematic research effort. Contemporaneously (and independently) an excellent synthetic effort of conceptually similar all-organic models was published (Kelly *et al.*, 1994; 1997).

The discovery of the clockwork analogy was followed by a systematic preparative work, aimed at the synthesis and characterization of similar molecules (Galamb *et al.*, 1981; Galamb and Pályi, 1986), as well as by a quantum chemical study which provided important elements for the understanding of the function of these molecular-level clockworks (Bencze *et al.*, 2002a). It was one of the most significant results of these theoretical studies, that *asymmetric rotation barriers* (that is: different height of the activation energy of rotations

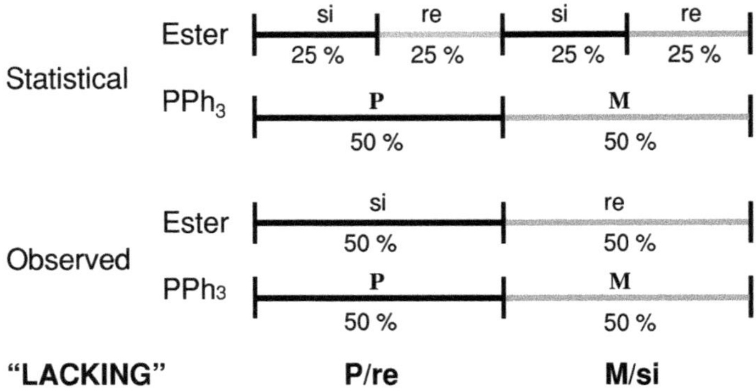

Scheme 2. Correlation of statistically possible and *de facto* observed isomers of [(alkoxycarbonyl)methyl] cobalt tricarbonyl triphenylphosphine compounds.

towards left and right) govern the development and selection of the resulting conformers. This is the key element in the chiral selection of some conformers, reducing radically the number of statistically possible isomers, as shown in Scheme 2.

Later, in course of these studies we found that the selection of chiral conformations was aided also by inter- and intramolecular H-bonds (Bencze, 2001; Zucchi *et al.*, 2001b), autosolvation (intramolecular charge transfer (Pályi and Varadi, 1975; Pályi, 1977; Pályi *et al.*, 1978; Szabo *et al.*, 2002) and the generation of new centres of chirality or rings of chiral conformation by these interactions (Szabo *et al.*, 2000; Zucchi *et al.*, 2001a) (Figure 4).

The "effectiveness" of the selection of these chiral conformers depends on the strength of these interactions, or in other words: on the (relative) height of the asymmetric potential barriers towards left or right (Bencze *et al.*, 2002a). If the asymmetry of the rotation barriers is sufficiently great (more than the Brownian motion energy at the given temperature), the chiral selection becomes *quantitative* (Pályi *et al.*, 1992/93; 1993; 1996; Szabo *et al.*, 2000; Bencze *et al.*, 2001; Zucchi *et al.*, 2001a; 2001b), while at lower (or less different) potential barriers it is only partial (Pályi and Zucchi, 1999; Zucchi, *et al.*, 1999).

Interestingly, the chiral self-organization of the conformers can be observed also at achiral molecules, if the constituents are of suitable geometry and/or capable of the "organizing" intramolecular interactions (Pályi, 1992/93; 1993; 1996; Zucchi *et al.*, 2001a).

Introduction of centers of configurational chirality into one (Szabo *et al.*, 2000) or more (Zucchi *et al.*, 2001b) segments (groups) of these molecules makes the chiral selection of conformers more efficient (Pályi *et al.*, 1996; Alper *et al.*, 2002), as it is expected on the basis of the well-documented phenomenon of chiral (asymmetric) induction (Janoschek, 1991; Mezey 1991). If, however, more configurational chiral centers are introduced into similar molecules, these do not generate "more chiral" conformations (Kajtar *et al.*, 1995; Bencze and Kurdi, 2000; Zucchi *et al.*, 2001a).

The structural interactions and selection rules identified in course of these studies can be regarded as **code letters of chirality information** in these model molecules. The present goals of this research project include the problem of generalization of the observed rules as well as a more general study of the role of transition metal ions in prebiotic/early biotic chemistry.

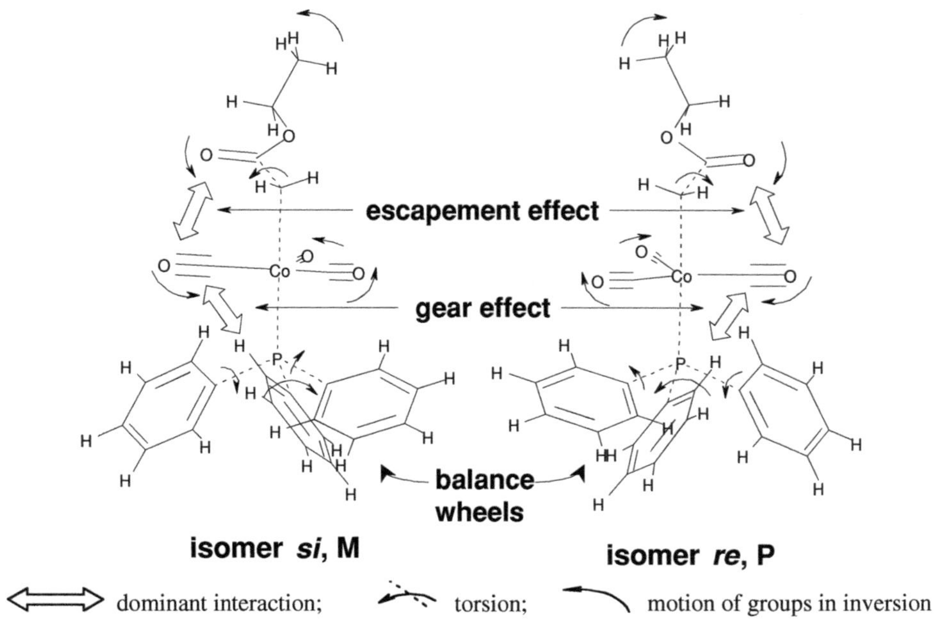

Figure 4. Clockwork analogy of the concerted intramolecular motions in organocobalt complexes.

It is a highly challenging aspect of these studies that the results obtained with these relatively small molecules could be useful in the construction of "molecular computers" (Reif, 2002). This goal has been successfully approached by attempts at using (polymeric) DNA for molecular computation (Braich *et al.*, 2002).

7. Summary

The article analyses the possibilities of viewing molecular structural parameters and reactivity as a kind of information carrier codes. These codes could be decoded by special physical and chemical methods. A field of utilizing these codes is the construction of "molecular devices", which is demonstrated by the example of molecular clockworks. Chirality is a special code, which might be very useful in digital (configurational chirality) or in analog (conformational chirality) molecular computation.

8. References

Alper, H., Bencze, L., Boese, R., Caglioti, L., Kurdi, R., Pályi, G., Tiddia, S., Turrini, D. and Zucchi, C. (2003) Intermediates of Cobalt-Catalysed PTC Carbonylation of Benzyl Halides. J. Mol. Catal. A: Chemical, **204/205**, 227–233.

Bencze, L. and Kurdi, R. (2000) Nanomachines: Concerted Development of Chiral Conformations in an Oxo-Catalyst Intermediate. In: K.S. Lakshiminarayanan, U. Devi, R. Bhavani Shankar and T.V. Gopal (eds.) *Nanocomputing Technology Trends*, Allied Publ. Ltd., New Delhi, pp. 27–34.

Bencze, L., Boese, R., Pályi, G., Szabo, M.J., Szilagyi, R.K. and Zucchi, C. (2001) A Királis Információ Terjedése Molekulán Belül (Intramolecular Transfer of the Chiral Information). Magyar Kem. Lapja, **56**, 215–219.

Bencze, L., Szabo, M.J., Szilagyi, R.K., Boese, R., Zucchi, C. and Pályi, G. (2002a) A Molecular Clockwork: Intramolecular Transfer of Chiral Information in [(Alkoxycarbonyl)methyl]cobalt Tricarbonyl Triphenylphosphine Complexes. In: G. Pályi, C. Zucchi and L. Caglioti (eds.) *Fundamentals of Life*, Elsevier, Paris, France, pp. 451–471.

Bencze, L., Pályi, G. and Kurdi, R. (2002b) Molecular-Level Machines: the Clockwork Model. NATO ASI Ser. accepted

Bengtson, S. (ed.) (1994) *Early Life on Earth*, Columbia University Press, New York, USA.

BioEssays, (2000) 22, (1), Special Issue: Biological Timing Mechanisms.

Bolli, M., Micura, R. and Eschenmoser, A. (1997) Pyranosyl-RNA: Chiroselective Self-Assembly of Base Sequences by Ligative Oligomerization of Tetranucleotide-2',3'-Cyclophosphates (with a commentary concerning the origin of biomolecular homochirality). Chem. Biol., **4**(4), 309–320.

Braich, R.S., Chelyapov, N., Johnson, C., Rothemund,, P.W.K. and Adleman, L., (2002) Solution of a 20-Variable 3-SAT Problem an a DNA Computer. Science, **296**, 499–502.

Cline, D.B. (ed.) (1996) *Physical Origin of Homochirality in Life*, AIP Press, Woodburg (NY), USA.

Crowe, K., Duclos, J., Fiorentini, G. and Torelli, G., (eds.) (1980) *Exotic Atoms '79: Fundamental Interactions and Structure of Matter*, Vol. 57. Plenum, New York, USA.

Dodziuk, H. (1995) *Modern Conformational Analysis*, VCH, New York, USA.

Dyson, F. (1985) *Origins of Life*, Cambridge University Press, Cambridge (UK).

Elliot, W.H. and Elliot, D.C. (1997) *Biochemistry and Molecular Biology*, Oxford University Press, Oxford, UK.

Emmons, T.P., Reeves, J.M. and Fortson, E.N. (1983) Parity-Nonconserving Optical Rotation in Atomic Lead. Phys. Rev. Lett., **51**, 2089–2092.

Feynman, R.P., Leighton, R.B. and Sands, M. (1964) *Feynman's Lectures on Physics*, Vol. I. Addison-Wesley, Reading (MA), USA, p. 52.

Fortson, E.N. and Wilets, L. (1980) Parity Nonconservation in Atoms: Status of Theory and Experiment. Adv. At. Mol. Phys., **16**, 319–373.

Galamb, V., Pályi, G., Cser, F., Furmanova, M.G. and Struchkov, Yu.T. (1981) Stable Alkylcobalt Carbonyls: [(Alkoxycarbonyl)methyl] Cobalt Tetracarbonyl Compounds. J. Organomet. Chem., **209**, 183–195.

Galamb, V. and Pályi, G. (1986) Alkylcobalt Carbonyls. In: J.J. Eisch and R.B. King (eds.) *Organometallic Syntheses*, Elsevier, Amsterdam, NL, pp. 142–146.

Ganti, T. (1984) *Chemoton Elmelet*, Vol. I. OMIKK, Budapest, Hungary.

Ganti, T. (1989) *Chemoton Elmelet*, Vol. II. OMIKK, Budapest, Hungary.

Ganti, T. (1997) Biogenesis Itself. J. Theoret. Biol., **178**, 583–593.

Gilat, G. (1997) The Concept of Structural Chirality. In: D.H. Rouvray (ed.) *Concepts in Chemistry. A Contemporary Challenge*, Research Studies Press/Wiley, London/New York, pp. 325–351.

Green, M.M., Peterson, N.C., Sato, T., Teramoto, A., Cook, R. and Lifson, S. (1996) A Helical Polymer with a Cooperative Response to Chiral Information. Science, **268**, 1861–1866.

Green, M.M., Park, J.-W., Sato, I., Teramoto, A., Lifson, S., Selinger, R.L.B. and Selinger, J.W. (1999) The Macromolecular Route to Chiral Amplification. Angew. Chem., Int. Ed., **38**, 3138–3154.

Holland, D.H. and Schidlowski, M. (eds.) (1982) *Mineral Deposits and the Evolution of the Biosphere*, Springer, Berlin, Germany.

Janoschek, R. (ed.) (1991) *Chirality*, Springer, Berlin, Germany.

Kajtar, M., Kajtar-Miklos, J., Giacomelli, G., Gaal, G., Varadi, G., Horvath, I.T., Zucchi, C. and Pályi, G. (1995) Dicobalt Hexacarbonyl Derivatives of Chiral Acetylenes. Tetrahedron Asymm., **6**, 2177–2194.

Kauffmann, S.A., (1993) *The Origin of Order: Self-organization and Selection in Evolution*, Oxford University Press, Oxford (UK).

Kelly, T.R., Bowyer, M.C., Bhaskar, K.V., Bebbington, B., Garcia, A., Lang, F., Kim., M.H. and Jette, M.P. (1994) A Molecular Brake. J. Am. Chem. Soc., **116**, 3657–3658.

Kelly, T.R., Tellitu, I. and Sestelo, J.P. (1997) In Search of Molecular Ratchets. Angew. Chem., Int. Ed. Engl., **36**, 1866–1868.

Kelvin, Lord (Thompson, W.) (1904) *Baltimore Lectures on Molecular Dynamics and the Wave Theory of Light*, Cambridge University Press, London, UK.

Keszthelyi, L. (1995) Origin of the Homochirality of Biomolecules. Quart. Rev. Biophys., **28**, 473–507.

Kondepudi, D. and Prigogine, I. (1998) *Modern Thermodynamics*, Wiley, New York, USA.

Kreil, G. (1999) Occurrence and Biosynthesis of Animal Peptides Containing a D-Amino Acid. In: G. Pályi, C. Zucchi, L. Caglioti, (eds.) *Advances in Biochirality*, Elsevier, Amsterdam, NL, pp. 297–304.

Kuhn, H. (1972) Self-Organization of Molecular Systems and the Evolution of the Genetic Apparatus. Angew. Chem., Int. Ed. Engl., **11**, 798–820.

Kuhn, H. (1976) Evolution Biologischer Information. Ber. Bunsenges. Phys. Chem., **80**, 1209–1223.

Kuhn, H. (1988) Origin of Life and Physics: Diversified Microstructure-Inducement to Form Information-Carrying and Knowledge-Accumulating Systems. IBM J. Res. Develop., **32**, 37–46.

Landauer, R. (1971) Stability and Instability in Information Processing and in Steady State Dissipative Systems. *Plenarvortrage Physik.* B.G. Teubner, Stuttgart, Germany, pp. 286–298.

Landauer, R. (1987) Computation: a Fundamental Physical View. Phys. Scripta, **35**, 88–95.

Lehn, J.-M. (1988) Supramolecular Chemistry—Scope and Perspectives Molecules, Supermolecules, and Molecular Devices (Nobel Lecture). Angew. Chem., Int. Ed. Engl., **27**, 89–112.

Lehn, J.-M. (1990) Perspectives in Supramolecular Chemistry—From Molecular Recognition towards Molecular Information processing and Self-Organization. Angew. Chem., Int. Ed. Engl., **29**, 1304–1319.

Lehn, J.-M. (1995) *Supramolecular Chemistry*, VCH, Weinheim, Germany.

Majer, Z., Lang, E., Vass, E., Szabo, S., Halgas, B. and Hollosi, M. (1999) Racemization-Induced Defolding and Aggregation of Segments of β-Amyloid Protein: An Early Step in the Formation of Amyloid Plaques. In: G. Pályi, C. Zucchi and L. Caglioti, (eds.) *Advances in Biochirality*, Elsevier, Amsterdam, NL, pp. 285–295.

Maynard Smith, J. and Szathmary, E. (1995) *The Major Transitions in Evolution*, W.H. Freeman/Spectrum, Oxford/New York.

Mecozzi, S. and Rebek, J., Jr. (1998) The 55% Solution: A Formula for Molecular Recognition in the Liquid State. Chem.—Eur. J., **4**, 1016–1022.

Mezey, P.G. (ed.) (1991) *New Developments in Molecular Chirality*, Kluwer, Dordrecht, NL.

Mezey, P.G. (1999a) Theory of Biological Homochirality: Chirality, Symmetry Deficiency, and Electron-Cloud Holography in the Shape Analysis of Biomolecules. In: G. Pályi, C. Zucchi and L. Caglioti (eds.) *Advances in BioChirality*, Elsevier, Amsterdam, NL, pp. 35–46.

Mezey, P.G. (1999b) The Holographic Electron Density Theorem and Quantum Similarity Measures. Mol. Phys., **96**, 169–178.

Mezey, P.G. (1999c) Holographic Electron Density Shape Theorem and Its Role in Drug Design and Toxicological Risk Assessment. J. Chem. Inf. Comp. Sci., **39**, 224–230.

Mezey, P.G. (2002) Theory and Detailed Computer Modeling of Biomolecules. In: G. Pályi, C. Zucchi and L. Caglioti (eds.) *Fundamentals of Life*, Elsevier, Paris, France, pp. 401–416.

Mojzsis, S.J., Arrhenius, G., McKeegan, K.D., Harrioso, T.M., Nutman, A.P. and Friend, R.L., (1996) Evidence for Life on Earth Before 3,800 Million Years Ago. Nature, **384**, 55–59.

Nagata, Y. (1999) D-Amino Acids in Nature. In: G. Pályi, C. Zucchi, L. Caglioti, (eds.) *Advances in Biochirality*, Elsevier, Amsterdam, NL, pp. 271–283.

Orgel, L.E. (1992) Molecular Replication. Nature, **358**, 203–209.

Pályi, G. (1977) Autosolvation. Violation of the 18-electron Rule Via Intramolecular Donor-Acceptor Interactions. Transition Met. Chem., **2**, 273–275.

Pályi, G. and Varadi, G. (1975) Methylidynetricobalt Nonacarbonyl Compounds, III. Evidence for Co-CO (Organic) Interaction in $CO_3(CO)_9CCH{=}CRCOOR'$ (R=H, alkyl, Ph; R'=H, Me) Derivatives. J. Organomet. Chem., **86**, 119–125.

Pályi, G., Kovacs-Toplak, M. and Varadi, G. (1978) Complessi del Cobalto-Carbonile con Derivati Propargilici (Cobalt Carbonyl Complexes with Propargylic Derivatives). Atti Accad Sci. Bologna, Rend. Cl. Sci. Fis., **266**, (13/5), 139–146.

Pályi, G., Zucchi, C., Bartik, T., Herbrich, T., Kriebel, C., Boese, R., Sorkau, A. and Frater, G. (1992/93) Induzione Chirale in fase Ordinata: il Cristallo e la Struttura Molecolare del $iPrOC(O)CH_2Co(CO)_3PPh_3$ (Chiral Induction in Ordered Phase: the Crystal and Molecular Structure of $iPrOC(O)CH_2Co(CO)_3PPh_3$). Atti Accad. Sci. Bologna, Rend. Cl. Sci. Fis., **281**, (14/10), 159–167.

Pályi, G., Zucchi, C., Bartik, T. and Boese, R. (1993) Conformational Asymmetric Induction in Solid Phase at Chiral Alkylcobalt Carbonyls. *J. Organomet. Chem. Conf., 1st*, (Nov. 4–5, Munchen, Germany), Abstr. p. 183.

Pályi, G., Alberts, K., Bartik, T., Boese, R., Frater, G., Herbrich, T., Herfurth, A., Kriebel, C., Sorkau, A., Tschoerner, C.M. and Zucchi, C. (1996) Intramolecular Transmission of Chiral Information: Conformational Enantiomers in Crystalline Organocobalt Complexes Generated by Self-Organization. Organometallics, **15**, 3253–3255.

Pályi, G. and Zucchi, C. (1999) The Rôle of Carbon Monoxide in Possible Prebiotic Reactions. *12th Internat. Conf. Origin of Life* (ISSOL-99), (July 11–17, San Diego, CA, USA) Abstr. p. 77.

Pályi, G., Zucchi, C. and Caglioti, L. (1999) Dimensions of Biological Homochirality. In: G. Pályi, C. Zucchi and L. Caglioti (eds.) *Advances in BioChirality*, Elsevier, Amsterdam, NL, pp. 3–12.

Pályi, G., Bencze, L., Micskei, K. and Zucchi, C. (2001) *Biological Chirality*: an Approach Through Coordination Chemistry. Atti Accad. Nazl. Sci. Lett. Arti (Modena), **317** (8/3) 457–477.

Pályi, G., Zucchi, C., Caglioti, L. (2002) Dimensions of Life. In: G. Pályi, C. Zucchi and L. Caglioti (eds.) *Fundamentals of Life*, Elsevier, Paris, France, pp. 1–13.

Pasteur, L. (1848a) Mémoire sur la relation qui peut exister entre la forme cristalline et la composition chimique, et sur la cause de la polarisation rotatoire C. R. Acad. Sci., **26**, 535–538.

Pasteur, L. (1848b) Recherches sur les relations qui peuvent exister entre la forme cristalline et la composition chimique, et le sens de la polarisation rotatoire Ann. Chim. Phys., **24**, 442–459.

Pasteur, L. (1922) *Dissymetrie Moleculaire*, Oeuvres de Pasteur, Vol. I., Mason et Cie, Paris, France.

Ponnamperruma, C. and Chela-Flores, J. (eds.) (1993) *Chemical Evolution: Origin of Life*, Deepak Publ., Hampton (VA), USA.
Rebek, J., Jr. (1994) Synthetic Self-Replicating Molecules. Sci. Am., 271 (July), 34–40.
Rebek, J., Jr. (2000) Host-Guest Chemistry of Calixarene Capsules. Chem. Commun., 637–643.
Rebek, J., Jr. (2002) Molecular Recognition, Replication and Assembly Through Chemical Synthesis. In: G. Pályi, C. Zucchi and L. Caglioti (eds.) *Fundamentals of Life*, Elsevier, Paris, France, pp. 417–426.
Reif, J.H., (2002) Computing: Successes and Challenges. Science, **296**, 478–479.
Rowan, A.E. and Nolte, R.J.M. (1998) Helical Molecular Programming. Angew. Chem., Int. Ed., **37**, 63–68.
Schidlowski, M. (1987) Application of Stable Carbon Isotopes to Early Biochemical Evolution on Earth. Ann. Rev. Earth Planet. Sci., **15**, 47–72.
Schidlowski, M., Golubic, S., Kimberley, M.M., McKirdy, D.M. and Trudinger, P.A. (eds) (1992) *Early Organic Evolution: Implications for Mineral and Energy Resources*, Springer, Berlin, Germany.
Schidlowski, M. (1998) Beginnings of Terrestrial Life: Problems of the Early Record and Implications for Extraterrestrial Scenarios. Instruments, Methods and Missions for Astrobiology. *Proc. Int. Soc. Opt. Eng. (SPIE)* 3441. Bellingham (WA), USA.
Schidlowski, M. (2002) Sedimentary Carbon Isotope Archives as Recorders of Early Life: Implications for Extraterrestrial Scenarios. In: G. Pályi, C. Zucchi and L. Caglioti (eds.) *Fundamentals of Life*, Elsevier, Paris, France, pp. 305–329.
Segre, D. and Lancet, D. (1999) Statistical Chemical Approach to the Origin of Life. Chemtracts—Biochem. Mol. Biol., **12**, (6) 382–397.
Shannon, C.E. (1948) A Mathematical Theory of Communication. Bell Syst. Tech. J., **27**, 379–424, 623–656.
Shannon, C.E. (1951) Prediction and Entropy of Printed English. Bell Syst. Tech. J., **30**, 50–64.
Shannon, C.E. and Weaver, W. (1949) *The Mathematical Theory of Communication*. University of Illinois Press, Urbana (IL), USA.
Szabo, M.J., Szilagyi, R.K., Bencze, L., Boese, R., Caglioti, L., Zucchi, C. and Pályi, G. (2000) Diastereoselection through Chiral Conformations. Enantiomer, **5**, 215–219.
Szabo, M.J., Szilagyi, R.K. and Bencze, L. (2003) Density Functional Studies of [Alkoxycarbonyl)methyl]cobalt tricarbonyl Triphenyl Phosphine Complexes: an α-Ester η^3-Coordination. Inorg. Chim. Acta, **344**, 158–168.
Szabo-Nagy, A. and Keszthelyi, L. (1999a) Demonstration of the Parity-Violating Energy Difference Between Enantiomers. Proc. Natl. Acad. Sci. USA, **96**, 4252–4255.
Szabo-Nagy, A. and Keszthelyi, L. (1999b) Experimental Evidences for Parity Violating Energy Differences Between Enantiomers. In: G. Pályi, C. Zucchi and L. Caglioti (eds.) *Advances in BioChirality*, Elsevier, Amsterdam, NL, pp. 367–376.
Voet, D. and Voet, J.G. (1990) *Biochemistry* (with 1991–1992 Supplements to Biochemistry), Wiley, New York, USA.
von Kiedrowski, G. (1986) A Self-Replicating Hexadeoxynucleotide. Angew. Chem., Int. Ed. Engl., **25**, 932–935.
von Neumann, J. (1966) *Theory of Self-Replicating Automata*, University of Illinois Press, London, UK.
Woese, C.R. (1987) Bacterial Evolution. Microbiol. Rev., **51**, 221–271.
Woese, C.R. (1998) The Universal Ancestor. Proc. Natl. Acad. Sci. USA, **95**, 6854–6859.
Woese, C.R., Kandler, O., Wheelis, M.L. (1990) Towards a Natural System of Organisms: Proposal for the Domains Archaea, Bacteria, and Eucarya. Proc. Natl. Acad. Sci. USA, **87**, 4576–4579.
Wolf, K.L., Frahm, H. and Harms, H. (1937) The State of Arrangement of Molecules in Liquids. Z. Phys. Chem., Abt. B, **36**, 237–287.
Yockey, H.P. (1992) *Information Theory and Molecular Biology*, Cambridge University Press, Cambridge, UK.
Yockey, H.P. (2000) Origin of Life on Earth and Shannon's Theory of Communication. Computers & Chemistry, **24**, 105–123.
Zucchi, C., Cornia, A., Boese, R., Kleinpeter, E., Alper, H. and Pályi, G. (1999) Preparation and Molecular Structures of Benzyl- and Phenylacetylcobalt Carbonyls. J. Organomet. Chem., **586**, 61–69.
Zucchi, C., Boese, R., Alberts, K., Herbrich, T., Toth, G., Bencze, L. and Pályi, G. (2001a) Concerted Development of Chiral Conformations in [(Alkoxycarbonyl)methyl]cobalt Tricarbonyl Triphenylphosphine Complexes. Eur. J. Inorg. Chem., 2297–2304.
Zucchi, C., Tiddia, S., Boese, R., Tschoerner, C.M., Bencze, L. and Pályi, G. (2001b) A Carbohydrate-Derived Alkylcobalt Carbonyl: {[(1,2:5,6-Di-O,O-isopropylidene-α-D-glucofuranos-3-yl)oxycarbonyl]-methyl}cobalt Tricarbonyl Triphenylphosphine. Chirality, **13**, 458–464.
Zurek, W.H. (1989) Thermodynamic Cost of Computation, Algorithmic Complexity and the Information Metric. Nature, **341**, 119–124.

IV. Genes & Genetics

Biodata of **Sven Degroeve** author (with coauthors Y. Saeys, B. Baets, Y. Van de peer and P. Rouzé) of the chapter "***Splice Site Prediction In Eukaryote Genome Sequences: The Algorithmic Issues***."

Sven Degroeve was born in Antwerp, Belgium, in 1977. He received his degree in Computer Science at R.U. Gent, Belgium, in 1999. The title of his graduate thesis is "Classification of Melanoma with a Neural Network". In 1999–2000, Sven has been affiliated with the "Center for Evolutionary Language Engineering" research group at Ieper, Belgium. Currently, Sven is working on a PhD. at the department of Bioinformatics at R.U. Gent, Belgium. The title of his PhD thesis is "Selecting Relevant Features for Biological Classification Models".

E-mail: **svgro@gengenp.rug.ac.be**

Biodata of **Yvan Saeys**, co-author (with Sven Degroeve *et al.*) of "***Splice Site Prediction In Eukaryote Genome Sequences: The Algorithmic Issues***."

Yvan Saeys was born in Sint-Niklaas, Belgium in 1977. He received his diploma of Master in Computer Science in 2000 at the university of Ghent, Belgium. His master's thesis was entitled "A study and enhancement of the genetic algorithm in the CAM-Brain Machine". After doing research on digital neural networks at the "Center for Evolutionary Language Engineering" (CELE) he started working towards a PhD. degree in Bioinformatics. Currently he is a PhD student in the Bioinformatics group of the Department of Plant Systems Biology (Ghent, Belgium) doing research on feature selection for nucleic acid classification.

E-mail: **yvsae@gengenp.rug.ac.be**

Biodata of **Bernard De Baets**, co-author (with Sven Degroeve *et al.*) of "***Splice Site Prediction In Eukaryote Genome Sequences: The Algorithmic Issues***."

Dr. Bernard De Baets received the MSc. degree in mathematics and computer science, the Postgraduate degree in knowledge technology, and the Ph.D. degree in mathematics, all summa cum laude from Ghent University, Belgium, in 1988, 1991 and 1995, respectively. Since 1999, he has been a Professor of applied mathematics at Ghent University, where he is leading the research unit. Knowledge-Based Systems (KERMIT). The activities of KERMIT concern the principles and practice of the extraction, representation and management of knowledge by means of intelligent technologies. His publications comprise chapters in various books and 60 papers in international journals. He serves on the editorial boards of Fuzzy Sets and Systems, 4OR, Intelligent Automation and Soft Computing, Mathware and Soft Computing, and Computing Letters, and he coordinates EUROFUSE, the EURO Working Group on Fuzzy Sets. Dr. De Baets is a Member of the Board of Directors of EUSFLAT, the Technical Committee on Artificial Intelligence and Expert Systems of IASTED, and of the Administrative Board of the Belgian OR Society.

E-mail: **Bernard.DeBaets@rug.ac.be**

J. Seckbach and E. Rubin (eds.), The New Avenues in Bioinformatics, 99–111.
© 2004 *Kluwer Academic Publishers. Printed in the Netherlands.*

Biodata of **Yves Van de Peer**, co-author (with Sven Degroeve *et al.*) of ***Splice Site Prediction In Eukaryote Genome Sequences: The Algorithmic Issues***."

Dr. Y. Van de Peer is a Full Professor in Bioinformatics and Genome Biology at the University of Ghent, Belgium. He obtained his PhD in 1995 on the studies of Ribosomal RNA as a tool in molecular evolution at the Department of Biochemistry, University of Antwerp (UIA), Belgium. Among his scientific activities he has been a Group Leader of Bioinformatics and Guest Professor at the Department of Plant Systems Biology at Ghent University (Belgium) and at the Department of Biomedical Sciences at the University of Antwerp (Belgium). During 1999–2001 he served as Assistant-Professor in the Research group "Evolutionary Biology" of Prof. Dr. Axel Meyer at the Department of Biology, University of Konstanz, Germany.

E-mail: **yves.vandepeer@gengenp.rug.ac.be**

Sven Degroeve

Yvan Saeys

Bernard De Baets

Y. Van de Peer

SPLICE SITE PREDICTION IN EUKARYOTE GENOME SEQUENCES
The Algorithmic Issues

SVEN DEGROEVE[1], YVAN SAEYS[1], BERNARD DE BAETS[2], YVES VAN DE PEER[1] and PIERRE ROUZÉ[3]
[1]Department of Plant Systems Biology, Ghent University, Flanders Interuniversity Institute for Biotechnology (VIB), K.L. Ledeganckstraat 35, 9000 Ghent, Belgium, [2]Department of Applied Mathematics, Biometrics and Process Control, Ghent University, Coupure links 653, 9000 Ghent, Belgium, and [3]Laboratoire associé de l'INRA (France), K.L. Ledeganckstraat 35, 9000 Ghent, Belgium

1. Introduction

Translating a gene into a protein starts by copying the part of the genome that codes for the protein on a primary transcript also called precursor RNA (or *pre-mRNA*). In eukaryotes, the primary transcripts of most protein-encoding nuclear genes are interrupted by *introns* that are removed by a process called splicing. The pre-mRNA serves as a code messenger between the cell nucleus that contains the DNA and the cytoplasm where the code (*mRNA*) is translated into a protein. Before leaving the nucleus, the pre-mRNA is spliced to obtain the mature mRNA. This splicing process identifies non-coding parts of the pre-mRNA transcript, the introns, and excises them out. The biological machinery that performs the actual splicing is called the *spliceosome*. This cellular machinery is a huge protein complex, which is formed through the dynamic association of smaller complexes called snRNP, embedding RNA components (*snRNAs*) labeled[1] U1, U2, U4, U5 and U6. The snRNP recognizes features in introns that are landmarks for splicing. The snRNAs are playing a central role in this process through base-pairing with specific binding sites located on the pre-mRNA and/or to each other. In a simplified splicing model, the transition from a coding (*exon*) to a non-coding (intron) part of the pre-mRNA, which is called the *donor site*, is identified by the U1 snRNP through base-pairing of the U1 snRNA with this donor site. The U2 snRNP recognizes a *branch* site that is located somewhere downstream the donor, through base-pairing of the U2 snRNA with the branch site. The U1 and U2 snRNP come together while the other snRNPs associate to form the fully functional spliceosome. The pre-mRNA is excised at the donor site and the free intron boundary loops with the branch

[1] The authors are aware of another class of spliceosomal splicing, involving other snRNAs besides U5, namely U11, U12, U4at and U6at. This class is nevertheless very rare (<0.1%), and for the sake of simplicity we will not include it in our presentation.

site, forming a lariat. Only then is the transition from the intron to the next exon, called the *acceptor* site, recognized. This acceptor site is usually the first AG dinucleotide located downstream the branch site. An intron can then in practice be defined as the part of the pre-mRNA located between a donor and an acceptor site. Conversely, locating the non-coding parts of the pre-mRNA involves the identification of the donor and acceptor sites in pairs on the pre-mRNA.

Computationally speaking, the location of the donor or acceptor sites can be reduced to a classification task by first defining the concept of a *candidate* binding site that can then be classified as either an *actual* or a *pseudo* site. It is observed that in eukaryotic organisms, the donor practically always contains the GT dinucleotide, sometimes GC. The donor site is indeed recognized by the U1 snRNA through base-pairing with an ACUUACCU motif, and should then ideally be AG/GTAAGT. Nevertheless, the base-pairing recognition is loose, especially in higher organisms, and tolerates many replacements in the motif, except for the border GT (or GC). The acceptor is observed to always contain the AG dinucleotide as the exact intron border. As such, all G{T,C}, resp. AG dinucleotides on the DNA are defined as candidate donor, resp. acceptor sites. Through the fast pace of sequencing of genes and their cognate transcripts, the number of experimentally identified eukaryotic donor and acceptor sites, i.e. GT or GC dinucleotides (resp. AG dinucleotides) known to be actual donor sites (resp. acceptor sites), has grown extensively in the last decade. This boost in the accumulation of publicly available biological data has boosted research in the field of Machine Learning and the classification of splice sites has become a popular task in that field.

The experimentally identified branch sites are very few, because they cannot be deduced from sequence data, contrary to acceptor and donor sites that are directly identified by comparing genomic DNA and mRNA sequences. Defining a candidate branch site is thus more difficult. Similarly to donor sites, branch sites are recognized through base-pairing by a snRNA, here U2 with the sequence GUAGUA, bringing the ideal branch site to be TACTA**A**C, in which **A** is the actual branch-point, i.e. the nucleotide where the lariat is formed. Again, the pairing can be very loose and the complementary motif very degenerated up to point that the only nucleotide to be always conserved is the nucleotide **A** at the branch point. Also, a candidate branch point site is therefore defined in terms of a "region on the DNA that probably contains a branch point" which we will refer to as a *branch point region*. This region differs slightly for different organisms but is typically assumed to fall between 15 and 60 nucleotides upstream the acceptor site. Since public domain databases contain very few data about branch points, except for yeast, the approach taken to classify a branch point region as actually containing a branch point site or not is an unsupervised approach, although the similarity to the consensus TACTAAC may be taken as an additional positive criterion.

The rest of this chapter is organized as follows: Section 2 describes the information sources used by the prediction methods to model the concept of a binding site. Section 3 describes the methods used for splice site and branch point prediction in some of the most popular DNA structural annotation systems that are publicly available for genetic research: GeneParser (Snyder *et al.*, 1995), SplicePredictor (Kleffe *et al.*, 1996), Genie (Kulp *et al.*, 1996; Reese *et al.*, 1997), Genscan (Burge *et al.*, 1997), NetGene2 (Hebsgaard *et al.*, 1996; Tolstrup *et al.*, 1997), SpliceView (Rogozin *et al.*, 1997), GlimmerM (Salzberg *et al.*, 1999), Genesplicer (Pertea *et al.*, 2001), and EuGène (Schiex *et al.*, 2001). Finally, Section 4 discusses some issues that are important for evaluating site prediction systems.

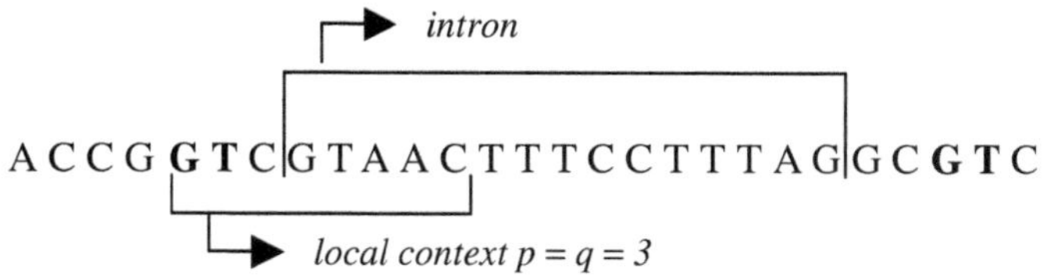

Figure 1. An intron in a pre-mRNA sequence. The GT dinucleotides in bold are pseudo donor sites, the GT dinucleotide at the start of the intron (in italic) is an actual site. The figure also illustrates the extraction of a local context $p = q = 3$ around the actual donor site.

2. Information Sources

In order to classify a GT dinucleotide as either an actual or a pseudo site, one needs discriminative information. For the pre-mRNA splicing task, information is extracted within the local context of the candidate site. In addition, sources of more global information, such as the distribution of intron lengths or the likelihood of a DNA sequence to be a genuine gene, could also be used as discriminative information for classifying splice sites. But, in the current practice of gene annotation, in which splice site prediction is one component, this is considered at another stage, as part of the integrative task of gene modeling.

The local context of a candidate splice site is a DNA subsequence that consists of p adjacent nucleotides upstream and q adjacent nucleotides downstream the candidate site. Figure (1) shows an example of local context $p = q = 3$ around an actual donor site. The information in the subsequence GTCGTAAC is used for classification. As we will see in the next section, the methods that use this local context to classify candidate splice sites can be divided into two classes, the *generative* and the *discriminative* methods.

In the generative approach, a probabilistic model is induced that generates subsequences of a certain class. This generative model can then be used to compute the probability that a local context subsequence belongs to the class of actual or pseudo sites, which can then be used to perform a classification. So, these methods exploit the 'sequential' property of sequences. Discriminative methods learn from a set of instances instead of sequences. Such an instance is a fixed length vector of features or attributes that represent the sequence. We denote the above subsequence as

$$z = s_1^z s_2^z \dots s_n^z = GTCGTAAC.$$

Extracting features $f_i^z = s_i^z$ that denote the nucleotide at position i in the local context results in an instance

$$\vec{z} = (f_1^z, f_2^z, \dots, f_n^z) = (G, T, C, G, T, A, A, C)$$

that does not imply a certain ordering between the features. The discriminative approach computes a relation between the values of these features and the expected prediction.

Other types of information sources can be found in the local context subsequence. For instance, we know that in most donor sites the upstream region in the local context subsequence contains protein-encoding DNA while the downstream region contains non-coding DNA. In the discriminative approach, this type of information could for instance be represented into the features of an instance that represents the subsequence, e.g. oligonucleotide frequencies (codon or hexamer), G+C richness, or local compositional complexity. The coding property could also be incorporated into a generative model as described in the next section. More information about information sources for splice site prediction can be found in Fickett (1996) and Mathé *et al.* (2002).

3. Binding Site Identification Methods

We will refer to a *classifier* as a *decision rule* that maps a candidate binding site on a class. In case of site prediction, the classes are *actual* or *pseudo*. This decision rule defines the structure of the *model* used for classifying candidate binding sites, e.g. a model can be a tree structure that tests symbolic features, or a non-linear function working on real valued features. The decision rule has a number of parameters that are optimized by a *learning method* to fit the data. The complexity of a decision rule is measured in terms of number of parameters that need to be optimized. Both the structure and the complexity of the model define what can be inferred from the data.

All methods described below are supervised, i.e. they compute a classifier referred to as *pred* and denoted as

$$pred(z) = c, \tag{1}$$

from a training set T that contains l labeled subsequences. This means that for each subsequence in T, we know whether it is an actual or a pseudo site. Generative methods induce a model of the joint probability distribution $p(x, c)$. This means that for every possible subsequence x and for every class c, the model represents the probability that x belongs to class c. Classification is performed by choosing the most likely class. Discriminative methods model the decision rule in Eq (1) directly, i.e. they actually learn the mapping between the feature representation of the subsequence and the expected class.

3.1. GENERATIVE LEARNING

In generative (or *Bayesian*) learning, a model is a joint probability distribution $p(x,c)$. If this distribution is known exactly, the Bayes optimal decision rule can be applied to compute the class distribution for a candidate subsequence z. Classification within the probabilistic framework is then performed by choosing the most likely class for z given the model $p(x,c)$. Using Bayes' formula we can write this as:

$$pred(z) = \arg\max_c \frac{P(s_1^z \dots s_n^z|c)P(c)}{P(s_1^z \dots s_n^z)} \tag{2}$$

which computes a weighted probability of observing subsequence z for each class c and outputs the class with maximum probability. The unconditional probability $P(s_1^z \dots s_n^z)$ is the same for every class and does not need to be computed.

3.1.1. Markov Chains

In practice, the space of possible subsequences is very large, making the estimation of $P(s_1^z \dots s_n^z|c)$ in Eq (2) impractical. A solution lies in making the model less complex. This can be achieved by reducing the number of parameters. Exploiting the knowledge that DNA sequences are transcribed directionally (from one nucleotide to the adjacent nucleotide) leads to the approximation of the optimal decision rule by means of a kth-order Markov Chain method. This method assumes dependencies between $k + 1$ adjacent nucleotides only:

$$P(s_1^z \dots s_n^z|c) = \prod_{k+1 \leq j \leq n} P(s_{j-k}^z \dots s_{j-1}^z s_j^z|c) \tag{3}$$

Typically, a generative model is computed for each class and the log-odds ratio is used as a prediction for z, i.e.

$$pred(z) = t\left(\log \frac{p\,(actual|z)}{p\,(pseudo|z)}\right),$$

with t a function that outputs a class based on a predefined threshold.

Choosing smaller values for k will decrease the complexity of the model, ending up with less word probabilities (the parameters of the model) that need to be reliably estimated from T. For $k = 5$, the model contains $4^{k+1} = 4^{5+1} = 4096$ parameters. So, the number of instances in T needed to make reliable estimates grows exponentially with k.

In the case of splice site prediction the U1 snRNP base-pairs with the donor site. The nucleotide pattern that discriminates between candidate donor sites is, relative to this donor site, fixed in location. To model this, the words $s_1 \dots s_k$ used for computing $P(s_{j-k}^z \dots s_{j-1}^z s_j^z|c)$ in Eq (3) are only those words that appear at position j in the subsequences in T. This special kind of Markov Chain, called the Weight Matrix Method (or WMM), where k was set to zero, was introduced for classifying donor and acceptor sites (Staden, 1984). As more biological data became available, Zhang *et al.* (1993) applied a first-order Markov Chain that they called the Weight Array Method (or WAM). The recent boost in the accumulation of biological data and improvements in methodology have made higher order models possible, but they have not shown further improvement in the classification of splice sites.

Higher order Markov Chains did prove to be very useful for predicting coding sequences or branch points. In this case, the discriminating pattern is invariant in sequence position and as such, words for estimating the parameters in Eq (3) are all words found in the subsequences of T. These models perform well by averaging the data available for estimating a parameter (the word probabilities), or by interpolating different order Markov Chains such that parameters for which there is less data have a decreased influence on the generation of sequences.

In the Windowed Markov Chain method the data available for estimating the conditional probability $P(s_{j-k}^z \dots s_{j-1}^z s_j^z|c)$ from Eq (3) is increased by averaging the conditional probabilities observed at position j over adjacent positions $j - w, \dots, j, \dots, j + w$ for some small value of w. The data available for estimating a parameter increases by a factor of $2w + 1$. This method was introduced by Burge *et al.* (1997) as a Windowed WAM to induce a model that generates Human DNA subsequences that contain the branch point. This model was then used for classifying candidate regions as either containing the branch point (somewhere in the subsequence) or not. The branch point region was defined as all regions between 21 and 38 nucleotides upstream an acceptor site. The value for w was 2, which increased the data available for estimating a parameter by a factor of 5. A somewhat similar probabilistic approach was used to model *Arabidopsis thaliana* branch point regions (Tolstrup *et al.*, 1997). Both sources report significant improvements in splice site identification performance when incorporating the branch point predictions.

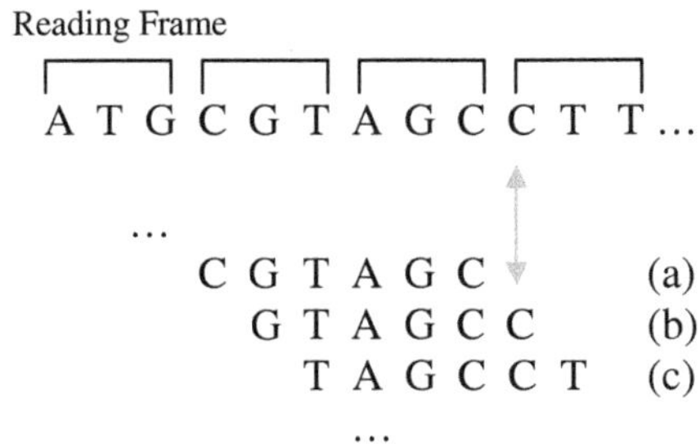

Figure 2. Word extraction for the 3 periodic IMM. Word CGTAGC (a) is used in the IMM associated with phase 3, GTAGCC (b) in the IMM associated with phase 1 and TAGCCT (c) in the IMM associated with phase 2.

Another improvement is the Interpolated Markov Model (or IMM) introduced by Salzberg *et al.* (1998), that generalizes the kth-order Markov Chain by defining a weighted linear combination of the parameters $P_0(s^z_j|c)$, $P_1(s^z_{j-1}s^z_j|c)$, $P_2(s^z_{j-2}s^z_{j-1}s^z_j|c)$ till $P_k(s^z_{j-k}\dots s^z_{j-1}s^z_j|c)$ for the estimation of the parameter $P(s^z_{j-k}\dots s^z_{j-1}s^z_j|c)$ in Eq (3). Each term is weighed proportionately to the amount of data available for estimating the term, i.e. unreliable estimates will have a decreased influence on the generation of sequences.

The IMM has proven to be a good method for modeling coding/non-coding sequences because higher-order Markov Chains can capture codon biases (or even hexamer biases using a fifth-order Markov Chain) that are very discriminative in terms of classifying coding sequences. Today's probabilistic coding models are 3-periodic. This means that the IMM is decomposed based on the phase of position k (in a word $s_1s_2\dots s_k$ that is used for estimating parameter $P(s^z_{j-k}\dots s^z_{j-1}s^z_j|c)$) in the coding sequence. For each of the three possible phases, a separate IMM is constructed that uses only those words $s_1s_2\dots s_k$ for which s_k is in the corresponding phase. Figure (2) illustrates how the words in a coding sequence are distributed over the three IMMs.

Markov Chain models are used to model donor and acceptor sites in SpliceView, GlimmerM and GeneParser, and to model acceptor sites in Genscan. Interpolated Markov Models are used to model coding sequences in GlimmerM and EuGène.

3.1.2. Bayesian Decision Trees

Although the Markov Chain method was biologically motivated, nucleotide dependencies studies by Burge (1998) and Vignal (1999) have shown that a large number of dependencies exist between non-adjacent nucleotides too. To model the most significant of these dependencies, the Maximal Dependence Decomposition method (or MDD) was introduced by Burge *et al.* (1997, 1998) for splice site prediction. This is a probabilistic decision tree method based on the χ^2 dependency test between all nucleotides in the local context subsequence. The method tries to capture the most significant of these dependencies, while keeping the complexity of the model moderate. It induces a model that is represented by a tree topology in which the local context subsequences x_i from T are stored as paths of connected nodes ending in leafs that represent subsets of T. The method begins by defining, for each nucleotide position in the subsequence, the most frequent base as the consensus

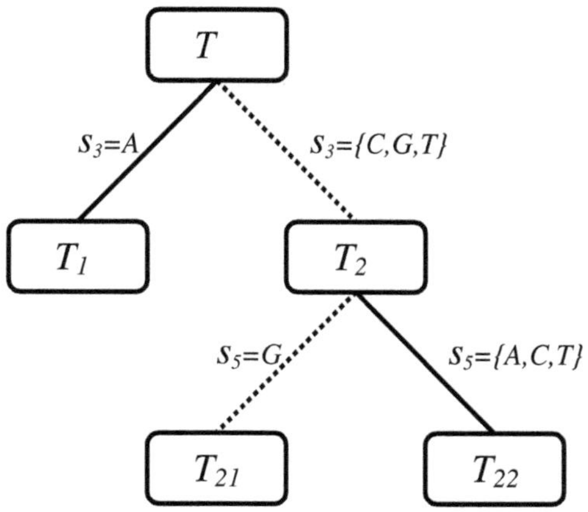

Figure 3. A simple tree-structure computed by the MDD method. See the text for more details.

of position i. Let variable C_i be 0 by default, but 1 if s_i is the consensus. Then, recursively repeat steps

(i) compute $m = \arg\max_i \sum_{j \neq i} \chi^2(C_i, s_j)$ for each nucleotide position i, and

(ii) partition T into T_1 containing all x_i that have the consensus at position m and T_2 containing all x_i from T not in T_1,

until a stopping criterion is fulfilled. This results in a tree structure that partitions the local context subsequences in T, i.e. the union of all sequences in the leaf is equal to T. For each leaf, a zero-order Markov Chain is computed form the instances associated with the leaf.

The parameter $P(s^z_{j-k} \ldots s^z_{j-1} s^z_j \mid c)$ is then estimated by following the correct path in the tree structure of the model, i.e. the path defined by the word $s^z_{j-k} \ldots s^z_{j-1} s^z_j$. Figure (2) shows a simplified tree structure that could be the result of applying the MDD method on a data set T. The path in the tree followed for a word $s_1 s_2 \ldots s_6 =$ ACTGGC is marked with dotted lines. First, s_3 is checked and the path $s_3 = \{C, G, T\}$ is followed. Then, in node T_2 the method checks s_5 and the path $s_5 = G$ is followed. This path ends in the leaf T_{21} and the zero-order Markov Chain associated with that leaf makes the prediction for z.

The stopping criterion controls the complexity of the model, which is typically defined in terms of the number of nodes in the tree. Tree construction is halted when the tree has reached a certain depth or when the number of instances associated with a leaf is below a certain threshold or when no significant dependencies exists between sequence positions. The MDD method predicts donor sites in Genscan and predicts both donor and acceptor sites in GeneSplicer.

The Glimmer 2.0 system (Delcher *et al.*, 1999) uses a somewhat different Bayesian decision tree method for classifying candidate coding sequences. Their method, the Interpolated Context Model (or ICM) is a generalization of the IMM by considering dependencies between position k in a word $s_1 s_2 \ldots s_k$ and all other positions i with $i = 1 \ldots (k-1)$. Recall that the IMM considers dependencies between adjacent positions only.

Dependencies between positions are measured by their mutual information. Position m with the highest mutual information with position k is used to split T in four disjoint sets T_A, T_C, T_G and T_T where set T_X contains all words $s_1 s_2 \dots s_k$ that have nucleotide X at position m. This procedure is repeated recursively on T_A, T_C, T_G and T_T until a stopping criterion is met, as in the MDD method. For each leaf in the tree, a zero-order Markov Chain is computed from the instances associated with that leaf.

3.2. DISCRIMINATIVE LEARNING

In discriminative learning, the decision rule is inferred directly from the data. The objective is to optimize the parameters of the rule such that some criterion is reached. The most common optimization criterion is to minimize the number of errors made by the decision rule on the instances in T (known as *empirical risk minimization*). As described in Section 2, discriminative methods represent a sequence in a vector of features. Some of these methods only work with numerically valued features and a mapping between symbolic and numerical values is needed. For instance, each symbol (nucleotide) can be mapped into four binary features: A as (1,0,0,0), C as (0,1,0,0), G as (0,0,1,0) and T as (0,0,0,1). The distance between any of these symbolic features is the same, so no information is lost or added.

In the most general case of discriminative learning, a classifier models linear dependencies between features in a space that consists of all possible combinations of all features. As was the case with the Bayes optimal decision rule (Section 3.1), this space of feature combinations is impractically large. Again the solution lies in making the decision rule less complex. This has led to different representations of the model, different objective functions and different learning methods that gave rise to a broad set of classifiers that solve the classification task differently and, as such, produce different predictions.

Currently the only discriminative method used for predicting splice sites is the *Multi-Layered Perceptron* (MLP) as introduced by Brunak *et al.* (1991). It is implemented in for instance the NetGene2 system to classify both splice sites and coding/non-coding nucleotides. But other discriminative methods such as rule-based, nearest-neighbor and kernel-based approaches have been evaluated (Rampone, 1998; Sonnenburg *et al.*, 2002) with promising results, indicating that discriminative methods construct decision rules that perform significantly better for site prediction than generative methods (Chuang *et al.*, 2001).

3.2.1. *Multi-layered perceptron*

The decision rule (or function) of a feedforward MLP is a linear combination

$$pred(\vec{z}) = \sum_{i=1}^{q} \vec{w}_i h_i(\vec{z}) \tag{4}$$

of q processing units (simple neurons):

$$h_i(\vec{z}) = f(\vec{w}_j \vec{z} - b_j) \quad (j = 1 \dots n). \tag{5}$$

In Eq (5), f is a sigmoid activation function that allows the induction of non-linear dependencies between features. All parameters $\vec{w}_i$, $\vec{w}_j$ and b_j in Eqs (4) and (5) are optimized from T, typically using a greedy search algorithm that minimizes the error on T (Hertz *et al.*, 1991). The parameter q controls the complexity of the model and is not automatically optimized. Choosing a small value for q results in a less complex decision rule that might not be able to capture the "true decision function" very well, while choosing large values can cause the method to capture patterns that are present in T, but are irrelevant for the "true decision function". The latter is known as *data-overfitting* and is one of the main problems with discriminative methods.

4. Evaluation Issues

The performance of a site classifier is usually measured as a sensitivity (Se) and specificity (Sp) ratio on a set of instances Z not used for training. The Se measure computes the probability that an actual site in Z is predicted as *actual* and the Sp measure computes the probability that a site from Z that is predicted as *actual* is an actual site.

When prediction is imperfect, there is a trade-off between the Se and Sp ratio's. Site classifiers that output a probability or a distance (as all methods described in Section 3 do) can be set to either "highly sensitive" or "highly specific" just by changing the threshold value used for mapping these probabilities or distances to a class. As such, evaluating a site classifier depends on their anticipated use. When used as standalone, e.g. to help molecular biologists to verify individual genes and plan experiments, classifiers should have high Sp at the expense of missing some actual sites, since the bioinformatics tool would then be used by biologists as a way to secure their experiments and their costs by reducing the chance of failure. Conversely, a site classifier that is integrated in the more global framework of gene modeling should have a high Se probability at the cost of classifying more pseudo sites wrongly. The rationale for this is that a site classifier makes predictions based on local features that might not be sufficient to perform perfect classification. Important dependencies between local site features and global gene features can easily correct misclassified pseudo sites. But sites that are missed cannot be recovered by most gene prediction systems. To illustrate this, Table 1 shows the result of a gene-prediction benchmark study by Pavy *et al.* (1999). Their set Z of evaluation sites was compiled from 168 carefully curated genes containing 859 intron sequences. Additionally, we added the results on the same set Z of a WMM and WAM trained on a set of 1486 *Arabidopsis* genes with no homologues in Z.

The methods used for predicting splice sites in the prediction systems listed in Table 1 are described in Section 3. Clearly the WMM and WAM methods that use local information only perform significantly less then SplicePredictor that uses additional global information such as splice-site-coupling and NetGene2 that also incorporates branch point prediction and many other post-processing rules. The last two systems perform a complete gene-prediction that can use gene structure features to eliminate false site predictions. For instance, the EuGène system uses the predictions of both SplicePredictor and NetGene2 for locating splice sites and improves on these by eliminating predicted sites that do not fit a correct gene-structure. Although these absolute values give some insight in the performance of the site classifiers, one has to be aware that factors other than the type of method used in the classifier can produce significantly different performance values. For instance, different training sets T used for training the methods as well as different source of information (features), e.g. different context lengths, have an influence on the performance of a classifier.

TABLE 1. Site prediction performance of several methods on an evaluation set of 168 *Arabidopsis thaliana* genes. Source: Pavy *et al.* (1999), unpublished results (WMM, WAM, Eugène)

	Donor sites		Acceptor sites	
	Se	Sp	Se	Sp
WMM	0.95	0.11	0.93	0.09
WAM	0.96	0.13	0.96	0.11
SplicePredictor	0.83	0.35	0.68	0.36
NetGene2	0.91	0.47	0.85	0.4
GENSCAN	0.77	0.82	0.73	0.78
EuGène	0.86	0.93	0.85	0.92

5. Summary and Limits

The identification of introns in pre-mRNA sequences is described as a search for donor and acceptor sites. This search is then formulated as a classification task and a detailed discussion about the methods used to solve this classification task is presented. The issues used for evaluating such a splice site identification system are also discussed.

There are further issues in splice site prediction which have not been addressed in this manuscript. First, our analysis is focusing on protein-encoding genes. There are indeed genes that are not encoding proteins, the end-products of which being various kind of RNAs, some well-known (e.g., rRNAs, tRNAs, snRNAs) some much less (snoRNAs, siRNAs) or not at all. While some of these genes are easy to find (e.g. the tRNA-encoding genes) most of them are not. Finding them is currently the subject of cutting-edge research. If spliceosomal introns do occur in some of these genes (other kinds of introns occur as well), the *in silico* search for their splice sites is a particular task, since the exon background at the intron borders differs for these genes compared to the protein-encoding ones. In a similar way, splice site prediction as we discussed in Section 3 applies well to introns that fall in the protein-encoding portion of genes (coding sequence, CDS). The region outside, in the untranslated regions (UTR) of genes upstream (5' UTR) and downstream (3' UTR) the CDS, also contains introns. The search for the splice sites of such introns is again a specific and more difficult task, for which—to the best of our knowledge—no algorithm has yet been developed. Last and not least, in most eukaryotes, a given gene may give rise to more than one mature transcript. One cause of such a biologically important phenomenon, the main one probably, is alternative splicing. The *in silico* prediction of the alternative splice sites (besides the comparative approach, when transcript information as cDNA & EST is available) is currently the object of research for several teams trying to decipher the biology behind alternative splicing by, for instance, using intron and exon-located specific signals, but which have not yet ended up in algorithms and software.

6. References

Brunak S., Engelbrecht J., and Knudsen S. (1991) *Prediction of Human mRNA Donor and Acceptor Sites from the DNA Sequence.* Journal of Molecular Biology, 220:49–65

Burge C., Karlin S. (1997) *Prediction of Complete Gene Structure in Human Genomic DNA.* Journal of Molecular Biology, 268:78–94

Burge, C.B. (1998) *Modeling dependencies in pre-mRNA splicing signals,* In: S. Salzberg, D. Searls, and S. Kasif (eds.) *Computational Methods in Molecular Biology*, Elsevier Science, Amsterdam, pp. 127–163

Chuang J., Roth D. (2001) Splice site prediction using a sparse network of Winnows. Technical Report UIUCDCS-R2001-2199, UIUC Computer Science Department

Delcher, A. L., Harmon, D., Kasif, S., White, O., Salzberg, S. L. (1999) *Improved microbial gene identification with GLIMMER.* Nucleic Acids Research, 27:4636–4641

Fickett J.W. (1996) *Finding genes by computer: the state of the art.* Trends in Genetics, 12(8):316–320

Hebsgaard S.M., Korning P.G., Tolstrup N., Engelbrecht J., Rouzé P., Brunak S. (1996) *Splice site prediction in Arabidopsis thaliana DNA by combining local and global sequence information.* Nucleic Acids Research 24, 17:3439–3452

Hertz J., Krogh A., Palmer R.G. (1991) *Introduction to the theory of neural computation.* Addison Wesley, Readwood City

Kleffe J., Hermann K., Vahrson W., Wittig B., Brendel V. (1996) *Logitlinear models for the prediction of splice sites in plant pre-mRNA sequences.* Nucleic Acids Research, 24:4709–4718

Kulp D., Haussler D., Reese G.M., Eeckman F.H. (1996) *A Generalized Hidden Markov Model for the Recognition of Human Genes in DNA.* Proceedings of ISMB-96:134–142, AAAI Press.

Mathé C., Sagot M.F., Shiex T., Rouzé P. (2002) *Current methods of gene prediction, their strengths and weaknesses.* Nucleic Acids Research, 30:4103–4117

Pertea M., Lin X., Salzberg S.L. (2001) *GeneSplicer: a new computational method for splice site prediction.* Nucleic Acids Research, 29(5):1185–1190

Rampone S. (1998) *Recognition of splice junctions on DNA sequences by BRAIN learning algorithm.* Bioinformatics, 14:676–684

Reese M.G., Eeckman F.H., Kulp D. Haussler D. (1997) *Improved splice site prediction in Genie.* Proc. IEEE 77(2):257–286

Rogozin I.B. and L. Milanesi. (1997) *Analysis of donor splice signals in different organisms.* Journal of Molecular Evolution, 45:50–59

Salzberg S., Delcher A., Kasif S., White O. (1998) *Microbial gene identification using interpolated Markov models.* Nucleic Acids Research, 26(2):544–548

Salzberg S.L., Pertea M., Delcher L.A., Gardner J.M., Tettelin H. (1999) *Interpolated Markov Models for Eukaryotic Gene Finding.* Genomics, 59(1):24–31

Schiex T., Moisan A., Rouzé P. (2001) *EuGène: an eukaryotic gene finder that combines several sources of evidence.* In: O. Gascuel and M.F. Sagot (eds.) *Lecture Notes in Computer Science 2006*, First International Conference on Biology, Informatics, and Mathematics, JOBIM 2000. Springer-Verslag, Germany, pp. 111–125

Snyder E.E., Stormo G.D. (1995) *Identification of Protein regions in Genomic DNA.* Journal of Molecular Biology, 248:1–18

Sonnenburg S., Ratsch G., Jagota A., Muller K.R. (2002) *New Methods for Splice Site recognition.* Proceed. of the International Conference on Artifical Neural Networks, February 2002

Staden R. (1984) *Computer methods to locate signals in nucleic acid sequences.* Nucleic Acids Research, 12:505–519

Tolstrup N., Rouze P., Brunak S. (1997) *A branch point consensus from Arabidopsis found by non-circular analysis for better prediction of acceptor sites.* Nucleic Acids Research, 25:3159–3163

Vignal L., Lisacek F., Quinqueton J., d'Aubenton-Carafa Y., Thermes C. (1999) *A multi-agent system simulating human splice site recognition.* Computer & Chemistry, 23(3–4):219–231

Zhang M.Q., Marr T.G. (1993) *A weight array method for splicing signal analysis.* Computer Applications in Biosciences, 9(5):499–509

Zhang C-T, Zhang R. (2002) *Evaluation of gene-finding algorithms by a content-balancing accuracy index.* Journal of Biomolecular Structure and Dynamics, 19:1045–1052

Biodata of **Hadi Quesneville** author (with co-author D. Anxolabéhère) of the chapter entitled: "***Object-Oriented Modeling in Genetics***".

Dr. **Hadi Quesneville** is assistant Professor at the Pierre and Marie Curie University (Paris 6). He carries out research at the "*Institut Jacques Monod*" (CNRS-Universities Paris 6 & 7) in the "*Laboratoire Dynamique du Génome et Evolution*" (Genome Dynamics and Evolution Laboratory). He obtained his PhD in 1996 from the University of Paris 6. Dr. Quesneville's main interests are bioinformatics, modeling, population genetics and transposable elements.

E-mail: **hq@ccr.jussieu.fr**

Biodata of **Dominique Anxolabéhère** co-author (with author H. Quesneville) of the chapter entitled: "***Object-Oriented Modeling in Genetics***".

Dr. **Dominique Anxolabéhère** is Professor at Pierre and Marie Curie University in Paris. He is currently head of the "*Laboratoire Dynamique du Génome et Evolution*" of the "*Institut Jacques Monod*" at the French National Research Council (CNRS). His PhD, obtained in 1979, concerned theoretical and experimental approaches to frequency-dependent selection. His current work focuses on interactions between transposable elements and host genomes. His group studies the mechanisms and regulation of transposition and their consequences for populations and evolution.

E-mail: **anxo@ccr.jussieu.fr**

Hadi Quesneville

Dominique Anxolabéhère

J. Seckbach and E. Rubin (eds.), The New Avenues in Bioinformatics, 113–134.
© 2004 *Kluwer Academic Publishers. Printed in the Netherlands.*

OBJECT-ORIENTED MODELING IN GENETICS

HADI QUESNEVILLE AND DOMINIQUE ANXOLABÉHÈRE
Laboratoire de Dynamique du Génome et Evolution Institut Jacques Monod, 2 place Jussieu, 75251 Paris Cedex 05, France

1. Introduction

Scientists and engineers use models as tools for studying systems. Many models are informal representations of a system and such models are referred to as *conceptual models*. This type of model is limited in terms of the ways in which it can be studied. The dynamics of such models and the predictions they allow depend on reasoning and are therefore tainted with a certain level of subjectivity. A formal description of the conceptual model makes it possible to eliminate this subjectivity. The formal description could take the form of a set of equations, a formal specification in simulation programs, or code in a computer programming language. The principal task in modeling is the translation of the conceptual model into this formal representation, which can be investigated objectively. Note that a very large number of factors are involved in biological systems. Consequently some of them are neglected when the system is modeled. Thus there is of course some subjectivity in these choices, but the validation of the model in some sense validates the choices made on the way.

Biologists use models to help them to understand biological systems. They have many motivations for adopting a modeling strategy: (i) to increase their understanding of a phenomenon or a system, (ii) to describe a system to other people, (iii) to measure performance, (iv) to test the fidelity of the model and to assess its appropriateness, (v) to select, to adjust or to estimate various parameters, or (vi) to validate the model empirically on the basis of its observed behavior. Unfortunately, biological systems are often complex by nature. The lack of detailed knowledge of some biological processes, the large number of missing, inaccurately or wrongly estimated parameters, and the need to model in multiple dimensions (space and time) make the modeling of biological systems a very challenging task. In most cases, they are complex aggregates of subsystems that may in themselves be complex aggregates. Thanks to developments in chaos theory, we now know that non-linear interactions almost always make the behavior of an aggregate more complicated than would be predicted by summing or averaging the behavior of its parts. It is essential to take this non-linearity into account if we want to understand most of the dynamics of complex systems. When modeling non-linear systems, the system must be formalized such that the non-linear interactions between the entities are dealt with. Unfortunately, most of our mathematical tools, from simple arithmetic to differential calculus, are based on an assumption of linearity. This lack

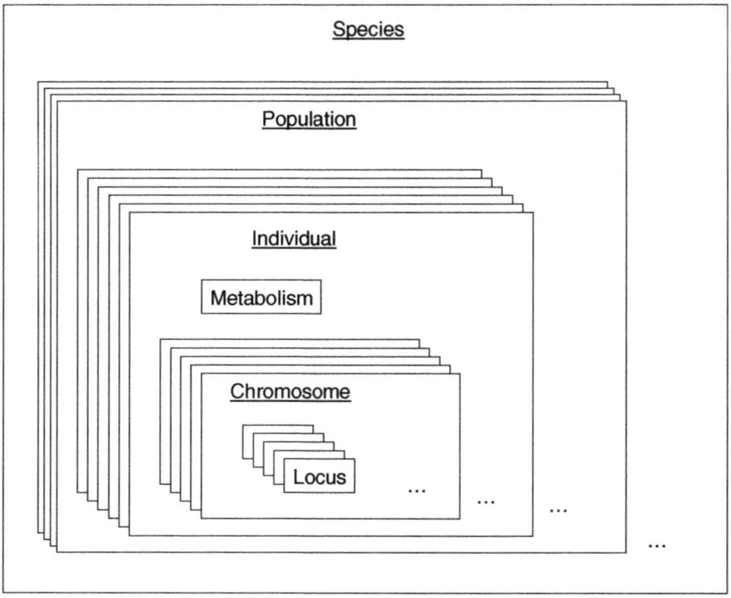

Figure 1. Example of the nested structure of a biological system. At each abstraction level, *Locus, Chromosome, Individual, Population* and *Species*, several objects of the same class may coexist. Modelers can navigate through these levels to understand the behavior of its system.

of appropriate mathematical tools renders the modeling of non-linear systems particularly difficult.

Object-oriented modeling provides a convenient framework for the modeling of complex systems. The formalism used supports the representation of aggregates of entities and their behavior. Moreover, this approach also allows to integrate all the steps in the modeling process, from description of the model to production of the executable code of a simulation program. This approach is also very intuitive because it involves the construction of a virtual world in which virtual representations of the entities are managed. The model deals with the entities and their interactions directly as processes rather than as synthetic variables and equations. The physical and logical constraints between the components can be conserved to improve the coherence of entity representation. The behavior of each simulated digital object can be followed, and the model explored by navigating through the various levels of abstraction, passing from one aggregate of objects to another. Figure 1 shows a complex aggregate for a model in genetics. Model validation is achieved by confronting its predictions with experimental data. This is simplified because *in silico* experiments can provide sets of simulated objects that can to be directly compared with real ones. Moreover, in the real world, events appear asynchronously. Sometimes, as in the examples presented in this review, the asynchronous nature is not relevant, and events can be modeled synchronously without changing the outcomes of the model. For the remaining cases, asynchronous modeling is achieved, usually by introducing a "scheduler" object that generates timed events to which the other objects react. Objects are also highly parallelizable. While the behavior of each object depends on its interactions with the other objects, this dependence can be asynchronous. Hence, the object-oriented systems produce objects that are relatively autonomous entities. Their codes can execute themselves in parallel, yielding to efficient distributed systems that allow managing a huge amount of objects on a computers cluster.

To further demonstrate the usefulness of object oriented modeling, we will describe some of the basic aspects of object-oriented modeling, and will then discuss three applications of this process in the modeling of genetic systems: the modeling of complex systems, knowledge representation, and the modeling of emergent behavior. These applications will be illustrated by three examples from our work.

2. The Basics of Object-Oriented Modeling

The goal of object-oriented modeling is to identify and to represent all the relevant entities of a system as *objects* belonging to a *class*.

2.1. THE CLASS OF AN OBJECT

Each entity of the modeled system must belong to a *class*. The *class* defines the set of states that an entity may take and the activities that it may perform. Each entity belonging to a *class i* is referred to as an *object* of *class i*. An *object* is thus one manifestation of a *class*. For example, two people would both be considered to belong to the *class Human*, but they are different manifestations of what a human may be. In the *object* paradigm, they are *objects* of the *class Human*.

The *attributes* and *methods* of a *class* determine its states and activities, respectively. *Attributes* are variables, the values of which determine the state of the *object*. The *methods* implement the activities that an *object* can perform. An encapsulation principle guides the design of a *class*. *Objects* interact with their environment by means of their *methods* (i.e. only the *methods* of the *object* can read or write *attribute* values and change the state of the *object*). In other words, the attributes are internal to the object, and are only visible to other objects through the methods. In the case of the *class Human*, the attribute family name can only be accessed, for example, through the method “talk”.

2.2. HIERARCHICAL RELATIONSHIPS BETWEEN CLASSES

Classes are organized with hierarchical relationships of two kinds: *aggregation/composition* and *generalization/specialization*. A *class* composed of one or several *subclasses* shares an *aggregation/composition* relationship with these *subclasses*. For example, the *class Human*, corresponding to a person could be represented as an *aggregation* of the class *Cell.* The relationship between the *classes Cell* and *Human* is an *aggregation/composition* relationship.

A *generalization/specialization* relationship characterizes *classes* with similar natures but different degrees of specialization. For example, the classes *Human* and *Mammal* display such a relationship as a *Human* is a *specialization* of a *Mammal*, and a *Mammal* is a *generalization* of a *Human* and a *Mouse*. Such relationships can be used to generate a taxonomy of *classes*, identifying their specific features and common characteristics.

2.3. COMMUNICATIONS BETWEEN OBJECTS

Objects interact in the system by exchanging messages. A message received by an *object* "A" activates one or several of its *methods*. The triggered *methods* of "A" can execute tasks, change the state of *object* "A" by modifying its *attribute* values, and/or send a message to another *object* "B", which in its turn reacts by triggering its *methods*, and so on. In the example above, if a message is passed to the *Human* "talk" method, he will pass back a message to the same object with the content of the "full name" attribute.

3. Modeling Complex Systems

Object-oriented formalism can be used to represent the entities of a system, together with their physical structures and their behavior. This makes it possible to describe a system with objects directly representing the real entities to be modeled.

The behavior of an entity can be described by interactions between its parts (i.e. the components of a lower abstraction level). It may also result from interactions (conflicts, collaborations, trades, etc) with other entities of the same abstraction level. Finally, higher levels define the environment in which these entities dwell, interacting with the lower levels by setting constraints. Consequently, the dynamics at one level of abstraction may be expressed as the result of interactions both within and outside that level. This facilitates the representation of high degree of non-linearity in the modeled systems. For example, several carnivores living in the same area interact, through conflicts or collaborations, to catch their preys. The population dynamics of the preys is the result of these interactions. The description of this dynamics at a population level is thus the result of interactions at the individual level (a lower abstraction level) of the carnivores and the preys. The environment where these species live (upper abstraction level) set also constraints on this dynamics through for example the food availability for the preys.

The behavior of each simulated entity can be followed, and the model explored by navigating through the various levels of abstraction, passing from one aggregate of objects to another. This offers two main advantages. Firstly, validation is made easier than with an analytical model (a set of equations) because it consists of comparing real entities with their abstract computer representations; thus, data can be extracted from this *in silico* world for comparison with the same kind of data from the real world. Secondly, questions concerning the dynamics of a system may be answered by considering several abstraction levels. The analyst is free to choose the abstraction level to be studied to investigate a given phenomenon. For example, the invasion dynamics of a new mutant allele in a species can be investigated by analyzing (i) linkage disequilibrium between loci at the level of the chromosome, (ii) allele frequencies at the level of the population, or (iii) indices of genetic differentiation between populations at the species level.

3.1. FROM THE MODEL TO THE SIMULATION SYSTEM

We skip here many issues of modeling that make the translation of the conceptual model to the object-oriented formalism a non-trivial step. This is out of the scope of this review, but we can mention key difficulties: how to deal with the non-synchronized nature of real world? How to deal with merging or disintegrating objects? How many objects can we represent to have an operational model?

Once the representation of the system has been formalized by means of object-oriented concepts, an object-oriented programming language can be used, almost directly, to produce an executable code. This enables the computer to generate its own representation of the model by simulation. All the steps, from identification of the abstractions which have to be modeled to the production of a simulation program, via formalization and translation into a programming language, can be managed by the use of methods developed for software production such as *The Unified Modeling Language for Object-Oriented Development* (UML; http://www.uml.org/). Software engineering methods, such as typing and assertion checking, may be used with this approach and are extremely useful in the production of reliable simulation systems.

Objects developed in one modeling project may also be re-used as building blocks in the construction of other object-oriented models. Modelers then build a true modeling framework, increasing the efficiency of production of new models. The work presented here illustrates this point. Historically, the simulation tool GENOOM (heading 4.2) was built by extending a simulation program produced for the modeling of *P* transposable element invasion in *Drosophila* (heading 5.1). This model was in turn extended by implementing the genetic algorithm-based model of the evolutionary dynamics of class II transposable elements (heading 6.1).

3.2. GENOOM: A SIMULATION TOOL FOR GENETIC OBJECT-ORIENTED MODELING

3.2.1. Motivation

Simulation approaches play an important role in the development of new statistical methods in genetics. Programs currently available include SIMLINK (Boehnke, 1986), SLINK (Weeks *et al.*, 1990), SIMULATE (Terwilliger and Ott, 1994), and GASP (Wilson *et al.*, 1996). These programs simulate the genotypes of families under a defined structure and linkage parameters. However, in the search for multifactorial disease susceptibility genes, methods make use of population features such as founder effects, small population sizes, or consanguinity. The programs currently available, such as POPSIM (Hampe, 1998), specifically designed for population studies, cannot simulate populations with such features.

Such simulation programs require modeling of a population, taking into account interactions between individuals, such as marriages or matings between members of the same family, but also complex genetic maps, and multifactorial determinism of phenotypes that involve interactions between susceptibility genes and environmental exposure.

We have developed a simulation package for genetics studies that we call GENOOM, for GENetic Object-Oriented Modeling (Quesneville and Anxolabéhère 1997, http://dynagen.ijm.jussieu.fr/equipes/bioinfo/software/genoom/). Based on an object paradigm, this tool implements a virtual computer world in which biological entities are digital objects. Each individual of a population is represented in this world. This makes it possible to study complex genetic models by means of simulations performed according to a genetic map with various types of genetic marker, and parameters such as penetrance matrix (phenotype probabilities for each genotype), exposure, reproductive rate, inter-relative mating probabilities, and the probability of migration in a two-dimensional space.

TABLE 1. Main Classes in GENOOM

Classes	Descriptions
Genetic elements	GENOOM manages four types of genetic element: microsatellites, RFLPs (restriction fragment length polymorphisms), qualitative trait loci (genes), and quantitative trait loci (QTL). Alleles are identified by an integer. Each element is mutated via a specific mutational process with a specific rate. Microsatellites increase or decrease in number by one; the allele numbers of RFLP markers, genes and QTL change randomly between two limits. An artificial tag locus, called MARKER, can be inserted at any map position for the tracing of chromosomal regions. It behaves like a true locus (but cannot be mutated) and each allele is unique in the starting population (initial frequency = 1/2N, where N is the initial effective population size).
Chromosomes	Chromosomes are composed of two chromatids. Chromatid length is expressed as a genetic distance, numbered from one extremity to the other to make it possible to position the genetic elements at a particular location. Crossover events may occur between two homologous chromatids of a chromosome, depending on genetic distance.
Individuals	Individuals are represented by their chromosomes, but also by attributes such as sex, age, phenotype, fertility status, vital status, and exposure. Methods are used to determine phenotype (using an exposure probability and a penetrance matrix), sterility status (according to phenotype and associated sterility probabilities), and migration.
Populations	Individuals are placed on an infinite two-dimensional grid. All individuals sharing a common location define a population.

Real data in this field is generally obtained from a population sample. As with the object-oriented approach we where able to compare directly simulated to real objects, GENOOM is supplied with a program that can be used to sample individuals in the simulated populations. This sampler program can randomly select individuals from a given geographic position or pedigrees with individuals sharing a particular phenotype. It is possible, for example, to sample nuclear families for sib-pair analysis, pedigrees with first cousins, or larger families over several generations. The data files generated can then be analyzed directly with various packages for statistical or genetic analysis, such as LINKAGE (Terwilliger and Ott, 1994), GENEHUNTER (Kruglyak, *et al.*, 1996, Kruglyak and Lander 1998) and MAPMAKER/SIBS (Kruglyak and Lander, 1995).

3.2.2. GENOOM

Based on an object paradigm, this program is characterized by its classes, which represent the structure of the system, and scenarios, which describe its dynamics. Table 1 describes the main classes, figure 2 shows the class diagram of GENOOM, and Table 2, the main scenarios.

3.2.3. Implementation

GENOOM is written in the C++ programming language. It can be compiled with the GNU gcc compiler under most UNIX systems. It takes a few minutes to run a simulation with 10 genetic elements and 10000 individuals, over 100 generations. GENOOM can be used with multi-processor architectures: several repeats of a population simulation can be run in parallel.

TABLE 2. Main scenarios in GENOOM. Time increments are generations; at each generation the simulation proceeds according to these scenarios

New generation	The age (in terms of generations) of each individual increases. Individuals who exceed the life span (a model parameter) die. Other individuals mate to produce new individuals.
Reproduction	1. For each individual born at the previous generation, the probabilities of inter-relative mating are checked. 2. If no inter-relative mating occurs, random mating is performed by randomly choosing the partner from the individuals of the opposite sex, born in the same generation, sharing the same geographic position, who have not already been involved in a previous mating. If no partner is available, no mating occurs. 3. If an inter-relative mating occurs, an individual is randomly chosen from all the possible relatives with the given level of kinship. If there is no such relative, random mating occurs (step 2). 4. If a partner is sterile (determined below in Birth, step 3), no offspring is produced (the other partner cannot be used in another mating). The number of offspring per couple is either constant (a simulation parameter) or determined at random according to a probability distribution (binomial, Poisson, or geometric). This number depends on reproductive rate and a maximum number of offspring per couple. The reproductive rate of a couple, r, is defined by: $r = (r_0 - 2)\left(1 - \frac{N}{K}\right) + 2 \quad (1)$ Where K is the maximum number of individuals sharing the same geographic position, N is the number of individuals of the parents' location, and r_0 is the maximum reproductive rate of the individuals.
Birth	1. A zygote is set, with two recombined gametes, one each selected randomly from the two parents. Crossovers between genetic elements are performed according to genetic distances. 2. Each allele of the zygote is checked for a mutation event. Mutation rates and processes are specific to the type of genetic element. 3. The phenotype (qualitative trait) is determined according to the genotype, the penetrance matrix, and the exposure. The associated probability of sterility is checked to set the fertility status of the individual. 4. The quantitative trait is calculated by summation over all QTL allele values. 5. The offspring is positioned according to its mother's coordinates.
Migration	Space is represented as an infinite two-dimensional grid. For each individual, the migration rate corresponds to the probability of moving to an adjacent site, randomly chosen from the 8 possible sites.

3.2.4. GENOOM as a simulation tool

A built-in model allows GENOOM to simulate basic genetic systems. Genetic markers and genetic determinants of quantitative or qualitative traits can be monitored. For the genetic determinism of the qualitative trait, a penetrance matrix can be used to model simple genetic determinisms such as dominance, and more complex determinisms such as additive or multi-factorial determinisms. With these genetic elements, inbred mating between relatives and in small populations can be simulated in spatially distributed populations. GENOOM has been

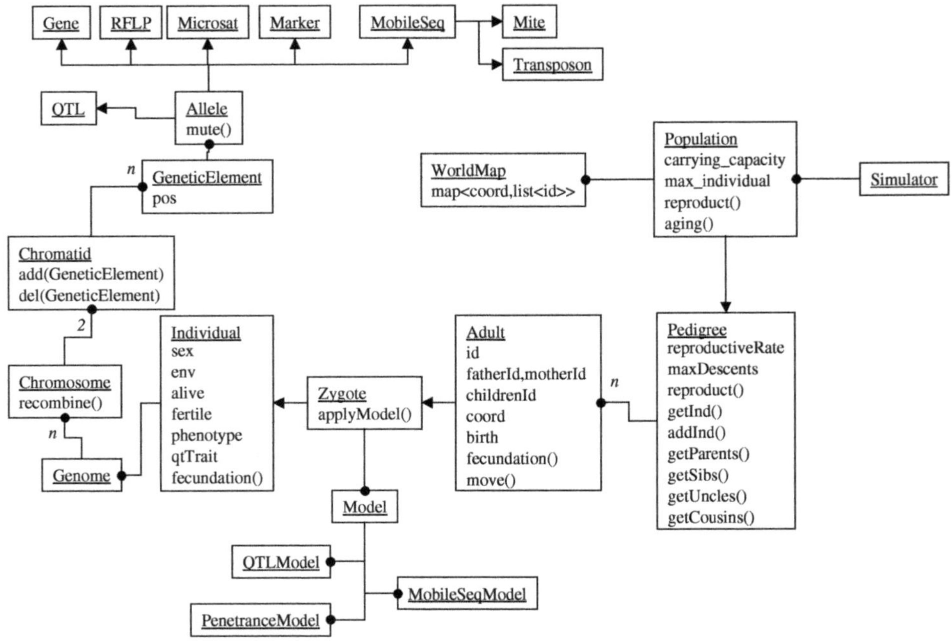

Figure 2. Diagram of the main classes of GENOOM. Generalization/specialization relationships are represented by arrows pointing towards the more general class. Aggregation/composition relationships are represented by a solid line ending in a circle indicating the contained classes. Near the circles, a number indicates the number of objects included in this relationship. If no number is given, then only one object is included in the relationship. The underlined names are class names, names ending in brackets are methods, and other names are attributes. For a description of the behavior of these classes see table 2.

used in various population genetic analysis and genetic epidemiology studies, in particular, in studies of founder population properties with a view to identifying multifactorial disease susceptibility genes (Bourgain, *et al.*, 2000).

This built-in model can be used as a tool for testing the robustness of methods for estimation from genetic markers (e.g. microsatellite sequences, RFLPs) or genes. The aim is to generate simulated populations closely resembling natural populations, in terms of the stochastic nature of sampled data distributions, as an *in silico* experimental study. These artificial populations may be sampled and analyzed like natural populations, giving sets of data comparable to experimental data sets, making it possible to use the same estimation methods: calculation of allele, genotype, and phenotype frequencies, population differentiation indices (F_{st}, F_{is}, F_{it}), population distances, linkage analysis, and so on. The robustness of the method in terms of its assumptions could be tested by simulating populations in which one or more of the assumptions does not apply, and by comparing estimates from a sample with those for the total population or with the simulation parameters used to produce the simulated populations.

This procedure is of great value for genetic epidemiologists, for the testing of various detection and estimation methods, for sampling families in the simulated population and for applying the tools used to analyze pedigrees, such as LINKAGE (Terwilliger and Ott, 1994), GENEHUNTER (Kruglyak, *et al.*, 1996, Kruglyak and Lander 1998), MAPMAKER/SIBS

(Kruglyak and Lander, 1995). As all the simulated objects are accessible and all the parameters are controlled, it is easy to check whether the predictions produced by an analysis method are accurate and to identify the cases in which they are not. Here, the object-oriented framework allows to produce highly accurate and detailed models of the reality in order to produce very realistic simulated data.

3.2.5. GENOOM as a modeling tool

The modeling concepts behind GENOOM demonstrate several important aspects of object oriented modeling. They make it possible for the user to represent experimentally deduced knowledge in a natural way, to model non-linear interactions in complex nested systems, and to build increasingly efficient modeling tools in an implicit manner. To do this, the program source codes can also be used as a framework of object classes into which modelers can integrate any new specific code. Hence, a simulation program has been developed from GENOOM for the study of transposable element dynamics (*see below*; Quesneville and Anxolabéhère, 2001).

4. Modeling and Knowledge Representation

Models are generally used as reasoning tools in elucidation of the behavior of dynamic systems. However, they may also be considered to be a language for exchanging information about a system and could thus be used as a tool facilitating communication between experts. In such an approach, models can be used to check whether knowledge about a system is consistent at all levels of description (i.e. from the molecular data to the species level). Is the knowledge about one of the levels of description of a system (e.g. a population) consistent with knowledge about lower levels (e.g. individuals)? In other words, can behavior at one abstraction level be accounted for by interactions between entities at lower levels? If the model is to be used in this way, then all the knowledge concerning the system must be represented as faithfully as possible in the model. Knowledge must be represented as specified by the expert of the domain, with minimal information loss (unless neglected details required achieving a useful model). Thus, although the models can be used in a predictive manner, the approach described here focuses on knowledge formalization and integration rather than predictive aspects, in order to check the coherence of the knowledge at several abstraction levels.

Modeling of this type requires a means of integrating various types of knowledge, heterogeneous in nature. The information to be modeled may concern physical structures, information flows, or processes. Their representations may adopt various forms, including: differential equations, Markov chains, cellular automata and finite state automata, etc.

Quesneville and Anxolabéhère (1997b, 1998) have demonstrated the application of these concepts to the modeling of *P* transposable element dynamics in *Drosophila* species. They demonstrated the integration of all available experimentally deduced knowledge into a single multi-model. Object-oriented methods were then used to formalize this knowledge, and to produce a simulation program for the multi-model. This made it possible to check the coherence of the knowledge represented at each abstraction level (molecular, organism, population, and species) and thus to determine whether experimentally identified mechanisms could account for the dynamics of *P* transposable element in an isolated population and through several populations via the exchange of migrants. However, it was also possible to explore the sensitivity of the model parameters, to estimate the parameters corresponding to real *Drosophila* species data, and to make predictions concerning the chromosomal distribution of insertions of the *P* transposable element, population equilibrium states, and the conditions necessary for successful invasion by this element.

4.1. DYNAMICS OF TRANSPOSABLE ELEMENTS IN METAPOPULATIONS: A MODEL OF *P* ELEMENT INVASION IN *DROSOPHILA*

4.1.1. Scientific context

Transposable elements (TEs) are mobile DNA sequences that may be present in multiple copies, dispersed throughout the genome. They may be classified according to the mechanism by which they move from one genomic site to another. Class I transposable elements use an RNA intermediate in their transposition; they are also called retrotransposons. Class II TEs, known as DNA-transposons or DNA transposable elements, use DNA. Within a class II TE family, the transposases produced by intact copies may mobilize defective elements.

Many transposable elements have been identified, but the *P* transposable one of the best studied at molecular, genetic, populational, and species levels. Genetic studies performed with *Drosophila melanogaster* have provided a lot of data concerning *P* element molecular structure, distributions of insertions, transposition and regulation mechanisms. The *P* transposable element invaded *D. melanogaster* during the second half of the Twentieth Century. Its spread has been described at the population level and at the molecular level (Kidwell *et al.* 1983, Anxolabéhère *et al.* 1988, 1990). These data show that there are various stages in the invasion process, some of which have been confirmed by experimental population studies (Anxolabéhère *et al.* 1987; Engels 1989; Preston and Engels 1989).

A model integrating the various mechanisms identified experimentally has been produced from the available data. These mechanisms have been formalized and connected to build a coherent global model. This model incorporates, at the individual level, transposition events and their regulation, as determined experimentally, in terms of changes in the chromosomal locations of the inserted copies. This model has been translated into a multilevel discrete-event simulation program, making it possible to follow *P* invasion dynamics at the molecular, chromosomal, population and metapopulation levels. It also makes it possible to simulate the various diagnostic crosses performed experimentally to study the regulatory properties of a population. The data obtained by our simulations can be compared directly with the data available for natural populations.

4.1.2. Types of P element copies

The *P* transposable element family is made of two types of copies: complete and deleted copies. The complete *P* transposable element is autonomous and encodes both an 87kDa transposase and a 66kDa repressor, produced by alternative splicing of the last intron. This splicing is restricted to the germ line, resulting in the synthesis of the transposase only in this tissue and the absence of transposition in somatic tissue. Deleted elements are non-autonomous: they can only be mobilized in *trans* by complete elements (for a review see Engels 1989, Rio 2002). Some deleted elements, such as the *KP* element, have regulatory properties (Black *et al.* 1987; Jackson *et al.* 1988; Rasmusson *et al.* 1993).

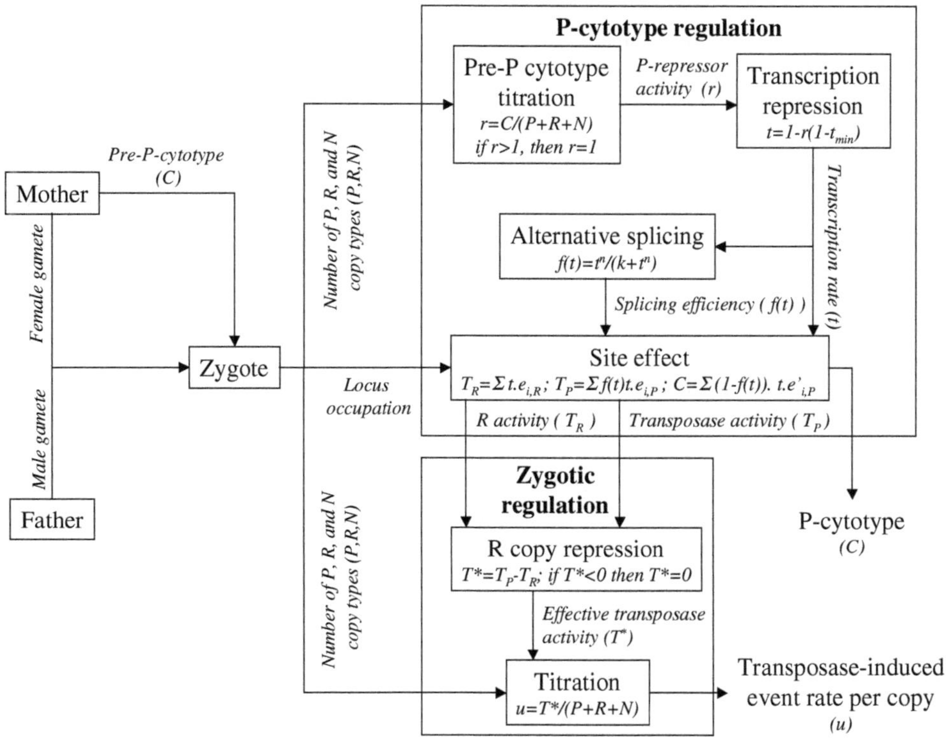

Figure 3. Diagram of information flow in the *P* regulation model. The correspondence between the variables (indicating the product effects in the equations) and the names of the product that they represent, can be found on the arrows (the variable name given between parentheses). For more details see Quesneville and Anxolabéhère 1998b.

In the model, three kinds of copy are considered: (i) P copies, corresponding to complete *P* elements. They produce the transposase and the *P*-repressor by alternative splicing. (ii) R copies, which produce another type of repressor, the R-repressor (*KP* repressor-like that probably play the role of a "poison" for the transposase). They are non-autonomous but can be mobilized by P copies in *trans*. (iii) N copies, which do not encode an active product. These copies are also non-autonomous and can be mobilized by P copies in *trans*.

4.1.3. Regulation

Three mechanisms of regulation have been modeled. The first is maternally determined and corresponds to P-cytotype regulation: maternal P elements lead to inactivation of early embryonic transposition as the repressor is accumulated in the eggs. The second is R copy-dependent and is zygotically determined. The third involves titration of the transposase by all the copies in the genome. Positional effects were modeled by considering insertion site affects on transposase expression levels. Figure 3 presents the flow of information in the regulation procedure described by Quesneville and Anxolabéhère 1998.

4.1.4. Transposition

For each insertion, u determines the probability of a transposition event and that of a sterilizing event. The number of elements transposed is determined by the binomial probability distribution $B(u; P + R + N)$, which gives the probability of n events.

Transposase activity is itself deleterious and decreases the fitness of individuals (Kidwell 1985, Engels 1989). Presumably, one of the major deleterious aspects is the possibility of generating rearrangements or chromosomal breakages, resulting in sterile individuals. If *PrSterile* is the probability of an event causing sterility, then the probability w of being fertile for an individual with m events assuming independence is:

$$w = (1 - PrSterile)^m \tag{2}$$

If the individual is fertile, n insertion sites are randomly selected. These sites correspond to the "donor sites" at which the transposition events take place. According to Engels' model (Engels *et al.* 1990), *P* element transposition is a "cut and paste" mechanism. The transposase "cuts out" the copy from the donor site and reinserts it elsewhere. A "target site" is randomly chosen from among the sites without an insertion, and this site receives the copy from the "donor site". This process leaves a gap (a DNA double strand break) at the donor site, which is repaired by the host, using a template. The probability of the sister chromatid acting as the template is 85%, and that of the homologous chromosome acting the template is 15% (with the exception of the male's X chromosome, for which the sister chromatid is always the template). This process repairs the gap at the insertion site according to the state of the site on the template chromosome used. If the template site has no inserted *P*element, then a precise excision occurs. If not, a copy identical to the template site insertion is inserted into the gap. However, this gap repair process may abort, with a probability*PrMut*. In this case, a defective copy is generated. This defective copy is of the N type unless the copy in the template site used is of the P type, in which case the probability of R type element mutation is *PrCopyR*.

4.1.5. Representation of individuals

Individuals are represented as three pairs of chromosomes. The Y and 4th chromosomes are not considered as they play only a minor role in *P* transposable element dynamics as they are much shorter than the other chromosomes. Males are hemizygous for the *X* chromosome. Crossovers between homologous chromosomes occur only during gamete formation in females, no recombination occurring in the *D. melanogaster* male germline. If transposition occurs, a new site is selected from the available *P*element-free sites. The probability of finding a new site is thus independent of the occupied sites, and the number of sites is not limiting. Females carry extrachromosomal information, the "pre-P-cytotype".

We treat transpositions that are harmful similarly to aborted transposition events because such insertions are rapidly eliminated by selection. The sites present in individuals in the model therefore corresponds to the subset of genome sites at which the insertion of a copy is not lethal. Simulations were thus carried out with chromatids with 96 insertion sites. Two adjacent sites are separated by 1 centimorgan: a chromosome is 96 centimorgans long. Females therefore possess 576 insertion sites and males 480. This number of sites is large enough to greatly exceed the total number of copies expected to be present in an individual from a natural *Drosophila* population, which ranges from 1 to around 100.

4.1.6. Population reproduction

The population reproduction procedure involves:

1. Randomly choosing two parents from the adult population.
2. The female brings a randomly recombined gamete and confers her own cytotype value on the egg. The male provides a non-recombined gamete to fertilize this egg.
3. Determining the probability of transposase-induced events according to the regulation model.
4. Returning to step 1 if the individual is sterile due to transposase action.
5. Carrying out the transposition and gap repair events according to the model.
6. Keeping this individual in a population of zygotes.
7. Returning to step 1 if the effective size of the population of zygotes is less than that of the adult population, or replacing the adult population with the population of zygotes if this is not the case.

4.1.7. Results and conclusions

Using object oriented modelling, it was relatively simple to integrate various experimentally identified mechanisms of regulation and transposition into a coherent global model. With this integrated model, we have tested the impact of these mechanisms at the population level. Data from experimental and natural populations are consistent with a model in which migrations associated with recurrent *P* invasion play an important role in the history of *P* invasions (Anxolabéhère *et al.* 1986, 1988, 1990). The three previously identified population equilibrium states (P, Q, M') were reconstructed in the simulations, and some new predictions were made. The simulations predicts, for example, a new equilibrium state (P', an equilibrium state with a P-cytotype regulation strong enough to repress intra-population transpositions but not invading *P* copies by migration) and that *P*element copies accumulate on the X chromosomes by contrast to the autosomes. The "molecular" and "genetic" simulation results are consistent with experimental and natural data, suggesting that the modeled molecular and genetic mechanisms are sufficient to account for the observed population data.

4.2. UNIVERSALITY OF OBJECT CONCEPTS

These previous studies involved the implementation of an object-oriented approach in a simulation program. However, the universality of object concepts in the computer science field makes it possible to envisage other types of implementation. For example, an object-oriented database could be used to construct a knowledge base for a system. It would be then possible to implement the same object model in several ways, and by questioning the model differently, to explore in more detail the outcomes of a set of knowledge. Hoogland and Biémont (1997) have constructed such an object-oriented knowledge base concerning the chromosomal distributions of transposable element insertions. The knowledge base approach is complementary to simulations for the study of transposable element dynamics. The knowledge base represents data concerning the distribution of insertions in natural populations, and the simulation program generates these distributions from underlying mechanisms. As both are based on object-oriented modeling, they identify classes corresponding to similar entities. The direct comparison of objects in the knowledge base (real data) and objects generated by the simulation program (simulated data) can be used in the testing of hypotheses. If temporal data are available, then the behavior of the system can be tested directly.

5. Modeling Emergent Behavior

One approach to investigating the global dynamics of complex systems is to describe each entity, looking for elementary phenomena, giving rise to a given pattern of global behavior. The dynamics at a given level of abstraction depend on interactions between components at lower levels, but also on interactions with components of the same level, in an environment determined by even higher levels.

In this last example, taken from Quesneville and Anxolabéhère (2001), we show how adaptive systems giving rise to emergent behavior (not predetermined by the model formalization) can be modeled. The conceptual framework underlying this type of modeling was inspired by approaches used to study "complex adaptive systems" (Holland, 1995; Forrest, 1993). Based on a specifically designed genetic algorithm, the emergence of a class II transposable element family can be studied as a self-organizing system.

5.1. GENETIC ALGORITHM-BASED MODEL OF THE EVOLUTIONARY DYNAMICS OF CLASS II TRANSPOSABLE ELEMENTS

5.1.1. Scientific context

Very little is known about the origin of TEs, but they may have originated during the putative transition from RNA-based genomes to DNA-based genomes (Jurka, 1998). It is clear that the genomes we observe today have evolved with the help of transposable elements. Many of these elements have evolved into parasites, but all seem to have retained their properties as "genome builders". Transposable elements are now thought of as essential participants in the evolution of genomes. What is known is that class II TE transposases have endonuclease properties: they cleave DNA at sequence-specific sites (Beall & Rio, 1997). Class II TEs may thus be thought of as endonuclease genes that recognize the extremities of their own nucleotide sequence. From this observation it is natural to hypothesize that TEs have originated from genes of this type. For such a primordial transposase to be maintained as a TE family, it needs to generate copies of itself that control their own mobility, In this study, we model the emergence of a TE family from a gene with basic endonuclease properties. We model the ancestral TE structure as the minimal organization common to all class II TEs. We also model the emergence of a mechanism for self-controlling mobility by means of protein-DNA interactions and DNA repair.

5.1.2. Structure and molecular activities of class II TEs

Most class II TEs have a sequence of about 30 to 250 nucleotides, present as two strictly identical copies in reverse orientation at each end of the element; these sequences are called terminal inverted repeats (TIR). Between the TIR, there is a gene that encodes a transposase

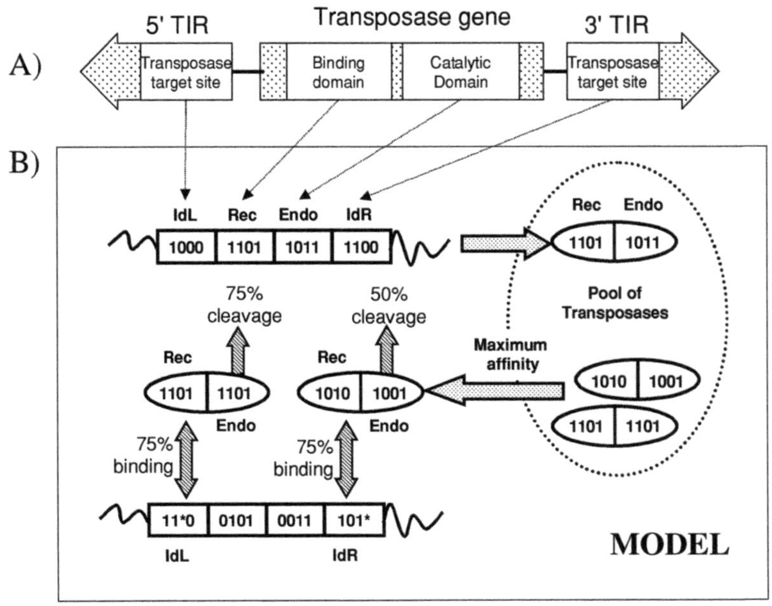

Figure 4. A) Schematic representation of a typical class II TE. The arrows indicate how the main functional regions are represented in the model. B) An example illustrating transposase synthesis, transposase recognition and the cleavage during transposition. A transposon is represented by 4 strings: the *IdL, IdR, Rec* and *Endo* strings. For *IdL* and *IdR*, the string contrains –, 0, 1, and *, with – and * characters representing a deleted or non-specific position accordingly. Each element produces a transposase, represented by its *Rec* and *Endo* domains. This transposase is stored in a pool of transposases away from the elements. For each element *IdL* and *IdR*, the sequences of the different transposase in the pool are tested for its binding to the element (recognition). This probability is given by the percent of matches between the two sequences, (see also table 3 below for a definition of matches). For example, if *Rec* is 1101 and *IdL* is 11*0, the table 3 indicates 3 matches. The resulting binding probability is $3/4 = 75\%$. If binding is successful, the transposase is tested for cleavage of the element, according to the probability encoded in the *Endo* sequence.

with two domains: a binding domain that binds to a target site within the TE sequence, within or close to the TIR region, and a catalytic domain that cleaves DNA in the TIR. Thus, a minimal TE would have a transposase with a binding domain that we call *Rec* (for recognition domain) and a catalytic domain that we call *Endo* (for endonuclease domain). Moreover, the sequence must also contain the two transposases target sites that we call *IdL* and *IdR* for identity left and right, because these sites determine which transposases recognize the sequence. As the recognition between the transposase binding domains and the transposase target sites is a key step in the transposition, we included this recognition in our model.

The sequence of each of these regions is modeled, with the properties of each region encoded by a string of four symbols: `-`, `0`, `1`, * (figure 4). The – and * symbol indicate a deleted and a non-specific matching position respectively, while `1` is a specific matching and `0` a non-matching position. For each target site, *IdL* or *IdR*, the probability of transposase binding is given by the matching percentage between the target site sequence and the *Rec* sequence of a candidate transposase. The matches between the symbols are summarized in Table 3. Note that if a sequence accumulates the '`*`' symbol, it evolves towards generalized

TABLE 3. Matches between TE sequence symbols. A '+' indicates a match whereas a '.' indicates no match

	*	1	0	-
*	+	+	+	.
1	+	+	.	.
0	+	.	.	.
-	.	.	.	.

recognition between the transposase target site (*IdL* or *IdR*) and the transposase recognition domain (*Rec*); if '1's accumulate, the *Rec* domain tends to specialize. The probability of cleavage depends on the percentage of '1's in the *Endo* sequence.

5.1.3. Host representation

Each individual host is represented as a set of chromosomes, with each chromosome consisting of two chromatids. Each chromatid contains a finite number of sites, each of which can contain several TE copies. Crossover occurs between two homologous chromatids, with 1 percent recombination between adjacent sites. Mutations occur in TE copy sequences during replication of the chromosome (independently of transpositions), with a probability m.

Populations of individuals are modeled according to standard population genetic hypotheses: random mating and constant population sizes (*i.e.* the parents of each zygote are selected randomly with replacement, zygotes are produced until the offspring population is the same size as the parent population, and sterile individuals (see below) are eliminated from the population).

5.1.4. Transposition

TE copies are inserted into the various sites on the chromatids of the host chromosomes. Each copy produces a transposase (*i.e.* the string of the element is reduced to the *Rec* and *Endo* domains; figure 4B). This transposase is stored in a position-free pool of transposases. For each element target site, *IdL* or *IdR*, the transposase of the pool with the maximum probability of binding is tested for its binding to the element. If binding is successful, the transposase is tested for cleavage of the element, according to the probability encoded in the *Endo* sequence. Finally, the element is transposed if binding and cleavage are successful on both sides of the element. The element then inserts itself into a new site chosen at random. If any one of these steps is unsuccessful, then the element remains in place. For each successful transposition, there is a probability s that the host becomes sterile. This sterility may be caused by chromosome breaks resulting from DNA cleavage during the transposition process.

When an element transposes, it leaves a gap at the donor site. This gap is repaired by a DNA "gap repair" mechanism. This process is modelled as two independent repair processes beginning simultaneously at each side and progressing towards each other. These processes restore the symbol present in the excised copy at each position in an imperfect way:

- The repair process may abort on one side, with a probability d. If it aborts on both sides, then an element with a deletion covering the unrepaired region appears. The symbol '-' is used to indicate a deleted position. An element is restored at the donor site if at least one repair process reaches the other on the sequence, even if the other process has aborted. Consequently if only one repair process abort no deletion occur.
- Ponctual mutations can occur, with a probability e. The original symbol is randomly replaced by another. Obviously no mutation event can cause a deletion or occur at a deleted position: a '-' symbol only appears by abortive gap repair and cannot be changed to another symbol.

According to the status of the four functional regions, various classes of copy with different properties may be produced:

- *Active copies:* If all regions contain at least one '1' or one '*', the transposase is functional and the copy is active. Its *Rec* and *Endo* domains are able to bind and cleave a particular copy.
- *Regulator copies:* If only the *Endo* domain is devoid of '1's, the encoded transposase can bind, but cannot cleave the element. When it binds, all other functional transposases cannot access to the target site and bind. Hence the product of this type of copy acts as a regulator.
- *Inert copies:* If the *Rec* domain is devoid of '1's and '*'s, the transposase is unable to bind to any element: the transposase is therefore inert.
- *Immobile copies:* Finally, if one transposase target site is devoid of '1's and '*'s, no transposase can bind at this site and thus no cleavage can occur: this copy is immobile.

With this model, non-autonomous copies are represented as *Regulator* or *Inert copies*.

5.1.5. Implementation

This model has been implemented in the simulation program GENOOM(Quesneville and Anxolabéhère, 1997a, see heading 4.2 in this text). To maximize speed and to minimize the amount of computer memory required, the four regions of the sequence were limited to four positions, making it possible to represent a TE copy with a computer word of 32 bits. Populations can be initialized with different kinds of copies and are simulated with the model. All probability tests are performed by means of a classical Monte-Carlo procedure. Note that due to the nature of object oriented modeling, little specific programming was required to produce a simulation from this model definition.

5.1.6. Results

We found that a transposable element could emerge from a single gene with basic endonuclease properties. Upon transposition, this new element can produce mutated and deleted copies. Some of these copies may interact by their product with other copies, reducing the capacity of the element to invade DNA. We observed the spontaneous formation of an organized molecular interaction network, resulting in control of mobility. The regulatory system displays auto-organization. The DNA double strand break repair process plays an important role in invasion dynamics, and appears to determine the success with which class II TEs emerge. The distribution of the deletions in the sequences retained by the selection is not centered on the middle of the copies as could be expected given the repair process, but it is shifted to the "endonuclease domain". The distribution of deletions also affects the rate of evolution of the sequence. Antagonism between two selective forces gives rise to heterogeneity both within TE sequences and between the different copy types.

The complexity of TE behavior results from the opposition of two driving forces acting on the copy mobility. The first selects against mobility because of its deleterious impact on host fitness. This "mobility disadvantage" is to the benefit of the host. The second selects mobility as a mechanism for the rapid multiplication of TE copies. This "mobility advantage" is to the benefit of TEs, and occurs during the initial phase of TE emergence. If the main driving force is the "mobility advantage", then the rate of transposition and copy number increase. However, if the level of mobility becomes too high, then there is selection against mobility and the "mobility disadvantage" becomes the main driving force. As a result, mobility is controlled by regulator copies. The outcome of this opposition is a state of dynamic equilibrium. Our ability to reconstruct the dynamic equilibrium between these opposing forces supports both the model validity describing the real world, and the current models of transposition evolution.

5.1.7. Discussion

The aim of this example was to demonstrate that object oriented modeling allows to easily and explicitly model emergent behavior. In this example, we demonstrate that TE family evolution can be reconstructed "mechanistically", given the structure and properties of the underlying class II TEs.

It is important to note that while this model proved useful in studying some behaviors, different or more sophisticated models are required to study others. For example, the coding of the four TE regions was modeled only for its functional properties, and not for its DNA sequence. Modeling the length of these regions on the DNA sequence was beyond the scope of this work. Moreover, the implementation of the model was constrained by the speed and memory requirements of the simulations. To ensure that dynamic equilibrium is reached in a reasonable simulation time, mutations are made to occur more frequently than would be expected in real life. Thus, only qualitative outcomes can be considered. It is therefore difficult to deduce the time required in real life to reach the observed dynamic equilibrium from the dynamics of the simulation.

This model should be considered as a null hypothesis. The results it produces depend on our knowledge of class II TE structures and transposition mechanisms. Discrepancies from real data concerning a particular TE family should be interpreted in terms of additional features to be added to the model, and to our knowledge of that TE. These features are very interesting, because they highlight specific features not present in other TEs.

6. Conclusion

Object-oriented modeling is an attractive approach to the modeling of complex genetic systems. It is very intuitive and makes it possible to model systems incorporating any desired level of complexity. Modeling is managed in the object-oriented framework, from the formalization of the conceptual model to the construction of a simulation program. Objects used in one modeling project can also be reused as building blocks in the construction of other models. This should make it possible to build up an increasingly complete toolbox, increasing the efficiency of new model production.

7. References

Anxolabéhère, D., Benes, H., Nouaud, D. and Periquet, G. (1987). Evolutionary steps and transposable elements in *Drosophila melanogaster:* the missing RP type obtained by genetic transformation. Evolution **41**: 846–853.

Anxolabéhère, D., Hu, K., Nouaud, D. and Périquet, G. (1990). PM system: a survey of *Drosophila melanogaster* strains from the People's Republic of China. Genet. Sel. Evol. **22**: 175–188.

Anxolabéhère, D., Kidwell, M.G. and Périquet, G. (1988). Molecular characteristics of diverse populations are consistent with a recent invasion of *Drosophila melanogaster* by mobile *P* element. Mol. Biol. Evol. **5**(3): 252–269.

Anxolabéhère, D., Nouaud, D., Périquet, G. and Ronsseray, S. (1986). Evolution des potentialités dysgénésiques du système P-M dans des populations expérimentales mixtes P, Q, et M' de *Drosophila melanogaster* . Genetica **69**: 81–95.

Beall, E. L. & Rio, D. C. (1997) *Drosophila P*-element transposase is a novel site-specific endonuclease. Genes Dev **11**, 2137–2151

Black, D. M., Jackson, M. S., Kidwell, M. and Dover, G. A. (1987). KP elements repress *P* induced hybrid dysgenesis in *Drosophila melanogaster*. EMBO J. **6**: 4125–4135.

Boehnke M. (1986) Estimating the power of a proposed linkage study: a practical computer simulation approach. Am. J. Hum. Genet., **39**, 513–527

Bourgain, C., Génin, E., Quesneville, H., and Clerget-Darpoux, F. (2000)—Search for multifactorial disease susceptibility genes in founder populations—Ann. Hum. Genet, **64**, 255–265

Engels, W. R. (1989). *P* elements in *Drosophila*, in *Mobile DNA*, edited by D. Berg and M. Howe. ASM Publication, Washington D.C., pp. 437–484.

Engels, W. R., Johnson-Schlitz, D. M., Eggleston, W. B. and Sved, J. (1990). High-frequency *P* element loss in *Drosophila* is homolog-dependent. Cell **62**: 515–525.

Forrest, S. (1993) Genetic algorithms : Principles of natural selection applied to computation. Science **261**, 872–878

Hampe, J., Wienker, T., Schreiber, S., and Nürnberg, P. (1998). POPSIM: a general population program. Bioinformatics, **14**, 458–464.

Holland, J. H. (1995) Addison-Wesley (ed) *Hidden Order, How adaptation builds complexity.*

Hoogland, C. and Biémont, C. (1997). DROSOPOSON: a knowledge base on chromosomal localisation of transposable element insertions in *Drosophila*. Comput. Appl. Biosci.**13**: 61–8

Jackson, M.S., Black, D.M. and Dover, G.A. (1988). Amplification of KP elements associated with the expression of hybrid dysgenesis in *Drosophila melanogaster*. Genetics **120**: 1003–1013.

Jurka, J. (1998) Repeats in genomic DNA: mining and meaning. Curr Opin Struct Biol **8**, 333–337

Kidwell, M. G. (1985). Hybrid dysgenesis in *Drosophila melanogaster:* nature and inheritance of *P* element regulation. Genetics **111**: 337–350.

Kidwell, M. G., Frydryk, T. and Novy, J.B. (1983). The hybrid dysgenesis potential of *Drosophila melanogaster* strains of diverse temporal and geographic origin. Drosophila Inf. Serv. **59**:63–69.

Kruglyak L., Daly, M.J., Reeve-Daly, M.P. and Lander, E.S. (1996). Parametric and nonparametric linkage analysis: A unified multipoint approach, Am. J. Hum. Genet. **58**: 1347–1363

Kruglyak, L. and Lander, E.S (1998). Faster multipoint linkage analysis using Fourrier transforms. J. Comput. Biol. 5:1–7

Kruglyak, L. and Lander, E.S. (1995). Complete multipoint sib-pair analysis of qualitative and quantitative traits, Am. J. Hum. Genet. **57**(2):439–54

Preston, C. R. and Engels, W. R. (1989). Spread of *P* transposable elements in inbred lines of *Drosophila melanogaster.* Proc. Nucleic Acid. Res. Mol. Biol. **36**: 71–85.

Quesneville, H. and Anxolabéhère, D. (1997a). GENOOM: a simulation package for GENetic Object Oriented Modeling.- In the Proceedings of the European Mathematical Genetics Meeting, Annals of Human Genetics **61**: 543.

Quesneville, H. and Anxolabéhère, D. (1997b). Simulation of *P* element horizontal transfer in *Drosophila*. Genetica **100**: 295–307

Quesneville, H. and Anxolabéhère, D. (1998). Dynamics of transposable elements in metapopulations: a *P* element invasion model. Theor. Popul. Biol **54**(2): 175–193.

Quesneville, H. and Anxolabéhère, D. (2001). Genetic algorithm-based model of evolutionary dynamics of class II transposable elements - J. Theor. Biol, **213**, 21–30

Rasmusson, K. E., Raymond, J. D. and Simmons, M. J. (1993). Repression of hybrid dysgenesis in *Drosophila melanogaster* by individual naturally occurring *P* elements. Genetics **133**:605–622.

Rio, D.C. (2002). *P* Transposable elements in *Drosophila melanogaster*, in N.L. Craig, R. Craigie, M. Gellert, and A.M. Lambowitz (eds) *Mobile DNA II*. ASM Press, Washington DC. pp 484–518

Terwilliger, J.D., Ott, J., (1994). Handbook of human genetic linkage. John Hopkins University Press, Baltimore.

Weeks, D.E., Ott, J, Lathrop, G.M. (1990) SLINK: a general simulation program for linkage analysis. Am. J. Hum. Genet., **59 (Suppl.)**, A204

Wilson A.F., Bailey-Wilson, J.E., Pugh, E.W., Sorant, A.J.M. (1996) The Genometric Analysis Simulation Program (G.A.S.P.): a software tool for testing and investigating methods in statistical genetics. Am. J. Hum. Genet., **59 (Suppl.)**, A193

Biodata of **Hanne Volpin** author (with co-author Hinanit Koltai) of "***Postgenomic Challenges in Plant Bioinformatics***".

Dr. Hanne Volpin is the Head of the Department of Genomics and Bioinformatics **at** The Agricultural Research Organization, The Volcani Center, Bet Dagan, Israel. She received her Ph.D. from the Microbiology, Faculty of Agriculture, Hebrew University of Jerusalem, Israel in 1995. Dr. Volpin served during 1999–2002 as the Head of Bioinformatics Services, Compugen Ltd, Tel Aviv. Her major research interests are in Data integration, Signal transduction, and Regulatory networks.

E-mail: **hanne@volcani.agri.gov.il**

J. Seckbach and E. Rubin (eds.), The New Avenues in Bioinformatics, 135–143.
© 2004 *Kluwer Academic Publishers. Printed in the Netherlands.*

POSTGENOMIC CHALLENGES IN PLANT BIOINFORMATICS

HANNE VOLPIN AND HINANIT KOLTAI
Department of Genomics and Bioinformatics, The Agricultural Research Organization, The Volcani Center, Bet Dagan 50250, Israel.
hanne@agri.gov.il

1. Introduction

There are more than 250,000 species of plants. They represent a wide variety of growth habits, adaptive responses, and useful traits. Plant genomics and post-genomics studies are expected to provide an unprecedented enhancement of our understanding of the biological events during plant growth, development and interactions with environmental factors. Plant diversity and complexity provide the opportunity to study about the similarities and differences among organisms and the basis of ecologic adaptations while taking advantage of large collections of cultivars and wild relatives with diverse life forms

Post-genomics approaches for plant biology studies include powerful techniques such as reverse genetics approaches involving genome-scale knockout mutations and, especially, T-DNA insertion mutants that have become a valuable resource for the study of gene function in Arabidopsis (Tax and Vernon, 2001; Thorneycroft *et al.*, 2001). In addition, proteomics studies are being performed using protein microarrays for the detection of immobilized antigen with antibodies or vice versa (reviewed by Kersten *et al.*, 2002), and metabolomics studies, aiming to examine the set of metabolites synthesized by a biological system (reviewed by Fiehn, 2002) are being integrated.

Facilitating plant genomics research is the combination of efficient of high throughput transformation systems, short generation times and high proliferation rates. Integrated genomics studies are likely to reinforce the ability not only to select the candidate genes most suitable for manipulation, in order to enhance or reduce biological processes, but also to provide a better and precise prediction of the outcome of such manipulation. The genetic pathways that typify a "plant" will be understood. Therefore, the influence of manipulation of one or more components of a certain signaling pathway, on other pathways that converge with the manipulated one, may be better predicted.

Using the growing collection of available sequence data, and combining bioinformatics and functional genomics should lead to a greater understanding of the genetic networks that are activated during various plant processes. Nevertheless, bioinformatics analysis presents some major intrinsic difficulties. Those include the variable quality of large sets of data and lack of standardization. For example, some EST libraries are redundant, some are not, and a gene may refer to the genomic DNA sequence, either to the transcribed mRNA, or only to the coding sequence. In addition, the computational tools for the mining of the data to converse it to usable knowledge are still immature. This is mainly because the mining becomes a significant statistical challenge once the high-throughput approach to data collection is not aimed at a specific biological question.

All the same, plant genomics is rapidly evolving. Currently, plant genomics is well behind human genomics, but the tools used in the Human Genome Project are being adapted and together with the obvious advantages of having high throughput transformation systems for plants, finding solutions for problems in food production and food quality will probably evolve much faster than solutions for parallel problems in drug target prediction.

Interdisciplinary research is constantly progressing towards the development of high-throughput laboratory techniques and the improvement of computational tools, for accurate data production and analysis. The correct interpretation of the data, and the correlation of the knowledge gained from many studies, both in genomics and in other fields, will pave the way towards the mapping of all genetic pathways, their interactions, divergences and convergences, and lead to the identification of the key genes that confer important traits. It is important to keep in mind that data-interpretation problems may distort the overall picture obtained by using genomics and bioinformatics approaches. This is mainly because the lack of the temporal and spatial cascades of events that take place during various developmental processes and high-throughput data concerning cell-specific reaction to external stimuli. Nevertheless, the elucidation of the larger, even ever so blurred, picture by computational means will increase our understanding of global phenomena in plant biology. This understanding will greatly facilitate the selection of candidates for the in-depth study of individual genes, proteins or processes. The combination of the correct global map and detailed information at the molecular level will eventually enable us to model the entire cellular program that governs plant growth, development and responses to environmental conditions and predict the outcome of hypothetical changes within the system.

2. Arabidopsis as a Model Plant

Arabidopsis is perhaps a good facilitator for understanding the functions of genes, proteins and genetic networks, due to the wide range of available Arabidopsis-related bioinformation. Nevertheless, extrapolating this knowledge to phylogenetic distinct families may be problematic. For example, only limited co-linearity (microsynteny), over small chromosomal segments exist between Arabidopsis and tomato (Reviewed by Mysore *et al.*, 2001) and Arabidopsis and maize genomes, and a significant proportion of the maize ESTs encode highly diverged or maize-specific proteins (Brendel *et al.*, 2002). Also, despite the high proportion of the predicted Arabidopsis proteins that are significantly homologous to those of rice (85%; draft of rice genome sequence released April, 2002; Goff *et al.*, 2002; Yu *et al.*, 2002), only 2% of the syntenic protein pairs in rice and Arabidopsis (two proteins found in close proximity in both species) on Arabidopsis chromosome 5 are adjacent to one another; most are separated by 1 to 150 intervening proteins (Goff *et al.*, 2002). Therefore, despite the conservation of the gene repertoire between Arabidopsis and other *Brassica* species (Paterson *et al.*, 2001) the establishment of synteny between genomes of species belonging to families as divergent as those of Arabidopsis and tomato, rice or maize may be difficult. In addition, many evolutionary homologues genes do not show identical expression patterns between different plant species and do not seem to be related to the same biological processes, although having the same biochemical function (Volpin unpublished).

Despite the limited similarities in genome organization and function between plants of economical importance and Arabidopsis, most of the high-throughput studies of plant processes have used Arabidopsis.

3. Molecular Sequence Characterization and Annotation

Several major genomics initiatives concerning both model plants and plants of economical importance, both with relatively small genomes are currently in progress, or they have led to the publication of genomes. The *Arabidopsis thaliana* genome has been completed (110 Mb; Arabidopsis Genome Initiative, 2000) and a draft of the sequence of the rice genome has recently been released. The target date for the final version is 2004 (Goff *et al.*, 2002; Yu *et al.*, 2002). Also in progress is the sequencing of the genome of the model legume *Medicago truncatula*. Genomics centers for plants include The Tomato Center (http://www.sgn.cornell.edu/), The Rice Center (http://www.tigr.org/tdb/e2k1/osa1/intro.shtml), The Legume *Medicago truncatula* Center (http://www.medicago.org/), The Banana Center (http://www.promusa.org/), Peach and other *Rosaceae* (http://www.genome.clemson.edu/gdr) and The Eucalypt Center (http://www.agrf.org.au/future_initiatives.html).

In contrast, the complete sequencing of plants with large genomes is yet impracticable, especially for those with genomes larger than the 3,000 Mb human genome, such as barley (5,000 Mb) and wheat (16,000 Mb). A more practical approach to gene discovery in such large-genome crop plants that provide a wealth of information in a relatively short time, is through the development of databases of expressed sequence tags (ESTs), a set of single-pass sequenced cDNAs from an mRNA population derived from cells of a specified tissue, organ, developmental state or environmental condition.

On March 26, 2004 the EST database (dbEST, GenBank, accessible at http://www.ncbi.nlm.nih.gov/dbEST/dbEST_summary.html) contained 20,442,611 public entries, and it continues to grow daily. The race between database growth and proficient query capabilities continues and computational improvements are continuously needed to support interpretation of the huge amounts of data for efficient biological discoveries. Plants for which there are major collections of ESTs include: *Triticum aestivum* (wheat), 549,926 ESTs; *Zea mays* (maize), 394,498; *Hordeum vulgare* ssp. *vulgare* (barley), 356,855 ESTs; *Glycine max* (soybean) 346,582 ESTs; *Oryza sativa* (rice), 283,989; *Arabidopsis thaliana* (thale cress), 204,396 ESTs; *Medicago truncatula* (barrel medic), 187,763; *Lycopersicon esculentum* (tomato), 150,519; *Solanum tuberosum* (potato), 144,730; and *Sorghum bicolor* (sorghum), 161,813 ESTs. Among the coniferophyta , *Pinus taeda* (loblolly pine) leads with a collection of 110,622 ESTs and the anthophyta, *Populus tremula x Populus tremuloides* heads the list with 65,981 ESTs.

Despite ESTs being typically short, with a majority representing only the 3' untranslated region (UTR) and of relatively low quality, which makes their functional annotation difficult, ESTs are useful molecular landmarks. They provide a profile of the mRNA population in a cell population, and those from groups of tissues can be used to answer questions about tissue-specific genes and to identify novel proteins (Allikmets *et al.*, 1995; Braren *et al.*, 1997). In particular, sequences that are longer and more accurate, and therefore better represent the underlying genes may be generated by clustering ESTs and mRNAs based on sequence overlaps. Among the public databases containing such putative transcripts are TIGR Gene indices (http://www.tigr.org/tdb/) and NCBI Unigene (http://www.ncbi.nlm.nih.gov/UniGene/). These transcripts can be annotated and otherwise analyzed. Multiple alignments of ESTs from a given gene may reveal polymorphism, and EST databases may be used for *in silico* expression profiling across multiple libraries (Volpin *et al.*, 2002).

A major challenge with respect to genome analysis is annotation—the elucidation and description of biologically relevant features in a sequence. The quality of the annotation will have direct impact on the value of the sequence. At a minimum, the data must be annotated to indicate the existence of gene coding regions and control regions. Further annotation activities that add value to a genome include finding simple and complex repeats, characterizing the organization of promoters and gene families, the distribution of guanine-cytosine (G + C) content, and tying together evidence for homologues and functional motifs (Head-Gordon and Wooley, 2001).

Large-scale genome sequencing projects depend greatly on gene finding to generate accurate and complete gene annotation. Automated methods for identifying protein coding regions in genomic DNA have progressed significantly in recent years. (reviewed by Pertea and Salzberg, 2002). Improvements in gene finding software are caused by a combination of better computational algorithms, a better understanding of the cell's mechanisms for transcription and translation, and the enormous increases in genomic sequence data.

Nevertheless, computational gene identification by sequence inspection remains a challenging problem. For a typical *Arabidopsis thaliana* gene with five exons, at least one of the exons is expected to have at least one of its borders predicted incorrectly by *ab initio* gene finding programs (Brendel and Zhu, 2002). The use of EST evidence or similarities to protein homologues often allows for accurate analysis of individual exon loci. Such methods are part of the routine annotation process. However, because the EST and protein databases are constantly growing, in many cases original annotation can be improved by the incorporation of updated input data.

Once the basic structure of genes has been modeled, similarity searches of new sequences against each other or an existing database is one of the most fundamental processes in computational sequence analysis. Such comparisons relate new sequences to archival sequences that may contain meaningful information about their biochemical function or biological processes. Multiple comparisons are the starting point for identification of protein motifs and evolutionary analysis of genomic/protein sequences.

Encoded in the DNA sequence is a protein's three-dimensional topography, which in turn determines function. Uncovering this sequence-structure-function-systems relationship is a core challenge of modern structural biology today.

Genomics has motivated a significantly increased effort in protein structure determination and structure prediction. The accumulation of 3D protein structures has increased dramatically in the past few years, bringing the total number in the Protein Data Bank (PDB, http://www.rcsb.org/pdb) to 24,908 in March 2004. However, many of these proteins are redundant, and the number of non-redundant structures is estimated to be about less than 25% of the total number (Virkup *et al.*, 2001)

The goal of fold assignment and comparative modeling is to assign each new genome sequence to the known protein fold or structure that it most closely resembles, using computational methods. The 3-D structures of proteins have been better conserved during evolution than their genome sequences (Head-Gordon and Wooley, 2001), and in cases where sequence identity dips below ~25 percent, the so-called "twilight zone", fold assignment algorithms can often be successful in functional assignments for a new sequence (Beamer *et al.*, 1998). Likewise, fold assignment and comparative modeling techniques can then be helpful in proposing and testing hypotheses in molecular biology, such as in inferring biological function, predicting the location and properties of ligand binding sites, and testing remote protein-protein relationships. (Head-Gordon and Wooley, 2001)

4. Genome-Scale Expression Analysis and Modeling of Biological Systems

Genetic networks comprising several or many genes virtually determine and affect every aspect of a living creature. Traditionally, molecular research has involved a "one gene at a time" approach, focused on understanding the influence of single genes. Plant genomics, that focuses on understanding processes occurring in plants by simultaneously studying all or most of its genes, proteins and the interactions of these (instead of a single gene at a time), is generating information about the functions of networks of genes, including processes that correspond to economically valuable traits, such as fruit quality, stress tolerance and disease resistance.

Transcriptomics and proteomics are among the powerful tools for the quantitative, real-time study of gene expression on the genome-scale, assigning products and functions to a greater part of the genes that comprise an organism. Transcriptomics is a mean to globally profile gene transcription in a cell or a tissue, and is mainly applied either by *in situ* synthesis of oligonucleotides ('oligonucleotide microarrays') or by deposition of pre-synthesized DNA fragments ('cDNA microarrays') on solid surfaces, followed by hybridization with labeled gene transcription products (reviewed by Aharoni and Vorst, 2002). Proteomics may offer direct approach of 2-dimensional gel electrophoresis (2DE) of proteins combined with mass spectrometry (MS), to generate a catalog of expressed proteins (reviewed by Kersten *et al.*, 2002); reverse proteomics globally map protein interactions utilizing yeast two-hybrid (Y2H) system (Fang *et al.*, 2002; Walhout and Vidal, 2001). However, besides initiatives like the Arabidopsis Information Resource (TAIR) (ftp://ftp.arabidopsis.org/home/tair/Microarrays/Datasets) that contains results from more than 500 Arabidopsis microarrays datasets, Nottingham Arabidopsis Stock Centre (http://affymetrix.arabidopsis.info/), and the Rice Microarray Opening Site (http://microarray.rice.dna.affrc.go.jp) with results from more than 600 rice microarrays, there is so far very limited public data available for plant expression analysis and plant proteomics.

In order to understand the functioning of organisms on the molecular level, we need to know which genes are expressed, when and where in the organism, and to which extent. The regulation of gene expression is achieved through genetic regulatory systems structured by networks of interactions between DNA, RNA, proteins, and small molecules. As most genetic regulatory networks of interest involve many components connected through interlocking positive and negative feedback loops, an intuitive understanding of their dynamics is hard to obtain. Consequently, formal methods and computer tools for the modeling and simulation of genetic regulatory networks will be indispensable. The formalisms that have been employed in mathematical biology and bioinformatics to describe genetic regulatory systems, in particular directed graphs, Bayesian networks, Boolean networks and their generalizations, ordinary and partial differential equations, qualitative differential equations, stochastic equations, and rule-based formalisms has recently been reviewed by de Jong (2002). In addition, the paper discusses how these formalisms have been used in the simulation of the behavior of actual regulatory systems. However, breakthroughs in experimental technologies, as well as advances in software, and analytical methods are required before the achievements of systems biology can be fully appreciated (Kitani, 2002).

5. Conclusions

The information age is revolutionizing the natural sciences. Genome sequencing and new techniques for high throughput measurements allow us to gather comprehensive data sets on biological system performance. The enormous amount of data obtained poses one of the major bioinformatics challenges of these days: that of interpreting the data in order to obtain useful knowledge. The obtained knowledge is likely to be implement in biological studies and lead to breakthroughs in all areas of biology, including plant sciences.

The first level of the computational challenge is analysis of the rapidly emerging genomic data at the sequence level. However, knowing the sequence of DNA only does not necessarily tell us about the position, structure or function of the genes, nor does it tell us about the combined action of their protein products, which is the essence of higher order biological function. Complete annotation will include the determination of structure and function of proteins. A major bioinformatics challenge is to move from analysis of the individual macromolecules and their interactions to the complex networks that make up the processes of cellular decisions.

Today, large-scale exploratory experiments are gathering as much data as possible, such as large collections of ESTs, the sequencing of entire genomes, extensive information on protein 3D structure etc. So now, when a biologist forms a hypothesis, the data may already be in such a collection, just a computer search away. However, currently, accessing genomics and proteomics data is a time consuming challenge of reformatting data, script writing and database querying. It is in great contrast to the ideal environment that would consist of a single location, that provides effective and perceptive access to a consistent view of data from many sources through an intuitive and useful interface. Strong emphasis should be placed on solid, user-friendly interfaces that will allow the biologist to perform both exploratory and predictive data mining as an integrated part of the biological experiment.

6. References

Aharoni A and Vorst O (2002) DNA microarrays for functional plant genomics. Plant Molecular Biology 48: 99–118

Allikmets R, Gerrard B, Glavac D, Ravnik-Glavac M, Jenkins NA, Gilbert DJ, Copeland NG, Modi W and Dean M (1995) Characterization and mapping of three new mammalian ATP-binding transporter genes from an EST database. Mammalian Genome 6: 114–117

Arabidopsis Genome Initiative (2000) Analysis of the genome sequence of the flowering plant *Arabidopsis thaliana*. Nature 408: 796–815

Beamer L, Fischer D, and Eisenberg D (1998) Detecting distant relatives of mammalian LPS-binding and lipid transport proteins. Protein Science 7:1643–1646

Braren R, Firner K, Balasubramanian S, Bazan F, Thiele HG, Haag F and Koch-Nolte F (1997) Use of the EST database resource to identify and clone novel mono(ADP-ribosyl) transferase gene family members. Advances in Experimental Medicine and Biology 419: 163

Brendel V, Kurtz S and Walbot V (2002) Comparative genomics of Arabidopsis and maize: prospects and limitations. Genome Biology 3: REVIEWS 1005

Brendel V, and Zhu W (2002) Computational modeling of gene structure in *Arabidopsis thaliana*. Plant Molecular Biology 48:49–58.

Fang Y, Macool DJ, Xue Z, Heppard EP, Hainey CF, Tingey SV and Miao GH (2002) Development of a high-throughput yeast two-hybrid screening system to study protein-protein interactions in plants. Molecular Genetics and Genomics 267: 142–153

Fiehn O. 2002. Metabolomics-the link between genotypes and phenotypes. Plant Molecular Biology 48: 155–171

Genome International Sequencing Consortium (2001) Initial sequencing and analysis of the human genome. Nature 409: 860–921

Goff SA, Ricke D, Lan TH, Presting G, Wang R, Dunn M, Glazebrook J, Sessions A, Oeller P, Varma H, Hadley D, Hutchison D, Martin C, Katagiri F, Lange BM, Moughamer T, Xia Y, Budworth P, Zhong J, Miguel T, Paszkowski U, Zhang S, Colbert M, Sun WL, Chen L, Cooper B, Park S, Wood TC, Mao L, Quail P, Wing R, Dean R, Yu Y, Zharkikh A, Shen R, Sahasrabudhe S, Thomas A, Cannings R, Gutin A, Pruss D, Reid J, Tavtigian S, Mitchell J, Eldredge G, Scholl T, Miller RM, Bhatnagar S, Adey N, Rubano T, Tusneem N, Robinson R, Feldhaus J, Macalma T, Oliphant A and Briggs S (2002) A draft sequence of the rice genome (*Oryza sativa* L. ssp. *japonica*). Science 296: 92–100

Head-Gordon T and Wooley JC (2001) Computational challenges in structural and functional genomics. IBM Systems Journal 40:265–296

de Jong H (2002) Modeling and simulation of genetic regulatory systems: a literature review. Journal of Computational Biology 9: 67–103.

Kersten B, Burkle L, Kuhn EJ, Giavalisco P, Konthur Z, Lueking A, Walter G, Eickhoff H and Schneider U (2002) Large-scale plant proteomics. Plant Molecular Biology 48: 133–141

Kitano H (2002) Systems biology: a brief overview. Science 295:1662–1664.

Mysore KS, Tuori RP and Martin GB (2001) *Arabidopsis* genome sequence as a tool for functional genomics in tomato. Genome Biology 2: REVIEWS 1003

Paterson AH, Lan TH, Amasino R, Osborn TC and Quiros C (2001) Brassica genomics: a complement to, and early beneficiary of, the Arabidopsis sequence. Genome Biology 2: REVIEWS 1011.

Pertea M and Salzberg SL (2002) Computational gene finding in plants. Plant Molecular Biology 48:39–48

Tax FE and Vernon DM (2001) T-DNA-associated duplication/translocations in Arabidopsis. Implications for mutant analysis and functional genomics. Plant Physiology 126: 1527–1538

Thorneycroft D, Sherson SM and Smith SM (2001) Using gene knockouts to investigate plant metabolism. Journal of Experimental Botany 52: 1593–1601

Virkup D (2001) Completeness in structural genomics. Nature Structural Biology 8:559–566.

Volpin H, Kahana A, Bendov R, Jaffee M, Koltai H, Kapulnik Y (2002) EST Analysis of Gene Expression in Host-Microbe Interactions (abs). 10th New Phytologist Symposium, Nancy, France. October

Walhout AJM and Vidal M (2001) High-throughput yeast two-hybrid assays for large-scale protein interaction mapping. Methods 24: 297–306

Walker MG, Volkmuth W, Sprinzak E, Hodgsdon D and Klinger T (1999) Prediction of gene function by genome-scale expression analysis: Prostate cancer-associated genes. Genome Research 9:1198–1203

Yu J, Hu S, Wang J, Wong GK, Li S, Liu B, Deng Y, Dai L, Zhou Y, Zhang X, Cao M, Liu J, Sun J, Tang J, Chen Y, Huang X, Lin W, Ye C, Tong W, Cong L, Geng J, Han Y, Li L, Li W, Hu G, Huang X, Li W, Li J, Liu Z, Li L, Liu J, Qi Q, Liu J, Li L, Li T, Wang X, Lu H, Wu T, Zhu M, Ni P, Han H, Dong W, Ren X, Feng X, Cui P, Li X, Wang H, Xu X, Zhai W, Xu Z, Zhang J, He S, Zhang J, Xu J, Zhang K, Zheng X, Dong J, Zeng W, Tao L, Ye J, Tan J, Ren X, Chen X, He J, Liu D, Tian W, Tian C, Xia H, Bao Q, Li G, Gao H, Cao T, Wang J, Zhao W, Li P, Chen W, Wang X, Zhang Y, Hu J, Wang J, Liu S, Yang J, Zhang G, Xiong Y, Li Z, Mao L, Zhou C, Zhu Z, Chen R, Hao B, Zheng W, Chen S, Guo W, Li G, Liu S, Tao M, Wang J, Zhu L, Yuan L and Yang H (2002) A draft sequence of the rice genome (*Oryza sativa* L. ssp. *indica*). Science 296: 79–92

Biodata of **Hanqing Xie** author (with R. Gill-More) of the chapter ***"Transcriptome Analysis Through Expressed Sequences."***

Dr. Xie obtained his B.Sc. from Department of Biochemistry Nanjing University in Nanjing People's Republic of China in 1987 and his Ph.D. Degree from Department of Biochemistry and Molecular Biology from The George Washington University in 1995. After finishing postdoctoral training in University of Alabama and the Johns Hopkins University, he became a computational biology researcher in Compugen Inc. in 2000. While interested in developing algorithms for a variety of biological problems, he currently works on transcriptome analysis and application Gen Ontology.

E-mail: **han@cgen.com**

J. Seckbach and E. Rubin (eds.), The New Avenues in Bioinformatics, 145–162.
© 2004 *Kluwer Academic Publishers. Printed in the Netherlands.*

TRANSCRIPTOME ANALYSIS THROUGH EXPRESSED SEQUENCES

HANQING XIE[1] and RAVEH GILL-MORE[2]
[1]*Compugen, Inc., 7 Center Drive, Suite 9, Jamesburg, NJ 08831 and*
[2]*Compugen, Ltd. 72 Pinchas Rosen St. Tel Aviv 69512, Israel*

1. Transcriptome Analysis

Recent advances in sequencing, computational biology, and information science have brought great new opportunities in biological research. One of the opportunities is that researchers are able to apply a systematic approach towards understanding some of the objects or phenomena in molecular biology/genetics and modern medicine. Not long ago, such a systematic approach was restricted to very few sub-disciplines in biological and medical sciences, such as anatomy or taxonomy, where scientists and researchers have been trying to catalogue and investigate every component of a given system. This new opportunity specifically refers to current research interest on various "-omes," including genome, transcriptome, proteome, metabolome, and others. Genome research addresses the physical sequence, the alternation, and the evolution of genetic materials, which are composed of DNA in most species. Transcriptome research (Velculescu *et al.*, 1995) focuses on the investigation of the expression of genetic materials and its regulation. A transcriptome is a collection of expressed sequences in a particular cell at a specific time point. A related concept is the "whole transcriptome," which refers to the complete set of RNA molecules derived from the genome during the lifetime of an organism. The expressed sequences could be mRNA, tRNA, rRNA, or other RNA species. While current transcriptome analysis focuses on mRNA molecules, which code for proteins, transcriptome is much more than proteome (the collection of all proteins in an organism), especially since more and more non-protein coding RNA molecules, besides tRNA and rRNA, have been identified recently (International Human Genome Sequencing Corsortium, 2001, Storz, 2002, Mouse Genome Sequencing Consortium, 2002). For instance, there has been a great interest in the significant role of small interference RNA in gene silencing (Plasterk, 2002). Furthermore, a large percentage of mouse genes are likely to be non-protein coding (The FANTOM Consortium, 2002) and the non-coding genes are likely to be conserved across different species and unspliced (The FANTOM Consortium, 2002).

Genome and transcriptome are two related, yet different concepts. In a particular cell, the genome remains stable. It undergoes little change during transitions, such as cell division and cell differentiation. A transcriptome in a cell, on the other hand, is dynamic and is changed both spatially and temporally. While different tissues or cell types in an organism have the same genome (except for somatic mutations), their transcriptomes could be vastly different. This dynamic and transitive nature of a transcriptome requires unique technologies and computational consideration. Various experimental and computational techniques have been employed to investigate the spatial and temporal patterns of gene expression in healthy and diseased states. These techniques contribute immensely to further understanding of both the biological fundamentals and disease mechanisms.

This chapter focuses on computational and technological aspects of transcriptome research through expressed sequences. It covers computational considerations and a brief discussion of experimental methods used for data acquisition. Alternative splicing modeling, especially on human expressed sequence data, is emphasized. A very schematic introduction to splicing and alternative splicing is included for those with limited exposure to molecular biology. Excellent reviews on the topics of splicing, alternative splicing, and transcriptome have been published recently (Burge, 2001, Strausberg *et al.*, 2001, Maniatis *et al.*, 2002, Modrek *et al.*, 2002), and it is recommended that the readers consult them for in-depth content. Gene prediction software and algorithms (Burge *et al.*, 1997, Lukashin *et al.*, 1998, Salamov *et al.*, 2000, Yeh *et al.*, 2001, Meyer *et al.*, 2002), which identify exons and introns through various computational algorithms (including analysis of expressed sequences), are not discussed here, and interested readers can refer to other chapters in this book and related literature.

Two additional techniques, SAGE and microarrays, have been used for investigating transcriptomes in specific cell types, tissues, or particular processes. These techniques have been extensively reviewed and are discussed here very briefly. SAGE (Serial Analysis of Gene Expression) is an innovative technology that has been used widely for transcriptome profiling (Valculescu *et al.*, 1995). In this method, a sequence tag of 10 or more bases is generated for each transcript in the cell or tissue of interest, and those 'unique' tags thereby constitute a SAGE library. Sequencing of the SAGE library creates a transcript profile. The resulting sequence tags are subsequently mapped to the genes, and the expression levels of the genes can be obtained through the counting of the SAGE tags, although it happens quite often that a SAGE tag may correspond to several genes, and some of the genes may not have a meaningful SAGE tag. An improved version of SAGE, called long-SAGE, has been devised, in which the tag length reaches 20–23 base pairs (Saha *et al.*, 2002), and therefore, the SAGE tags can be mapped more accurately to unique genes. SAGE has been used to construct the yeast whole transcriptome (Velculescu *et al.*, 1997), a human transcriptome (Caron *et al.*, 2001), and a non-small cell lung cancer transcriptome (Fujii *et al.*, 2002), and to understand the treatment-induced changes in transcriptomes (Robert-Nicound *et al.*, 2001, Menssen *et al.*, 2002). Modified SAGE has been used to investigate the transcriptome changes during monocytic leukemia cell differentiation (Piquemal *et al.*, 2002).

Microarray technology holds great potential in investigating transcriptomes. Microarray studies with high-density oligonucleotide probes have been used to discover new transcripts in the intergenic region of *E. Coli* (Tjaden *et al.*, 2002), circadian cycling of the mouse liver transcriptome (Akhtar *et al.*, 2002), transcriptome changes during mouse placental development (Hemberger *et al.*, 2002), the alteration of budding yeast transcriptome during meiosis (Primig *et al.*, 2000), and *Arabidopsis thaliana* transcriptome during systemic acquired resistance (Maleck *et al.*, 2000). In yeast, microarray probes have been designed and used for identifying splicing events (Clark *et al.*, 2002).

Other more 'traditional' techniques for transcriptome analysis, such as *in situ* nucleic acid hybridization analysis (see Carson *et al.*, 2002 for an interesting discussion), Northern hybridization, differential display (Liang *et al.*, 1992), and real-time PCR, investigate transcriptomes from different perspectives, and they all contribute to the understanding of transcriptomes. High throughput processes for those techniques are not established or widely adopted yet, and their results remain to be linked with genomic sequences. Therefore, their contributions toward transcriptome analysis on a large scale are yet to be seen and they are not discussed here.

2. Transcription and Alternative Splicing

In the process of transcription, RNA molecules are synthesized according to DNA templates through the actions of DNA-dependent RNA polymerases. Several major RNA species have been characterized—mRNA, rRNA, tRNA, and other small RNA molecules. Transcription is accomplished through a transcription complex where probabilistic and sequential assembly of proteins and small nuclear RNA (snRNA) likely occurs (Dundr *et al.*, 2002). So far, transcriptome research has been mainly concerned with mRNA molecules, which are translated into proteins and constitute the starting point of the proteome. It is hard to estimate the total number of transcripts present in a cell, or in the whole transcriptome, especially since the number of human genes is still controversial, with estimates ranging from 30,000 to 80,000 genes (Ewing *et al.*, 2000, Venter *et al.*, 2001, International Human Genome Sequencing Consortium 2001). In addition, the percentage of the human genome being expressed still lacks systematic assessment. The exons in the human genome, as derived from expressed sequence information and from gene prediction, cover slightly more than 1% of the genome (Venter *et al.*, 2001). On the other hand, a recent study using microarray techniques indicates that at least 10% of the genome is likely expressed (Kapranov *et al.*, 2002).

Transcription of protein-coding sequences (mRNA) by RNA polymerase II requires the participation of a diverse array of proteins, including one or more transcription activators, chromatin-regulating factors, and general initiation and elongation factors. More recently, a protein thought to be mainly involved in the proteolytic pathway, ubiquitin, has been shown to be involved in the general transcription process (Conaway *et al.*, 2002). One of the important features of the transcription process in eukaryotes is pre-mRNA splicing. During splicing, the introns from a pre-mRNA molecule are excised, and the exons are joined together to form mature RNA in a ribonucleoprotein complex called splicesome. Splicing occurs before translation of the vast majority of human protein-coding genes. A recent survey of human Refseq sequences (Pruitt *et al.*, 2001) through alignment to the genome sequence indicates that only 8.5% of Refseq sequences are unspliced (Xie *et al.*, submitted). Specific base pair consensi (for example, the canonical splicing site consensus exon...GT...intron...AG...exon) exist in the exon-intron boundaries and are mechanistically important to the splicing process. These consensus signals are routinely used in the computational transcriptome analysis to aid the identification of intron-exon boundaries and transcription directionality.

Differential joining of exons in a pre-mRNA molecule produces several different mRNA molecules (called splice variants) from the same gene. This process is known as alternative splicing. Alternative splicing occurs to more than 50% of human genes and most alternative splicing events result in altered protein products (Brett *et al.*, 2000, International Human Genome Sequencing Consortium, 2001, Modrek *et al.* 2001. The FANTOM Consortium,

2002). Thus, alternative splicing has been regarded as one of the major means of achieving protein diversity in higher organisms. A preliminary study indicated that the percentage of alternatively spliced genes is likely to be comparable across different species (Brett et al, 2002), although separate studies indicated that human may have a higher incidence of alternative splicing among the analyzed genes than mouse (Modrek *et al.*, 2001, The FANTOM Consortium 2002). Individual alternative splicing events have been customarily grouped into four categories: intron retention, exon skipping, alternative 5' splicing, and alternative 3' splicing. In addition, some of the exons are mutually exclusive; that is, those exons do not appear in same mature transcripts. The specific alternative splicing pattern in a pre-mRNA molecule is mainly determined by the short base pair sequences in exons, introns, and neighboring sequences (Maniatis *et al.*, 2002). Changes in splicing sites and these splicing regulating elements cause erratic splice transcripts and have been implicated in various diseases (Krwczak *et al.*, 1992), Ars *et al.*, 2000, Teraoka *et al.*, 2000, Liu *et al.*, 2001, Stella *et al.*, 2001). Some of the disease-causing mutations, including synonymous mutations, may reside in the splicing regulating elements or the splicing sites, and thus alter the splicing pattern in the mutated genes.

Transcriptome analysis also addresses other phenomena that increase the diversity of transcripts and proteins in many species, such as alternate transcription initiation, alternate transcription termination/polyadenylation, interleaving genes, overlapping genes, and antisense/sense pairing genes. Many of these phenomena have been investigated through the analysis of expressed sequences. Trans-splicing, where the exons from different pre-mRNA molecules are joined together, has been described in vivo and in vitro; however, its involvement in cellular functions in mammalians and mechanistic details remain to be investigated (Maniatis *et al.*, 2002 and references therein).

3. EST Data

In the last two decades, molecular biologists and genome researchers have deposited sequence data, including both genomic and expressed sequences, into International Nucleotide Sequence Database Collaboration, comprised of DNA DataBank of Japan (DDBJ), the European Molecular Biology Laboratory (EMBL), and GenBank at NCBI, NIH. Expressed sequences are composed of RNA sequences and ESTs (Expressed Sequence Tags). In GenBank release version 131, there are slightly over 100,000 human mRNA sequences, including those in the gbpri dataset and those in high throughput cDNA sequencing (gbhtc dataset), and over 4.7 million human EST sequences (Adams *et al.*, 1991). The expressed sequences dataset includes entries from the MGC project for human (Strausberg *et al.*, 1999) and the FANTOM project (The FANTOM Consortium, 2001 and 2002) for mouse, which have sought to obtain complete lists of full-length mRNA sequences. Such full-length sequences are of high quality and maintain the contiguity of many exons of individual genes, and are therefore invaluable in analyzing transcriptomes (Haas *et al.*, 2002). The same can be said about the earlier mRNA sequences, especially those submitted individually by the investigators. They were usually examined, analyzed, and annotated by the investigators, and their quality tends to be higher than that of ESTs. Such a distinction in quality commonly translates to different weights being applied to different data types during various computational transcriptome analyses.

The vast majority of expressed sequences are ESTs, resulting from single-pass and high throughput sequence projects. Briefly, mRNA is extracted from tissue sources or cultured cells, and first strand cDNA synthesis through reverse transcriptase and second strand syntheses through DNA polymerase produce double-stranded cDNA. These cDNA are subsequently ligated with cloning vectors to produce cDNA libraries. Clones selected from these libraries are then sequenced from either end. Each sequencing reaction generates on average 300 base pairs representing a sequence tag for a particular transcript. An EST sequencing project, though expensive, is technically simple to execute since all it requires is a cDNA library, automated DNA sequencing capabilities, and standard bioinformatics protocols.

Different ways in which cDNA libraries are generated result in different portions of expressed sequences or differentially enriched expressed sequences. The first strand cDNA synthesis can be based either on random priming or on poly d(T) priming. Random priming is likely to generate ESTs that correspond to the central portion or the 5' end of a transcript, although it more likely introduces sequences from pre-mRNA molecules (immature RNA). Poly d(T) priming tends to generate 3' ESTs. The cDNA library can be normalized or subtracted in order to obtain sequences that are expressed at low levels, or differentially expressed in one particular tissue or cell source (Bonaldo et al, 1996). In addition, various cDNA libraries have different levels of quality due to a variety of reasons (Sorek *et al.*, in press)

Several large-scale human EST projects have been undertaken in the past decade. Among the 4.7 million human ESTs in GenBank version 131, one quarter came from the Mammalian Gene Collections projects in the National Institute of Health (http://mgc.nci.nih.gov), which aims "to provide a complete set of full-length (open reading frame) sequences and cDNA clones of expressed genes for human and mouse." Another quarter of these ESTs originated from the Cancer Genome Anatomy Project (CGAP) (http://www.ncbi.nlm.nih.gov/CGAP/) whose aim is "to determine the gene expression profiles of normal, precancer, and cancer cells." The ORESTES project, using thousands of mini-libraries, produced more than 700,000 ESTs (Camargo *et al.*, 2001). An earlier EST sequencing effort called I.M.A.G.E. (Integrated Molecular Analysis of Genomes and their Expression) contributed more than half a million ESTs between 1995 and 1997 (Hillier et al, 1996). A few hundred other projects account for the rest of the human EST collections. EST sequences from well-defined tissue sources have been used as a BodyMap of expression information (Okubo *et al.*, 1992). Analysis of EST information also led to the identification of putative single nucleotide polymorphisms (Garg *et al.*, 1999, Buetow *et al.*, 1999)

4. Computational Analysis of Expressed Sequences

An integral part of transcriptome analysis is the investigation of expressed sequences through computational means. The need for algorithmic development and software implementation for this kind of investigation is apparent, considering that in human, tens of thousands of genes, millions of sequences, and billions of expressed data bits are analyzed in the backdrop of more than 3 billion base pairs of genomic sequences. In addition, a myriad of factors complicate the analysis. ESTs and mRNA sequences have errors in base calling and in directionality annotation, may be contaminated with vectors and microbial sequences, and may be chimerical. The alignment of expressed sequences against genomic sequence must also resolve the complexity stemming from repeats, paralogues, pseudogenes, overlapping genes, interleaving genes, sense/anti-sense genes, satellite sequences, and nucleotide polymorphisms, in addition to alternative splicing, alternate transcription initiation, and alternate transcription termination. It is thought that in the post-genomic era, data analysis rather than data collection presents the biggest challenge to biologists (Boguski, 1999). It is an extremely challenging task to derive the transcriptomes, or even to partially understand them, from expressed sequences in view of the multitudes of complicating factors and imperfections. A lot of works have been published on the computational analysis of expressed sequences. These efforts differ in scope and focus. A widely used public platform for human and other species is the UniGene system (http://www.ncbi.nlm.nih.gov/entrez/query.fcgi?db=unigene) (Boguski *et al.*, 1995, Schuler, 1997). UniGene is an automatic system that uses multiple steps to partition expressed sequences into unique gene-oriented clusters. While the system is able to cluster the majority of expressed sequences into different gene clusters, it does not align them to yield any prediction of transcripts. The most recent UniGene Build #156 for *Homo sapiens* includes 83,162 mRNAs, 1,377,792 3' ESTs, 1,778,818 5' ESTs, and 665,548 other/unknown ESTs in a total of 121,062 clusters, among which 22,360 clusters contain at least one mRNA and 109,603 contain at least one EST. There have been efforts to assemble UniGene clusters through genomic alignments (Zhuo *et al.*, 2001, Geier et al., 2001). Other published and/or online databases focus on specific aspects of expressed sequence analysis. For instance, many published works or databases focus on alternative splicing (Burke *et al.*, 1998, Mironov *et al.*, 1999, Brett et al, 2000, Modrek et al, 2001). PALS db (http://palsda.ym.edu.tw) uses UniGene clustering and the longest mRNA in each cluster to obtain alternative splicing information (Huang *et al.*, 2002). SpliceDB collects mammalian splice site information (Burset *et al.*, 2001). STACK (Sequence Tag Alignment and Consensus Knowledgebase) database (http://www.sanbi.ac.za/Dbases.html) uses tissue-level information during the clustering and assembly process (Christoffels *et al.*, 2001). TAP (Transcript Assembly Program) (http://stl.wustl.edu/~zkan/TAP/) clusters and assembles expressed sequences and predicts putative transcripts (Kan *et al.*, 2001). Intronerator (http://www.cse.ucsc.edu/~kent/intronerator/) includes a collection of software tools with an emphasis on alternative splicing for *C. Elegans* (Kent *et al.*, 2001). ASAP (http://www.bioinformatics.ucla.edu/HASDB/) presents extensive information about alternative splicing in human genes through a web interface. ACDB (http://cbcg.nersc.gov/asdb) collects annotations from SWISS-PROT and GenBank entries and assembles the genes accordingly (Dralyuk *et al.*, 1999). ISIS (http://isis.bit.uq.edu.au/) provides a variety of information on splicesomal introns (Croft *et al.*, 2000). In addition, computational analysis of expressed sequences is part of several integrated genome resources, such as NCBI Genome resources (http://www.ncbi.nlm.nih.gov), University of California, Santa Clara Genome Browser (http://genome.cse.ucsc.edu), Ensemble Project (http://www.ensembl.org) from the European Bioinformatics Institute and Sanger Institute, and Gene Indices (Quackenbush *et al.*, 2001) (http://www.tigr.org/tdb/tgi) in The Institute for Genome Research. One of the important tools is the SIM4 program (http://globin.cse.psu.edu), which is used for aligning expressed sequences to genomic sequences and for defining exon-intron boundaries (Kan *et al.*, 2001, The FANTOM Consortium, 2002).

4.1. LEADS PROCESS

ESTs are short, highly redundant, and error-prone. They constitute the majority of expressed sequences in the sequence depositories. A comprehensive computational approach that overcomes data quality issues is needed in order to extract valuable information from EST and mRNA data. The following discussion focuses on the LEADS platform, an EST clustering and assembly platform developed by Compugen. However, processes and issues discussed here are applicable to any computational platforms for transcriptome analysis based on expressed sequences. Whenever possible, publicly available software is cited instead of Compugen's proprietary system. During LEADS process, expressed sequences are initially cleaned and mapped to the genome. Expressed sequences mapped to the same genomic locus are clustered together and assembled into graph theory models where transcripts are predicted. The LEADS platform has been used to investigate various interesting phenomena. For example, Rotem Sorek and colleagues discovered that exonized *Alu* elements are likely to be alternatively spliced with a low retention ratio (Sorek *et al.*, 2002). Anat David and colleagues identified novel prostate-specific proteins from the first introns of human Kallikrein genes (David *et al.*, 2002). The LEADS platform was used to identify genes transcribed from different strands of the same genomic region, the so-called sense/antisense genes (Yelin *et al.*, submitted). In addition, the LEADS platform has been used for new gene discoveries (Matloubian *et al.*, 2001), tissue or cancer-specific alternative splicing events (Xie *et al.*, 2002), design of microarray oligonucleotide probes (Shoshan *et al.*, 2001, Hu *et al.*, 2002), and design of small interference RNA (siRNA) (Compugen, unpublished results).

4.1.1. Cleaning and masking

There are lots of artifacts and contaminants in EST data. For example, ends of ESTs may contain vector sequences, which should be removed. A software program Vecscreen from NCBI (http://www.ncbi.nlm.nih.gov:80/VecScreen/VecScreen.html) can be used to screen out vector sequences. Since ESTs result from single-pass sequencing, and the base calling on the ends of sequencing gel lanes tend to be error-prone, there is a high incidence of sequencing errors in ESTs. It is estimated that up to 3% of EST base calling might be erroneous (Wolfbergs *et al.*, 1997). Regions identified as having low sequence quality are marked as such. A small percentage of ESTs are apparently of microbial origins, either from the contamination of microbes during cDNA library preparation and EST sequencing process or from mRNA tissue sources infected with viruses (Weber *et al.*, 2002). BLAST analysis against microbial genome sequence databases can be used to identify those contaminating sequence portions. ESTs of highly abundant genes (such as some housekeeping genes) and genes that undergo rearrangement processes (such as immunoglobulin genes and T cell receptors) are removed because they impair the assembly process. A substantial number of ESTs and mRNAs contain whole or partial repeat sequences (ALU, L1, etc.) (Sorek *et al.*, in press), and their locations are marked using the REPEATMASKER software (http://ftp.genome.washington.edu/cgi-bin/RepeatMasker) or through simple BLAST analysis against repeat databases.

4.1.2. Genome mapping

EST and mRNA sequences are mapped to genomic locations through local alignments, which can be efficiently accomplished using MEGABLAST. An important step of the mapping process is to differentiate between alignments of an expressed sequence to several different genomic regions (finding potential paralogues and pseudogenes). In order to exclude pseudogenes and identify the actual genes, preference is given to alignments to the genome that include gaps (introns), preferably with the canonical splice site (GT..AG). Segments in expressed sequences that are aligned to many different genomic sequences are marked as potential repetitive elements and excluded. These repeats can be manually added to the repeat database, and this addition improves the genome mapping in the next run.

A substantial number of ESTs and mRNAs fail to align with genomic sequences completely. Many factors, besides EST sequencing artifacts and errors, contribute to this failure. The expressed sequences were derived from tens of thousands of tissues and cell samples from different people. Polymorphisms, especially those with significant inserts or deletions, or large stretches of substituted bases, exist between the sources of genomic sequence and the sources of expressed sequences, which prevent some of the expressed sequences from fully aligning with genomic sequences. Also, a large number of expressed sequences are from cancerous tissues and cells, and somatic mutations, especially those with defective DNA repair process (Lynch *et al.*, 1999), are likely to be more prevalent in cancer tissues and cells (although a recent study indicates that rates of somatic non-synonymous mutation events may be similar in both normal cell types and cancer cells (Wang *et al.*, 2002)). In addition, chromosomal aberrations are common in cancer cells. Some expressed sequences were from the affected chromosomal loci, and these expressed sequences can align with genomic sequences only partially. Furthermore, human genomic sequences have not been finished yet, and those finished could be misassembled; therefore, expressed sequences corresponding to the unfinished or misassembled genome portion cannot be readily aligned. In the end of the genome mapping process, each expressed sequence is aligned to a single location in the genome using a sequencing error model and an intron/exon model. The sequencing error model assumes a higher probability of sequencing errors at the beginning and end of expressed sequences, and the intron/exon model distinguishes between actual genes, pseudo-genes, and gene duplications and prefers alignments that show legal intron/exon splice sites.

4.1.3. Clustering

Overlapping expressed sequences in the same genomic location are clustered together to form a contig. Sequences that are not locally aligned to the genome (mainly sequences that did not have sufficient homology to genomic data) can be added based on partial homology. To make contig data more comprehensible, redundant EST data is removed from large contigs (for example, from those supported by more than 1,000 ESTs).

4.1.4. Assembly

The assembly stage builds a complete "gene picture" with intron/exon structure and splicing information. The multiple alignment process determines gene structure by evaluating all possible alignments between expressed sequences and the genomic location in a contig. It assembles ESTs into a continuous sequence, while modeling exons, introns, and their boundaries according to statistical models of splice sites. More accurate gene structure is obtained by taking into account the abundance and type of supporting expressed sequences (mRNAs are considered more reliable than ESTs). Transcripts with various confidence levels can be predicted. Transcripts that are fully supported by expressed sequences or are covered by mRNA sequences or contain clone mates (5 and 3 sequences that originate from the same clone) are considered to be of high quality. The output of the assembly stage is typically a set of assembled contigs aligned to the genome with predicted transcripts. This provides a comprehensive, high-quality view of the transcriptome and its alignment to the genome.

4.2. PROCESS CONSIDERATIONS

A myriad of problems complicate the clustering and assembly process. They arise both from imperfections of EST and mRNA data, and from a battery of biological phenomena involved in the process. The following discusses some of the most prevalent issues that affect the outcome of transcriptome analysis.

4.2.1. Chimerical sequences

During the genome mapping step, thousands of expressed sequences have perfect alignments with genomic segments from two different chromosomes, or from two distant portions of the same chromosome (for example, regions that are 400,000 bp apart on the same chromosome). Those expressed sequences are likely chimerical sequences, generated by mistake during EST sequencing or resulting from genomic translocations. There are well-documented fusion mRNA sequences, which resulted from translocations in tumor tissues; for example, AB000267 and AB000268, AB001342 and AB001343 (Arai *et al.*, 1997), AF060927 (Silliman *et al.*, 1998), HSCOLPFU1 and HSCOLPFU2 (Simon *et al.*, 1997), HSRLF (Makela *et al.*, 1991), and AF179280 (Lagerstedt *et al.*, 2000). The GenBank annotations of these mRNA sequences may sometimes reveal whether or not they are results of fusion events. On the other hand, those identified chimerical sequences from genomic mapping process could be used for identification of translocation events in normal or abnormal tissues (Xie *et al.*, unpublished result).

4.2.2. NotI sites in the expressed sequences

Many ESTs are obtained from directional cloning, where linkers containing recognition sites for rare cutting restriction enzymes, mostly *NotI*, are used. The recognition site of *NotI* is eight bases long (GC^GGCCGC) and possesses two CpG dinucleotides. CpG islands are rare in genome sequences, but appear in relatively high frequency in the 5 ends of genes (Zabarovsky *et al.*, 2000). A substantial fraction of cDNA clones that contain internal *NotI* sites are inserted into cloning vectors in the opposite orientation, resulting in incorrect annotation of directionality of EST sequences. The sequences derived from such clones usually align together with flush ends or flush starts in the vicinity of a *NotI* site, and their strand orientation should be adjusted (Yelin *et al.*, submitted).

4.2.3. Poly-A or Poly-T tracts

Most of mature RNA molecules contain poly-A tails. Many expressed sequences deposited in GenBank were derived from oligo (dT)-primed reverse transcription on the poly-A tails of mRNAs. This becomes a poly-T 'head' in the 3' ESTs. However, poly-A priming may also occur if RNA molecules have internal poly-A stretches. During genomic mapping, these poly-A or T stretches from the poly-A tails will not align with the genome, and these portions of sequences are not used during clustering and assembly. The existence of such poly-A signal is useful in determining directionality of aligned genes, and termination of transcripts for transcript prediction (Iseli *et al.*, 2002).

4.2.4. Sense/antisense genes

Sense and antisense transcripts represent two different genes that are encoded, most likely partially, by complementary DNA strands of the same genomic region (Lehner *et al.*, 2002, Shendure *et al.*, 2002). The occurrence of such sense/anti-sense pair seems to be more widespread than expected (the FANTOM Consortium, 2002, Yelin *et al.*, submitted). The sense/anti-sense pair poses a challenging problem during clustering and assembly, especially if the directionalities of expressed sequences are not known or are inaccurate. RNA sequences and canonical splice sites within the cluster serve as reliable markers of transcription direction. In addition, directionalities of ESTs as indicated in their annotation, as well as poly-A 'tails' and poly-T 'heads,' help determine sequence direction.

4.3. TARGETED SEQUENCING

An important aspect of expressed sequence data acquisition and transcriptome analysis is to obtain expressed sequences that are not redundant to the available collection of expressed sequences, and thus can be used to improve the transcriptome analysis. The following discusses an approach we have taken to obtain such significant expressed sequences (Xie et al., submitted). A typical EST sequencing project tends to select 5 or 3 portions of transcripts, while missing the central portions. Computational clustering and assembly of these ESTs separate a gene into separate contigs, and thus fail to yield accurate description of gene structures. We sought to obtain overlapping sequences for a selected group of such contigs through targeted cDNA sequencing. The chosen contig pairs include those linked by the EST clone information (clone mate-connected contig pairs) and those whose predicted proteins shared sequence similarity with different parts of the same known protein (homology-connected contig pairs). 948 contig pairs, including 651 clone mate-connected, 335 homology-connected, and 38 connected by both clone mates and protein homology, were selected from GenBank 121 LEADS production. More than 600 overlapping sequences for 363 contig pairs were obtained from reverse transcription and sequencing in different tissue samples. Inclusion of the new sequences in the LEADS clustering and UniGene mapping process allowed better characterization for a substantial number of UniGene clusters. Those sequences were submitted to GenBank. Although computational algorithms and software tools have been developed for identification of exons, their predictions need to be verified through experimental means. Both EST sequencing and full-length cDNA sequencing produce expressed sequences that greatly enhance the prediction of alternative splicing, and hence, the transcriptome; however, they become less and less efficient in obtaining those overlapping sequences due to their non-selective nature. The targeted sequencing approaches represent an efficient alternative.

5. Transcriptome Analysis in Combination with Tissue Information

One of the major challenges in transcriptome analysis is that gene expression is constrained spatially and temporally. Since expressed sequences are derived from specific tissues, the sources of expressed sequences have been used to infer expression patterns of genes, and to investigate the tissue-specific alternative splicing (Xie *et al.*, 2002, Xu *et al.*, 2002). The numbers of ESTs from different tissue types in UniGene clusters were used to indicate endothelium-specific genes (Huminiecki *et al.*, 2000), disease-specific or tissue-specific polyadenylation sites (Beaudoing, 2001), colon cancer-related genes (Brett *et al.*, 2001), and genes that are differentially expressed in normal or cancer tissues (Schmitt *et al.*, 1999), and to build tissue expression profiles for adult skeletal muscle (Bortoluzzi *et al.*, 2000) and retina (Bortoluzzi *et al.*, 2002).

We analyzed tissue-specific alternative splicing through the LEADS platform (Xie *et al.*, 2002) and found that very few alternative splicing events supported by ESTs from more than four libraries were restricted to a single tissue type, suggesting that tissue-specific alternative splicing might be rare among the genes analyzed. One example, among those identified, is the short form of a prostate-specific protein PSP57 (Xuan *et al.*, 1995), which results from the concatenation of exon 2 and exon 4 and is supported by ESTs from 11 prostate libraries; whereas in the long form PSP94, concatenations of exon 2, 3, and 4 were supported by ESTs from six types of tissues. Using more relaxed criteria, Xu and colleagues found that 10–30% of human alternatively spliced genes show evidence of tissue-specific splice forms (Xu *et al.*, 2002).

Such a tissue-linked expressed sequence analysis provides a road map into an expression atlas. However, the results might not link to specific cell physiology or pathology since the majority of EST libraries are from bulk preparations (although a few of them are from microdissected tissues), and most tissues are heterogeneous in their cell compositions (for instance, a liver is mainly composed of hepatocytes, but also of vessel cells, immune cells, etc.). Thus, cell type-specific transcriptome analysis remains a significantly unmet goal and endeavor.

6. Future Directions

6.1. INTEGRATION OF TRANSCRIPTOME ANALYSIS THROUGH EXPRESSED SEQUENCES WITH OTHER TECHNOLOGIES

Transcriptome analysis through integration of genomic and expressed sequences provides the best available gene structure description of individual genes in terms of alternative splicing, alternate initiation, and alternate poly-adenylation. Such a description is generally qualitative, although there have been some efforts to obtain semi-quantitative data on expression levels based on expressed sequences. On the other hand, high-throughput methodologies such as SAGE and microarrays, and more traditional techniques such as *in situ* nucleic acid hybridization analysis, Northern hybridization, differential display, and real-time PCR, provide quantitative assessments of the expression levels of specific genes.

Nevertheless, alternative splicing and other advanced biological phenomena associated with many genes complicate interpretations of data from most techniques. For instance, in microarray studies, a probe specific for a gene, but not for any of its transcripts, will indicate the sum of expression levels of all its transcripts. Such a result, while accurately revealing the total transcript level of the gene, misses important information about individual transcripts that likely have different (sometimes even opposing) cellular functions. In many cases, probes specific to each of the transcripts can be designed, based on the gene structure description from the transcriptome analysis of expressed sequences, and used in the microarray experiments. Such an integration of transcriptome analysis with microarray technologies and with other techniques enables more accurate qualitative and quantitative analysis of transcriptome.

6.2. NON-PROTEIN CODING RNA SEQUENCES

Current transcriptome analysis is focused on those of mRNA molecules that code for proteins. Two additional RNA classes, rRNA and tRNA, play important roles directly in protein synthesis, and have been studied extensively. Besides mRNA, rRNA and tRNA, several recent analyses point to the prevalence of other non-protein coding RNA molecules. These RNA molecules range from 21 nucleotides to longer than 10,000 nucleotides, and perform a variety of cellular functions (Storz, 2002). A recent study in human chromosome 21 indicates that a much broader region of the genome than currently derived from the expressed sequence database is expressed (Kaporov *et al.*, 2002), and in mouse, full-length cDNA sequence project reveals a substantial number of transcription units which are non-protein coding RNA (The FANTOM Consortium, 2002). These molecules likely constitute a major component of the transcriptome, and elucidation of their composition and expression profiles through experimental and computational approaches remains a challenging and worthy goal.

7. Acknowledgements

The authors wish to thank many colleagues, in particular, Sarah Pollock, Avner Magen, Eyal Fink, Ariel Scolnicov, Guy Kol, Eran Halperin, Eitan Rubin, Avi Rosenberg, Yuval Cohen, Ohad Shoshany, David Lehavi, Alex Golubev, Tomer Zecharia, Gil Dugon, Dror Dotan, Avishai Vaaknin, Iftach Nachman, Galit Fuhrmann, Ariel Farkash, Amit Gal, Dror Efrati, Mor Amitai, Ami Haviv, Zipi Fligelman, Moshe Havilio, Danny Schaffer, Alon Wasserman, Avi Shoshan, Brian Meloon, Vladimir Grebinskiy, and Andrew Olson who developed the LEADS platform. Thanks are also due to Brian Meloon for his comments and suggestions and to Elena Sinyavsky and Lital Asher for their help on manuscript preparations.

8. References

Adams, M.D., Kelley, J.M., Gocayne, J.D., Dubnick, M., Polymeropoulos, M.H., Xiao, H. *et al.* (1991) Complementary DNA sequencing: expressed sequence tags and human genome project, *Science* **252**, 1651–1656.

Akhtar, R.A., Reddy, A.B., Maywood, E.S., Clayton, J.D., King, V.M., Smith, A.G. *et al.* (2002) Circadian cycling of the mouse liver transcriptome, as revealed by cDNA microarray, is driven by the suprachiasmatic nucleus, *Curr. Biol.* **12**, 540–550.

Arai, Y., Hosoda, F., Kobayashi, H., Arai, K., Hayashi, Y., Kamada, N. *et al.*. (1997) The inv(11)(p15q22) chromosome translocation of de novo and therapy-related myeloid malignancies results in fusion of the nucleoporin gene, NUP98, with the putative RNA helicase gene, DDX10, *Blood* **89**, 3936–3944.

Ars, E., Serra, E., Garcia, J., Kruyer, H., Gaona, A., Lazaro, C., and Estivill, X. (2000) Mutations affecting mRNA splicing are the most common molecular defects in patients with neurofibromatosis type 1, *Hum. Mol. Genet.* **9**, 237–247.

Beaudoing, E. and Gautheret, D. (2001) Identification of alternate polyadenylation sites and analysis of their tissue distribution using EST data, *Genome Res.* **11**, 1520–1526.

Boguski, M.S. and Schuler, G.D. (1995) ESTablishing a human transcript map, *Nature Genet.* **10**, 369–371.

Boguski, M.S. (1999) Biosequence exegesis. *Science* **286**, 453–455.

Bonaldo, M.F., Lennon, G., and Soares, M.B. (1996) Normalization and subtraction: two approaches to facilitate gene discovery, *Genome Res.* **6**, 791–806.

Bortoluzzi, S., d'Alessi, F., and Danieli, G.A. (2000) A novel resource for the study of genes expressed in the adult human retina, *Invest. Ophthalmol. Vis. Sci.* **41**, 3305–3308.

Bortoluzzi, S., d'Alessi, F., Romualdi, C., and Danieli, G.A. (2000) The human adult skeletal muscle transcriptional profile reconstructed by a novel computational approach, *Genome Res.* **10**, 344–349.

Brett, D., Hanke, J., Lehmann, G., Haase, S., Delbruck, S., Krueger, S. *et al.* (2000) EST comparison indicates 38% of human mRNAs contain possible alternative splice forms, *FEBS Lett.* **474**, 83–86.

Brett, D., Kemmner, W., Koch, G., Roefzaad, C., Gross, S., and Schlag, P.M. (2001) A rapid bioinformatic method identifies novel genes with direct clinical relevance to colon cancer, *Oncogene* **20**, 4581–4585.

Brett, D., Pospisil, H., Valcarcel, J., Reich, J., and Bork, P. (2002) Alternative splicing and genome complexity, *Nature Genet.* **30**, 29–30.

Buetow, K.H., Edmonson, M.N., and Cassidy, A.B. (1999) Reliable identification of large numbers of candidate SNPs from public EST data, *Nature Genet.* **21**, 323–325.

Burge, C.B. (2001) Chipping away at the transcriptome, *Nature Genet.* **27**, 232–234.

Burge, C. and Karlin, S. (1997) Prediction of complete gene structures in human genomic DNA, *J. Mol. Biol.* **268**, 78–94.

Burke, J., Wang, H., Hide, W., and Davison, D.B. (1998) Alternative gene form discovery and candidate gene selection from gene indexing projects, *Genome Res.* **8**, 276–290.

Burset, M., Seledtsov, I.A., and Solovyev, V.V. (2001) SpliceDB: database of canonical and non-canonical mammalian splice sites, *Nucleic Acids Res.* **29**, 255–259.

Camargo, A.A., Samaia, H.P., Dias-Neto, E., Simao, D.F., Migotto, I.A., Briones, M.R. *et al.* (2001) The contribution of 700,000 ORF sequence tags to the definition of the human transcriptome, *Proc. Natl. Acad. Sci. USA* **98**, 12103–12108.

Caron, H., van Schaik, B., van der Mee, M., Baas, F., Riggins, G., van Sluis, P., (2001) The human transcriptome map: clustering of highly expressed genes in chromosomal domains, *Science* **291**, 1289–1292.

Carson, J., Thaller, C., and Eichele, G. (2002) A transcriptome atlas of the mouse brain at cellular resolution, *Curr. Opin. Neurobiol.* **12**, 562–265.

Chou, H.H. and Holmes, M.H. (2001) DNA sequence quality trimming and vector removal, *Bioinformatics* **17**, 1093–1104.

Christoffels, A., van Gelder, A., Greyling, G., Miller, R., Hide, T., and Hide, W. (2001) STACK: Sequence Tag Alignment and Consensus Knowledgebase, *Nucleic Acids Res.* **29**, 234–238.

Clark, T.A., Sugnet, C.W., and Ares, M., Jr. (2002) Genomewide analysis of mRNA processing in yeast using splicing-specific microarrays, *Science* **296**, 907–910.

Conaway, R.C., Brower, C.S., and Conaway, J.W. (2002) Emerging roles of ubiquitin in transcription regulation, *Science* **296**, 1254–1258.

Croft, L., Schandorff, S., Clark, F., Burrage, K., Arctander, P., and Mattick, J.S. (2000) ISIS, the intron information system, reveals the high frequency of alternative splicing in the human genome, *Nature Genet.* **24**, 340–341.

David, A., Mabjeesh, N., Azar, I., Biton, S., Engel, S., Bernstein, J. *et al.* (2002) Unusual alternative splicing within the human kallikrein genes KLK2 and KLK3 gives rise to novel prostate-specific proteins, *J. Biol. Chem.* **277**, 18084–19090.

Dralyuk, I., Brudno, M., Gelfand, M.S, Zorn, M., and Dubchak, I. (2000) ASDB: database of alternatively spliced genes, *Nucleic Acids Research* **28**, 296–297.

Dundr, M., Hoffmann-Rohrer, U., Hu, Q., Grummt, I., Rothblum, L.I., Phair, R.D., and Misteli, T. (2002) A kinetic framework for a mammalian RNA polymerase in vivo, *Science* **298**, 1623–1626.

Ewing, B. and Green, P. (2000) Analysis of expressed sequence tags indicates 35,000 human genes, *Nature Genet.* **25**, 232–234.

The FANTOM Consortium and the RIKEN Genome Exploration Research Group Phase I & II Team (2001) Functional annotation of a full-length mouse cDNA collection, *Nature* **409**, 685–690.

The FANTOM Consortium and the RIKEN Genome Exploration Research Group Phase I & II Team (2002) Analysis of the mouse transcriptome based on functional annotation of 60,770 full-length cDNA, *Nature* **420**, 563–573.

Fujii, T., Dracheva, T., Player, A., Chacko, S., Clifford, R., Strausberg, R.L. *et al.* (2002) A preliminary transcriptome map of non-small cell lung cancer, *Cancer Res.* **62**, 3340–3346.

Garg, K., Green, P., and Nickerson, D.A. (1999) Identification of candidate coding region single nucleotide polymorphisms in 165 human genes using assembled expressed sequence tags, *Genome Res.* **9**, 1087–1092.

Geier, B., Kastenmuller, G., Fellenberg, M., Mewes, H.W., and Morgenstern, B. (2001) The HIB database of annotated UniGene clusters, *Bioinformatics* **17**, 571–572.

Haas, B.J., Volfovsky, N., Town, C.D., Troukhan, M., Alexandrov, N., Feldmann, K.A. *et al.* (2002) Full-length messenger RNA sequences greatly improve genome annotation, *Genome Biol.* **3**, RESEARCH0029.

Hemberger, M., Cross, J.C., Ropers, H.H., Lehrach, H., Fundele, R., and Himmelbauer, H. (2001) UniGene cDNA array-based monitoring of transcriptome changes during mouse placental development, *Proc. Natl. Acad. Sci. USA* **98**, 13126–13131.

Hillier, L.D., Lennon, G., Becker, M., Bonaldo, M.F., Chiapelli, B., Chissoe, S. *et al.* (1996) Generation and analysis of 280,000 human expressed sequence tags, *Genome Res.* **6**, 807–828.

Hu, G.K., Madore, S.J. Moldover, B., Jatkoe, T., Balaban, D., Thomas, J., and Wang, Y. (2002) Predicting Splice Variant from DNA Chip Expression Data, *Genome Res.* **11**, 1237–1245.

Huang, Y.H., Chen, Y.T., Lai, J.J., Yang, S.T., and Yang UC. (2002) PALS db: Putative Alternative Splicing database, *Nucleic Acids Res.* **30**, 186–90.

Huminiecki, L. and Bicknell, R. (2000) In silico cloning of novel endothelial-specific genes, *Genome Res.* **10**, 1796–1806.

International Human Genome Sequencing Consortium (2001) Initial sequencing and analysis of the human genome, *Nature* **409**, 860–921.

Iseli, C., Stevenson, B.J., de Souza, S.J., Samaia, H.B., Camargo, A.A., Buetow, K.H. *et al.* (2002) Long-range heterogeneity at the 3' ends of human mRNAs, *Genome Res.* **12**, 1068–1074.

Kan, Z., Rouchka, E.C., Gish, W.R., and States, D.J. (2001) Gene structure prediction and alternative splicing analysis using genomically aligned ESTs, *Genome Res.* **11**, 889–900.

Kapranov, P., Cawley, S.E., Drenkow, J., Bekiranov, S., Strausberg, R.L., Fodor, S.P., and Gingeras, T.R. (2002) Large-scale transcriptional activity in chromosomes 21 and 22, *Science* **296**, 916–919.

Kent, W.J. and Zahler, A.M. (2000) The intronerator: exploring introns and alternative splicing in Caenorhabditis elegans, *Nucleic Acids Res.* **28**, 91–93.

Krawczak, M., Reiss, J., and Cooper, D.N. (1992) The mutational spectrum of single base-pair substitutions in mRNA splice junctions of human genes: causes and consequences, *Hum. Genet.* **90**, 41–54.

Lagerstedt, K., Carlberg, B.M., Karimi-Nejad, R., Kleijer, W.J., and Bondeson, M.L. (2000) Analysis of a 43.6 kb deletion in a patient with Hunter syndrome (MPSII): identification of a fusion transcript including sequences from the gene W and the IDS gene, *Hum. Mutat.* **15**, 324–331.

Lehner, B., William, G., Campbell, R.D., and Sanderson, C.M. (2002) Antisense transcripts in the human genome, *Trends Genet.* **18**, 63–65.

Liang, F., Holt, I., Pertea, G., Karamycheva, S., Salzberg, S. L., and Quackenbush, J. (2000) Gene index analysis of the human genome estimates approximately 120,000 genes, *Nature Genet.* **25**, 239–240.

Liu, H.X., Cartegni, L., Zhang, M.Q., and Krainer, A.R. (2001) A mechanism for exon skipping caused by nonsense or missense mutations in BRCA1 and other genes, *Nature Genet.* **27**, 55–58.

Lukashin, A.V. and Borodovsky, M. (1998) GeneMark.hmm: new solutions for gene finding, *Nucleic Acids Res.* **26,** 1107–1115.

Lynch, H.T. and de la Chapelle, A. (1999) Genetic susceptibility to non-polyposis colorectal cancer, *J. Med. Genet.* **36**, 801–818.

Makela, T.P., Saksela, K., Evan, G., and Alitalo, K. (1991) A fusion protein formed by L-myc and a novel gene in SCLC, *EMBO J.* **10**, 1331–1335.

Maleck, K., Levine, A., Eulgem, T., Morgan, A., Schmid, J., Lawton, K.A., Dangl, J.L., Dietrich, R.A. (2000) The transcriptome of Arabidopsis thaliana during systemic acquired resistance, *Nature Genet.* **26**, 403–410.

Maniatis, T. and Tasic, B. (2002) Alternative pre-mRNA splicing and proteome expansion in metazoans, *Nature* **418**, 236–243.

Matloubian, M., David, A., Engel, S., Ryan, J.E., and Cyster, J.G. (2000) A transmembrane CXC chemokine is a ligand for HIV-coreceptor Bonzo, *Nature Immunol.* **1**, 298–304.

Menssen, A. and Hermeking, H. (2002) Characterization of the c-MYC-regulated transcriptome by SAGE: identification and analysis of c-MYC target genes, *Proc. Natl. Acad. Sci. USA* **99**, 6274–6279.

Meyer, I.M. and Durbin, R. (2002) Comparative *ab initio* prediction of gene structures using pair HMMs, *Bioinformatics* **18**, 1309–1318.

Mironov, A.A., Fickett, J.W., and Gelfand, M.S. (1999) Frequent alternative splicing of human genes, *Genome Res.* **9**, 1288–1293.

Modrek, B., Resch, A., Grasso, C., and Lee, C. (2001) Genome-wide detection of alternative splicing in expressed sequences of human genes, *Nucleic Acids Res.* **29**, 2850–2859.

Modrek, B. and Lee, C. (2002) A genomic view of alternative splicing, *Nature Genet.* **30**, 13–19.

Mouse Genome Sequence Consortium (2002) Initial sequencing and comparative analysis of the mouse genome, *Nature* **420**, 520–562.

Okubo, K., Hori, N., Matoba, R., Niiyama, T., Fukushima, A., Kojima, Y., and Matsubara, K. (1992) Large scale cDNA sequencing for analysis of quantitative and qualitative aspects of gene expression, *Nature Genet.* **2**, 173–179.

Piquemal, D., Commes, T., Manchon, L., Lejeune, M., Ferraz, C., Pugnere, D. *et al.* (2002) Transcriptome analysis of monocytic leukemia cell differentiation, *Genomics* **80**, 361.

Plasterk, R.H. (2002) RNA silencing: the genome's immune system, *Science* **296**, 1263–1265.

Primig, M., Williams, R.M., Winzeler, E.A., Tevzadze, G.G., Conway, A.R., Hwang, S.Y. *et al.* (2000) The core meiotic transcriptome in budding yeasts, *Nature Genet.* **26**, 415–423.

Pruitt, K.D. and Maglott, D.R. (2001) RefSeq and LocusLink: NCBI gene-centered resources, *Nucleic Acids Res.* **29**, 137–140.

Quackenbush, J., Cho, J., Lee, D., Liang, F., Holt, I., Karamycheva, S. *et al.* (2001) The TIGR Gene Indices: analysis of gene transcript sequences in highly sampled eukaryotic species, *Nucleic Acids Res.* **29**, 159–164.

Robert-Nicoud, M., Flahaut, M., Elalouf, J.M., Nicod, M., Salinas, M., Bens, M. *et al.* (2001) Transcriptome of a mouse kidney cortical collecting duct cell line: effects of aldosterone and vasopressin, *Proc. Natl. Acad. Sci. USA* **98**, 2712–2716.

Saha, S., Sparks, A.B., Rago, C., Akmaev, V., Wang, C.J., Vogelstein, B. *et al.* (2002) Using the transcriptome to annotate the genome, *Nature Biotechnology* **20**, 508–512.

Salamov, A.A. and Solovyev, V.V. (2000) *Ab initio* gene finding in Drosophila genomic DNA, *Genome Res.* **10**, 516–522.

Schmitt, A.O., Specht, T., Beckmann, G., Dahl, E., Pilarsky CP, Hinzmann, B. *et al.* (1999) Exhaustive mining of EST libraries for genes differentially expressed in normal and tumor tissues, *Nucleic Acids Res.* **27**, 4251–60.

Schuler, G.D. (1997) Pieces of the puzzle: expressed sequence tags and the catalog of human genes, *J. Mol. Med.* **75**, 694–698.

Shendure, J. and Church, G.M. (2002) Computational discovery of sense-antisense transcription in the human and mouse genomes, *Genome Biology* **3**, 1–14.

Shoshan, A., Grebinskiy, V., Magen, A., Scolnicov, A., Fink, E., Lehavi, D., and Wasserman, A. (2001) Designing oligo libraries taking alternative splicing into account, In: M.L. Bittner, Y. Chen, A.N. Dorsel and E.D. Dougherty (eds.) *Proc. SPIE Microarrays: Optical Technologies and Informatics* **4266**, pp. 86–95.

Silliman, C.C., McGavran, L., Wei, Q., Miller, L.A., Li, S., and Hunger, S.P. (1998) Alternative splicing in wild-type AF10 and CALM cDNAs and in AF10-CALM and CALM-AF10 fusion cDNAs produced by the t(10;11)(p13–14;q14–q21) suggests a potential role for truncated AF10 polypeptides, *Leukemia* **12**, 1404–1410.

Simon, M.P., Pedeutour, F., Sirvent, N., Grosgeorge, J., Minoletti, F., Coindre, J.M. *et al.* (1997) Deregulation of the platelet-derived growth factor B-chain gene via fusion with collagen gene COL1A1 in dermatofibrosarcoma protuberans and giant-cell fibroblastoma, *Nature Genet.* **15**, 95–98.

Sorek, R., Ast, G., and Graur, D. (2002) Alu-containing exons are alternatively spliced, *Genome Res.* **12**, 1060–1067.

Sorek, R. and Safer, H. (in press) A novel algorithm for computational identification of contaminated EST libraries, *Nucleic Acids Research.*

Stella, A., Wagner, A., Shito, K., Lipkin, S.M., Watson, P., Guanti, G., Lynch, H.T., Fodde, R., and Liu, B. (2001) A nonsense mutation in MLH1 causes exon skipping in three unrelated HNPCC families, *Cancer Res.* **61**, 7020–7024.

Storz, G. (2002) An expanding universe of noncoding RNAs, *Science* **296**, 1260–1263.

Strausberg, R.L., Feingold, E.A., Klausner, R.D., and Collins, F.S. (1999) The mammalian gene collection, *Science.* **286**, 455–457.

Strausberg, R.L. and Riggins, G.J. (2001) Navigating the human transcriptome, *Proc. Natl. Acad. Sci. USA* **98**, 11837–11838.

Teraoka, S.N., Telatar, M., Becker-Catania, S., Liang, T., Onengut, S., Tolun, A. *et al.* (1999) Splicing defects in the ataxia-telangiectasia gene, ATM: underlying mutations and consequences, *Am. J. Hum. Genet.* **64**, 1617–1631.

Tjaden, B., Saxena, R.M., Stolyar, S., Haynor, D.R., Kolker, E., and Rosenow, C. (2002) Transcriptome analysis of Escherichia coli using high-density oligonucleotide probe arrays, *Nucleic Acids Res.* **30**, 3732–3738.

Velculescu, V.E., Zhang, L., Vogelstein, B., and Kinzler, K.W. (1995) Serial analysis of gene expression, *Science* **270**, 484–487.

Velculescu, V.E., Zhang, L., Zhou, W., Vogelstein, J., Basrai, M.A., Bassett, D.E., Jr. *et al.* (1997) Characterization of the yeast transcriptome, *Cell* **88**, 243–251.

Venter, J.C., Adams, M.D., Myers, E.W., Li, P.W., Mural, R.J., Sutton, G.G. *et al.* (2001) The sequence of the human genome, *Science* **291**, 1304–1351.

Wang, T.L., Rago, C., Silliman, N., Ptak, J., Markowitz, S., Willson, J.K. *et al.* (2002) Prevalence of somatic alterations in the colorectal cancer cell genome, *Proc. Natl. Acad. Sci. USA* **99**, 3076–3080.

Weber, G., Shendure, J., Tanenbaum, D.M., Church, G.M., and Meyerson, M. (2002) Identification of foreign gene sequences by transcript filtering against the human genome, *Nature Genet.* **30**, 141–142.

Wolfsberg, T.G. and Landsman, D. (1997) A comparison of expressed sequence tags (ESTs) to human genomic sequences, *Nucleic Acids Res.* **25**, 1626–1632.

Xie, H., Zhu, W.Y., Wasserman, A., Grebinskiy, V., Olson, A., and Mintz, L. (2002) Computational analysis of alternative splicing using EST tissue information. *Genomics.* **80**, 26–30.

Xie, H., Diber, A., Pollock, S., Nemzer, S., Safer, H., Meloon, B. *et al.* Bridging Expressed Sequence Alignments through Targeted CDNA Sequencing, submitted.

Xu, Q., Modrek, B., and Lee, C. (2002) Genome-wide detection of tissue-specific alternative splicing in the human transcriptome, *Nucleic Acids Res.* **30**, 3754–3766.

Xuan, J.W., Chin, J.L., Guo, Y., Chambers, A.F., Finkelman, M.A., and Clarke, M.W. (1995) Alternative splicing of PSP94 (prostatic secretory protein of 94 amino acids) mRNA in prostate tissue, *Oncogene* **11**, 1041–1047.

Yeh, R.F., Lim, L.P., and Burge, C.B. (2001) Computational inference of homologous gene structures in the human genome, *Genome Res.* **11**, 803–816.

Yelin, R., Dahary, D., Sorek, R., Levanon, E., Goldstein, O., Shoshan, A. *et al.* Widespread occurrence of antisense transcription in the human genome, *Nature Biotechnology*, submitted.

Zabarovsky, E.R., Gizatullin, R., Podowski, R.M., Zabarovska, V.V., Xie, L., Muravenko, O.V. *et al.* (2000) NotI clones in the analysis of the human genome, *Nucleic Acids Res.* **28**, 1635–1639.

Zhuo, D., Zhao, W.D., Wright, F.A., Yang, H.Y., Wang, J.P., Sears, R. *et al.* (2001) Assembly, annotation, and integration of UNIGENE clusters into the human genome draft, *Genome Res.* **11**, 904–918.

V. Glycoinformatics & Protein Data

Biodata of **Ofer Markman**, author of ***"Challenges in Glycoinformatics 2003: Glycoinformatics: Bioinformatics of Glycomolecules."***

Dr. Ofer Markman is the co-founder and VP New Technology Development of the company ProCognia (formerly Glycodata) Ltd. located in Ashdod, Israel. He obtained his MSc. in 1990 from the Feinberg Graduate School at the Weizmann Institute of Science, and his PhD in 1996 from Boston College, MA. USA.

E-mail: **ofer. Markman@procognia.com**

J. Seckbach and E. Rubin (eds.), The New Avenues in Bioinformatics, 165–174.
© 2004 *Kluwer Academic Publishers. Printed in the Netherlands.*

CHALLENGES IN GLYCOINFORMATICS 2003
Glycoinformatics: bioinformatics of glycomolecules

OFER MARKMAN
Procognia Ltd. Procognia (Israel) Ltd., 3 Habosem St., Ashdod, Israel 77610

1. Introduction

This chapter entitled Challenges in Glycoinformatics will review new themes in describing and archiving glycostructures, their interface with the analysis tools and how to extract relevant glycomic data out of those.

Glycoinformatics is the common technology and electronic language that connects the field of glycobiology. It allows the presentation, and the description of glycomolecules in a useful way that enables their utilization and manipulation. Glycoinformatics further permits pathway and systemic analysis of glycans to create diagnostic tools, aid in drug discovery and development, and develop biotechnological applications for food, medicine and industry. The technologies enabling this are now only emerging, and new abilities are now developed. The developments in technologies and in strategies for using data require progress and re-evaluation in the field of glycoinformatics.

2. Glycoinformatics is Dealing With Glycans Which are Highly Complex Molecules

Glycoinformatics complications are first due to the complexity of the structure/sequence of glycans. Glycans are carbohydrate polymers that include monosaccharide (sugar) units connected to each other via glycosidic bonds. These polymers have a structure, which is known as the two-dimensional structure of the glycan and can be described by the linear sequence of the monosaccharide subunits. Glycans can also be described by way of the structures formed in space by their component monosaccharide subunits.

A chain of monosaccharides that forms a glycan has two dissimilar ends. One end contains an aldehyde group and is known as the reducing end. The other end is known as the non-reducing end. Additional glycan chains may branch from any of the carbon moieties at C1, C2, C3, C4, or C6 of the carbohydrate ring of the monosaccharide as long as the connection is with a hexose unit. In addition, a given monosaccharide may be linked to more than two different monosaccharide units simultaneously. Moreover, the connection at the C1 position may be in either the α or β configuration. Thus, both the two-dimensional and three-dimensional structure of the carbohydrate polymer can be highly complex.

The structural determination of glycans is of fundamental importance for the development of glycobiology. Research in glycobiology relates to subjects as diverse as the identification and characterization of antibiotic agents that affect bacterial cell wall synthesis, plasma protein glycans, growth factor and cell surface receptor structures involved in viral disease, and autoimmune diseases such as insulin dependent diabetes. Rheumatoid glycans have also been used in the development of biomaterials for contact lenses, artificial skin, and prosthetic devices. Furthermore, glycans are used in a number of non-medical fields, such as the paper industry. Additionally, of course, the food and drug industry uses large amounts of various polysaccharides and oligosaccharides.

In all of the above fields, there is a need for improved saccharide analysis technologies. Saccharide analysis information is useful in, *e.g.*, quality control, structure determination in research, and conducting structure-function analyses.

The structural complexity of polysaccharides has hindered their analysis. For example, saccharides are believed to be synthesized in a template-independent mechanism. In the absence of structural information, the researcher must therefore assume that the building units are selected from any of the saccharide units known today. In addition, these units may have been modified, during synthesis, *e.g.*, by the addition of sulfate groups.

Second, saccharides can be connected at any of the carbon moieties, *e.g.*, a the C1, C2, C3, C4, or C6 atom as long as the connection is with a hexose unit. Moreover, the connection to the C1 atom may be in either α or β configuration.

Third, saccharides may be branched, which further complicates their structure and the number of possible structures that may arise from an identical number and kind of sugar units.

A fourth difficulty is presented by the fact that the difference in structure between many sugars is minute, as a sugar unit may differ from another merely by the position of the hydroxyl groups (epimers).

2.1. WHAT ARE **GLYCOMOLECULES** AND **GLYCOINFORMATICS**, AND WHAT ARE THEIR RELATIONS TO **GLYCOMICS AND** OTHER **OMICS**

"Glycomolecules" in this chapter is a general term describing sugars, oligosaccharides, polysaccharides and glyco-conjugates (glycoproteins, glycolipids and glycoseaminoglycans); the molecules that bind them (lectins); and the molecule that process them—e.g. glycosyltransferases and glycosidases.

Glycoinformatics is the general technology that store, process and analyze the information on glycomolecules.

Glycomics is the interface of wet-bench glycoanalytical technology and glycoinformatics to make sense of the glycome (the general glycan phenotype of an organism).

Glycogenomics is the connection of **genomics** to the field of glycobiology.

Glyco-proteomics is the analysis of glycans on distinct proteins, and by thus is a subdivision of glycomics.

While the field of glycoinformatics has a wide usage in all avenues of glycobiology, the massive use of it currently is in the field of glycoproteomics and the mass data analysis of glycans from glycoproteins. The description of the molecules and the handling of data, structure and sequence information are not only challenges by themselves, but also an opportunity of learning for the whole field of glycoinformatics.

3. Challenges in Glycoprotein Description

Once glycoproteins are in question, there are several theoretical considerations concerning the description of the molecule: The glycoprotein can be presented as a chain of amino-acids to which at certain positions glycan derivations are attached, or as the mixture of glycoforms *e.g.* a chain of amino-acids that holds a certain combination of glycans at each glycosylation site. Thus, a glycoprotein is an ensemble of one peptide chain and certain glycans distributed on the protein according to a set of probabilistic rules. This is the most common way of presenting a glycoprotein and the closest to what the analytical machinery can provide information on.

Little attention has been put to the analysis and presentation of the relationship of multiple structures within the molecule. This lack of formal description is inline with the ability of most analysis technologies.

A more accurate way of presenting the glycoprotein would be similar to the one used for multiple sequences of a protein—distinct structures with a set of combinations of glycans at exact positions. This is a detailed description of the molecules—rather then of the ensembles. Analytical methods that result in such a detailed description are non-existent.

The glycan synthesis mechanism is complex and involves the transfer of the glycoprotein through several compartments in the cell with a rigorous mechanism of editing and control. [Roth J. (2002)] Glycan synthesis in the cell occurs within a certain time frame (in fermentation we may look at full days of synthesis). While different glycoproteins may probe the cell in different time frames, two glycosylation sites on the same glycoprotein will always probe the cell at the same time. These are some of the mechanisms that balance conformity and diversity in glycoproteins.

Moreover, nature may have more then one way to create glycan diversity, and we may be surprised to find that, some proteins are found in mixtures that are in fact a "soup" of glycoforms (i.e. are a result of a probabilistic glycosylation mechanism and post-translational modification process), and some are a result of rigorous mechanistic pathways or time frame differences. For example, all glycans with terminal glucose are eliminated by the mechanism of fold editing.

Even if the production of the protein is an ensemble in nature, its function may not be necessarily so. Most of the knowledge we have is on the level of specific glycans, rather then their composition (either on all glycosylation sites or their distribution on one site), it is the role of modern glycoinformatics and glycomics to bring this knowledge to the ensemble and distinct molecule level and provide generalization and usable rules on these.

Most methods of representing glycans only represent the carbohydrate structure to a certain and very limited level. Glycoproteins are described to the level of present glycans (total glycans and type of glycan connectivity e.g. N-glycan or O-glycan). For a high percentage of glycoproteins this is the highest level of information available.

Information on abundance and distribution of glycans and on relationship between sites is not available in the format of the current databases.

Yet another level of complexity in describing the molecules is on the level of the glycoprotein mixture and the right way to describe distinct glycosylation and glycoform ensembles.

4. Distinct Glycosylation and Glycoform Ensembles

Glycoproteins are often an ensemble of glycoforms rather then a homogenous mixture. The biochemical mechanism is such that a certain glycoprotein-molecule is exposed in the course of its production and processing to a set of biochemical conditions and bio-compartmental conditions that are likely to impose conformity, even if the peptide backbone and folding are exact or even if they are produced by the same cells but in a different time frame [Roth J. (2002)].

The ensemble nature of the glycoprotein is an unexplored avenue, we do not have a clear answer to whether the glycosylation mechanism ensures uniformity, heterogeneity, or rather both are correct. Nor is the exact nature of such mechanism clear.

For the industrial glycoprotein producer this is a key issue in ensuring product consistency. Protein folding and glycan relationship are key issues in productivity. While glycans play an important role in glycoprotein folding-editing mechanism, the influence of the folded proteins or domains on glycosylation patterns and homogeneity have not been explored. The relationship of glycosylation and other post translation modifications (PTMs) have only been demonstrated in few cases. Nevertheless, it is well known that glycoproteins are exposed to other forms of PTM. Few examples of such relationship have been extremely valuable [Eisenhaber B., Bork P (1998) ; Kondo A (1996)]. The recent works on O-GlcNAc and phosphorylation are another example to the importance of the phenomena [Zachara NE. and Hart GW (2002)].

Recent analytical advances are promising in the field of glycomics and glycoprotein application—advances in whole organ [Sutton Smith M, Morris HR., Grewa PK., Hewitt JE., Bittner RE., Goldin E., Shiffmann R., Dell A. (2002)] and whole organism glycomics [Sachchter H. Chen S., Zhang W., Spence A M., Zhu S. Callahan JW. Mahuran DJ. Fan X., Bagshow RD., She YM., Rosa JC., Reinhold VN. (2002); Haslem SM., Gems D., Morris HR., Dell A. (2002)], and the interface between organism genetic databases and glycomic information [Comelli EM., Amado M., Head S., Paulson JC., (2002)], several of these have been documented. These can help decipher information such as genome—structure correlation, genomic—glycomic, glycan—cell line and glycoprotein-production-system interactions.

Information revealed on the function of glycans in glyco-conjugate trafficking (or actually glycans as reporters of trafficking events, [Helenius J, Aebi M (2002)]) holds promise to help decipher the fate or pathway history of glycoconjugates.

5. Glycoinformatics as a Mature Field of Bioinformatics—Current Status

There are several claims to be made from THE glycomics data base[1], yet the field is extremely young in professional terms: a lack of benchmarking[2] opens competition between

---

[1] http://www.glycosuite.com/; http://www.dkfz-heidelberg.de/spec2/sweetdb/; http://www.boc.chem.uu.nl/sugabase/sugabase.html; http://www.glyco.org/

[2] for example http://www-igbmc.u-strasbg.fr/BioInfo/BAliBASE/, http://www.csm.ornl.gov/evaluation/BIO/, http://www.smi.stanford.edu/projects/helix/psb02/genomepathwayIntro.pdf; http://bioinformatics.ljcrf.edu/liwz/research/benchmark/

databases and the lack of common gold standards for database quality, as well as the lack of common language and e-standards between the databases is a feature that prevents constant improvement of glycomic-databases. The accuracy of the information, the reference accuracy and quality of entries and benchmark data are currently the sole interpretation of the database makers. These databases are also in constant competition with internal databases of analytical glycobiology laboratories in industry and in academia, which still hold access to the majority of analyses of experimental data.

Tools for data mining, as well as tools for enhancing the ensemble nature of the glycan structure in the conformational space are starting to emerge [Bohne A and von der Lieth CW (2002); Frank M, Bohne A, Wetter T and von der Lieth CW (2002)]

6. Quality of the Input—Glycoanalytical Data

The way data is obtained in the glycomic databases is affecting the way the glycans are described. Most of the data on protein glycosylation is obtained by Mass Spectrometry (MS). The majority of MS experiments used are MALDI based methods. Other methods are picking up rapidly in the field [Mechref Y and Novotny V (2002)].

A short overview on current status of analytical methods has to start from sample preparation (which is the most critical and sensitive step in glycoanalysis): These can be divided to three categories, intact molecule analysis, partial release of glycans and chemo/enzymatic glycan release. Choosing any of these steps may result in major differences in analysis.

A most commonly used method is that of glycoprotein tryptic digestion followed by enzymatic removal of the glycans from the glycoprotein, and by peptide—glycan separation by chromatography—a highly sensitive method with three complicated steps resulting in the release and separation of the glycans.

Variations on this method are present to widen the scope of glycan release.
The capturing of the glycans is also a critical step (which is outside the scope of this chapter). Desialylated (sialidase treated) and native glycans are often analyzed separately although sometimes the total glycan content is analyzed.

Post release treatment by reduction, end-labeling for detection, per-methylation, as well as desialylation are often the practice.

Qualitative analysis—major advances are made in the field of identification of glycan structures. Today if needed glycans that are of interest can be identified with a high degree of accuracy using a variety of glyconalytic technologies. Yet, it is seldom that all glycans are identified or can be identified, for reasons such as degeneracy in mass, behavior in the HPLC or for lack of analytical tools (the scope of this chapter does not allow detailed analysis of the reasons for such phenomena). Moreover, the quantitative assignment of abundance weights to such glycans is in most cases impossible or possible to a very rough estimation. Combination of mass-spectrometry, chromatography and limiting exoglycosylation are commonly used and are often referred to as glycan sequencing.

Semi quantitative analysis—also known as profiling—includes several chromatographic methods or analysis of per-methylated glycans by MS and NMR. These will output the percentage of the glycans in the glycan mixture, at an accuracy still to be defined.

Fluorescence glycan tagging (such as in FACE) adds another chemical step to the sample preparation—end-labeling of the released glycans, and thus adds another source of variance and error depending on the exact protocol used, the reagents and the fluorophore used. The technology is used both for qualitative and semi quantitative analysis.

These methods employ the fact that most common glycans have only one reducing end open to reduction in the anomeric position (commonly position 1).

The current databases are limited in annotation of the sample preparation methods, and a systematic assesment to establish the implications of such variation on the data are needed.

The analytical stage of the process (e.g. by MS—MALDI, NMR, Capillary Electrophoresis, HPAEC-PAD—a chromatographic method, FACE etc.) is more established, but careful annotation is needed here as well. Other drawbacks of the current sequencing output relates to the final output of the analysis, that comes in one of the following forms:

1. Traditional chain description of the glycan/glycans;
2. Traditional binary "decorated chain" description of the glycoprotein (phosphorylated yes/no)—N-glycosylation and their positions, O-glycosylation and their positions;
3. N-glycans and their abundances vs. O-glycans and their abundances and
4. Positional sequencing—per peptide sequencing, per position sequencing.

7. Other Glycoinformatics Challenges

The relevant language used to describe glycans depends on the question asked. Often antigenic epitopes are family (O-linked/N-linked) independent and are dependent on a certain glycosyl-transferase e.g. fucosylation in lewis antigens. Yet, in many cases the biological nature of the glycan in context of the pathway that led to its formation is the most relevant (e.g. Asialylated N-Glycan, A-galacto-N-Glycan, Defucosylated glycan) and has a major impact on our understanding of the pathway leading to such glycans and our ability to understand phenotypes.

The current language that is commonly used is a combination of the biological glycan synthesis pathway and the analytical technology with which it was detected and is strongly biased toward the latter. (e.g. Mass-spec based tetra-antennary structure can be mistaken for three antennary in which one antenna is modified with a bi LacNAc (GlcNAc-Gal-GlcNAc-Gal) sequence.

Similarity in glycobiology is still a meaningless word, Tri-sialylated tri-antennary structures can be in some context more similar to tetra- antennary tetra-sialylated then to the monosialylated or bi-sialylated forms of this glycan. Thus the glycobiology similarity matrices are context dependent and need to be developed further.

Graphical needs of the analytical biochemist are starting to be addressed in recent years [Bohne A., Lang E. and. and von der Lieth CW (1999)]. Advances are not made in the same pace in the field of representing enzymes, lectins and their products. The interface between glycomics and glycogenomics has recently been given an innovative attention [Dirckamer K. and Dell A (2002)]. Advances in this important theme are largely dependent on a solid glycoinformatics ground.

8. Conclusions and Futuristic View

The emergence of glycomics emphasizes the need for a real glycoinformatics basis. Such is a language that could bind and allow interface between Glyco-conjugate data applications: these include the analysis of glyco-conjugates, their connection to synthesis pathways, the ability for *in-silico* manipulation and design of glycomolecules and follow-up of glycan differences in the production process and in disease. Effort is needed in integration tools, such as converters between various data formats, and tools for data integration. Community based tests for data—and database-quality has to be established and the issues that influence data quality and data annotation quality have to be clearly defined.

A more accurate description of glycomolecules and their ensembles is needed, including ways to annotate collected analytical data. Glycoinformatics will bridge between the sciences of carbohydrate glyco-molecules and of the carbohydrate modifying/binding molecules.

9. Glycan Presentation Glossary

Linear description—current standards are of linearizing the tree structures using several codes [Bohne-Lang A, Lang E, Forster T and von der Lieth CW, (2001); Loss A, Bunsmann P, Bohne A, Schwarzer E, Lang E, and von der Lieth CW (2002)]. This linear format is at disadvantage for several reasons: (1) It is fairly not straight forward to decipher the structure from it, (2) It is extremely complicated with regards to brackets and other codes and while tools to edit such writing are within reach today it is far from being an easy to use nomenclature.

The advantage of linear description: the ability to use complicated algorithms developed for linear molecules in languages known to the bioinformatics community from DNA and peptide sequence analysis.

Tree description—a rather graphic description with a complicated structure but with rather straightforward glycan to graphic representation structure.

Advantage—ability to use tools from graph mathematics to present, analyze and look at structure including higher algorithms which are developed for such problems which is the subject of many years of mathematical research in the field of operation research and graph theories. The ease of using the graphics and the linear code lands its advantage to many tools and algorithms developed in a field not quiet associated with bioinformatics before.

"Vectorial" description—describing a polysaccharide as a linear description of "vectors": each "vector" is further described by its mass and/or the exact monosaccharide structure. The advantage of vectorial description—describing the polysaccharide by formal computer language—may help in computer search of e.g. mass spectroscopy data.

10. Acknowledgements

Contributions and comments from the fellow scientists in Procognia (Israel) Ltd. are acknowledged. The author thanks Tamar Erlich for her help in editorial work and for her comments on the manuscript.

11. References

Bohne A, and von der Lieth CW (2002) Glycosylation of proteins: a computer based method for the rapid exploration of conformational space of N-glycans. Pac Symp Biocomput, 285–96

Bohne A., E. Lang and C.-W. von der Lieth (1999) SWEET—WWW-based rapid 3D construction of oligo-and polysaccharides Bioinformatics, 1999, 15, 767–768

Comelli EM., Amado M., Head S., and Paulson JC. (2002), Custom microarray for glycobiologists: considerations for glycosyltransferase gene expression profiling. Biochemistry Society Symposia 69, 135–142. Portland Press.

Dirckamer K. and Dell A (2002) Glycogenomics: the impact of genomics and informatics on glycobiology. Biochemistry Society Symposia 69, vii–viii. Portland Press

Eisenhaber B., and Bork P., (1998) Sequence properties of GPI-anchored proteins near the omega-site: constraints for the polypeptide binding site of the putative transamidase Protein Engineering 11, No.12, 1155–1161

Frank M, Bohne A, Wetter T, and von der Lieth CW (2002) Rapid generation of a representative ensemble of N-glycans conformations. Silico Biol., 2, 1–13

Haslem SM., Gems D., Morris HR., and Dell A. (2002) The glycomes of C. elegans and other model organisms. Biochemistry Society Symposia 69, 117–134 Portland Press.

Helenius J, and Aebi M (2002) Transmembrane movement of dolichol linked carbohydrates during N-glycoprotein biosynthesis in the endoplasmic reticulum. *Semin Cell Dev Biol.* (2002) 13:3 171–8

Helenius J, Ng DT, Marolda CL, Walter P, Valvano MA, and Aebi M (2002) Translocation of lipid-linked oligosaccharides across the ER membrane requires Rft1 protein. *Nature* 2002 415:6870 447–50

Kondo A. (1996) De-*N*-glycosylation Will Be the Novel Pathway for Processing of the Tumor Antigen Presented by MHC Class I. *Trends Glycosci.Glycotechnol.* **8**:299–300

Mechref Y and Novotny V (2002) Structural investigation of glycoconjugates at high resolution Chem. Rev. 102 321–369

Roth, J (2002) Protein glycosylation along the Secretory pathway: Relationship to Organelle Topography and function, protein quality control, and cell Interactions. Chem. Rev. 102 321–369

Sachchter H. Chen S., Zhang W., Spence AM., Zhu S. Callahan JW. Mahuran DJ. Fan X., Bagshow RD., She YM., Rosa JC., and Reinhold VN. (2002) Functional post-translational proteomics approach to study the role of N-glycans in the development of C. elegans, Biochemistry Society Symposia 69, 1–29 Portland Press.

Sutton Smith M, Morris HR., Grewa PK., Hewitt JE., Bittner RE., Goldin E., Shiffmann R., and Dell A. (2002) MS screening strategies: investigating the glucomes of knockout mice and myodystrophic mice and leukodystrophic human brain, Biochemistry Society Symposia 69, 105–115. Portland Press.

Zachara NE. and Hart GW (2002) The emerging significance of O-GlcNAc in cellular regulation *Chem Rev* **102:2** 431–8

Biodata of **Ron Unger** author of the chapter ***"The building block approach to protein structure prediction."***

Dr. **Ron Unger** is a senior lecturer in the faculty of Life Science in Bar-Ilan University, Israel. He earned his Ph.D. at the department of Computer Science in the Weizmann Institute of Science in 1990, and spent a post-doctoral period at the University of Maryland. His research interests are in the interface between computer science and biology, namely in the emerging fields of bioinformatics and computational biology. Dr. Unger is currently the head of the computational biology undergraduate program in Bar-Ilan University, and is involved with several national committees to advance Bioinformatics in Israel. Dr. Unger is a board member of the ISTMB (Israel Society of Theoretical and Mathematical Biology) and ISBCB (Israel Society of Bioinformatics and Computational Biology).

E-mail: **ron@biocom1.ls.biu.ac.il**

J. Seckbach and E. Rubin (eds.), The New Avenues in Bioinformatics, 175–188.
© 2004 *Kluwer Academic Publishers. Printed in the Netherlands.*

THE BUILDING BLOCK APPROACH TO PROTEIN STRUCTURE PREDICTION

RON UNGER
Faculty of Life Science, Bar-Ilan University, Ramat Gan, 52900, Israel

1. Introduction

1.1. PROTEIN STRUCTURE PREDICTION

The ability to predict the three dimensional structure of a protein from its linear sequence of amino acids is one of the major challenges of modern biology. In the several decades since the importance of this question was recognized, there has been progress on related issues, for example, building homology models for the structure of proteins when the structure of a related molecule is known. Progress was also achieved in fold-recognition where the goal is to gain information about the overall fold of a novel sequence based again on knowledge about the fold of similar proteins. Yet, the "hard core" of the problem, the ab-initio prediction of the structure of a protein from its amino acid sequence where no knowledge on the structure of related proteins is available, has had only limited success [for a review see Moult 1999]. Recent studies [Rubin, 2001] indicate that even as more sequences become known, the percentage of proteins that have low sequence similarity to any other proteins remains high. Since so many sequences with therapeutic and industrial potential are known, the need for a method enabling direct prediction of structure from sequence is greater than ever before.

The quality of a protein structure prediction is often judged by the RMS (Root Mean Square) deviation of the main chain (or C_α atoms) in Angstroms (Å) when the predicted structure is superimposed on the actual structure, solved by experimental means. While RMS measures are often criticized as inadequate, no other measure is commonly accepted. One of the problems with the RMS deviation measure, is the narrow range between values that are considered useful predictions versus those that are considered inaccurate. As a rule of thumb, for short proteins (up-to approx. 150 amino acids) RMS distances of less than 3 Å are considered very good predictions, RMS distances of less than about 4.5 Å are considered acceptable and useful, and predictions with deviations above 5 Å are considered to be uninformative.

In recent years, the performance of prediction schemes has been evaluated at CASP (Critical Assessment of methods of protein Structure Prediction) experiments. CASP is a community-wide blind experiment in protein prediction [Moult *et al.*, 1995]. In this test, the organizers collect sequences of proteins that are in the process of being experimentally solved, but whose structure is not yet known. These sequences are presented as a challenge to predictors who must submit their structural predictions before the experimental structures become available. Previous CASP meetings have shown progress in the categories of homology modeling and fold-recognition, but minimal progress in the category of ab-initio folding. However, in CASP 4 which was held in 2000, a method based on the building-block approach, presented by David Baker and his co-workers [Bonneau *et al.*, 2001b] was able to predict the structure of a small number of proteins with an RMS below 4 Å. The prediction success was still rather low and the method has significant limitations, yet it was the first demonstration of a successful systematic approach to protein structure prediction. For a recent general review of protein structure prediction methods see [Baker and Sali, 2001].

1.2. THE BUILDING BLOCK APPROACH TO PROTEIN STRUCTURE PREDICTION

The structure of proteins utilizes recurring motifs on many levels, starting from the very limited values for the dihedral angles of the main chain [Ramakrishnan and Ramachandran, 1965], and a relatively small set of rotamers for sides chain conformations [Ponder and Richards, 1987]; the repeated patterns of hydrogen bonding that lead to the formation of a small number of possible secondary structure elements (namely helices, sheets and turns) [Kabsch and Sander, 1983]; and the observation that the overall number of protein folds is rather limited [Chothia, 1992].

Early studies of Unger [Unger *et al.*, 1989] and of Wodak [Rooman *et al.*, 1990] showed that the structure of short fragments (say 5 to 9 residues) also tend to cluster into specific conformations, which are repeatedly utilized in the structure of proteins. Furthermore, it was shown that these building blocks carry a sequence signal, in the sense that the sequences associated with occurrences of a particular structural motif tend to be somewhat similar. This and other studies [Jones and Thirup, 1986; Claessens *et al.* 1989] have also shown that proteins can be re-constructed using "replacement parts", i.e. fragments taken from other proteins. The finding that the structure of proteins can be re-constructed by using a library of standard building blocks, has been validated in many subsequent studies [for example see: de Brevern at al, 2000, Micheletti *et al.*, 2000, Bystroff and Baker, 1998, Lessel and Schomburg, 1997].

These observations led to the building-block approach to protein structure prediction that, to the best of our knowledge, was first described in [Unger at al 1989]. The approach is based on the simple idea that, to a certain extent, local sequence determines local structure. Thus, each fragment of a protein can be replaced by a standard building block of defined structure, where the selection of the building block is guided by the sequence preference that each building block carries. Note that it is not realistic to expect to find a single building block that suits a given sequence fragment. But, the method may suggest a small set of building blocks that includes the correct one. Since multiple choices of building blocks are suggested for each fragment, a global optimization algorithm is used to select the optimal building blocks and how to concatenate these overlapping fragments to form a complete structure.

The physical logic behind the scheme suggests that in the actual folding process, each fragment of the chain tends to occupy only a small subset of possible conformations. Occasionally, neighboring fragments will occupy compatible conformations that can be considered as extensions in a nucleation-like model, [e.g. Moult and Unger, 1991] and thus will form longer, semi stable structures. Eventually, these conformations will interact with other semi stable structures being formed in other parts of the chain [as described by the diffusion-collision model, see Karplus and Weaver, 1994] and become further stabilized until the entire structure is formed.

In spite of the simple logic behind this scheme, early efforts using such an approach did not yield reasonable predictions for proteins with unknown structure. This is mainly due to the fact that while sequence-based restriction of the number of relevant candidate building blocks in each position is quite effective, concatenating such building blocks to a full structure is an under-defined problem and too many solutions are possible. The situation has somewhat changed recently with the work of Baker's group presented at the CASP4 meeting [Bonneau *et al.*, 2001b]. Their method is based on using two libraries of building blocks (size 3 and 9 amino-acids) [Simmons *et al.*, 1999]; selecting possible structural building blocks compatible with a sequence signal enhanced by multiple sequence alignments [Bonneau *et al.*, 2001a]; concatenating building blocks to create a large ensemble of candidate conformations; and clustering these conformations to select conformations from highly populated clusters as the final predictions. The major innovations in the method are the use of multiple sequence alignments to amplify the sequence signal, and the clustering of conformations to achieve the final selection. Using this approach, the Baker group was able to produce good predictions (less than 4 Å deviation from the experimental structure for a fragment of more than 60 residues) for a large fragment of the structure for three small proteins. In about a dozen other cases the predictions were less good (4.5 to 6.5 Å. from the correct structure for the best fragment) [Bonneau *et al.*, 2001b]. Initial accounts from the very recent CASP5 meeting report that while other groups, notably that of Skolnik's, are closing the gap with Baker's group, further improvements beyond the CASP4 achievements were limited.

The term "building blocks" was recently used by Tsai and Nossinov [2001] to describe structural elements that play a key role in the folding process. This notion is somewhat similar to the notion of early folding units [Moult and Unger, 1991] or to the concept of foldons [Panchenko *et al.* 1997]. These terms usually denote identifying selected parts of the protein that are critical during the folding process. Our subject here is different: Describing protein structure in terms of short structural motifs that span the entire sequence, and using this description to facilitate protein structure prediction.

2. The Basic Technique

The building block approach to structure prediction is schematically described in Figure 1: It includes three main components. (A) First, preparation of a library of "standard" building blocks. (B) Second, for each fragment of the sequence, the method should come up with several building blocks that might be compatible with the sequence of that fragment. Note that it is not realistic to expect to identify a single building block that suits a given sequence fragment. Rather one can expect that the method will select a small set of building blocks that includes the correct answer. (C) Third, an optimization algorithm is used to select one building block from each selection in such a way that all the selected building blocks can be "fused" together, in an overlapping manner, to form the predicted structure. Alternatively,

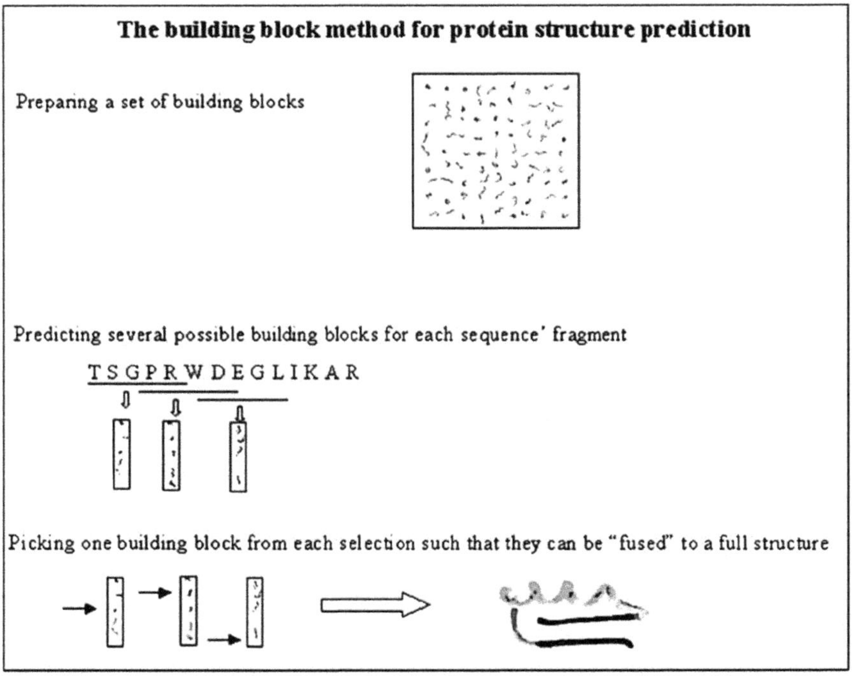

Figure 1. A schematic view of the building block prediction approach. A: A library of standard building blocks is prepared based on their frequent occurance in the structural database. B: based on the sequence preference attached to each building block, a small number of building blocks are suggested for each sequence fragment. C: An optimization algorithm is used to select one building block for each sequence fragment and to concatenate them to a complete structure. Alternatively, the algorithm can produce a large ensemble of possible conformations, from which the correct structure should be further selected by additional algorithmic step.

by selecting different building blocks from the candidate sets, a large ensemble of possible conformations for the complete sequence is created. Then, an appropriate scoring function must be used to select the best overall structure.

Several aspects of the building block approach will be discussed in this review: For each subject we survey the current state of the art and also point to directions in which we believe the current methods can be extended.

1. The completeness of libraries of structural building blocks.
2. Libraries of "natural" building blocks with variable sizes.
3. Libraries containing longer fragments (super secondary structures)
4. Enhancing the sequence signal to identify the appropriate building block to be used.
5. Algorithms and score functions to concatenate building blocks into a complete structure.
6. Using sparse experimental data as a compliment to the computational approach.

2.1. EXPLORING THE COMPLETENESS OF THE LIBRARY OF BUILDING BLOCKS

Many studies have demonstrated that a library of about hundred building blocks of length six or seven residue is enough to represent most protein fragments. For example in [Unger at al, 1989] it was shown that a library of 81 hexamers can represent 78% of all hexamer structures in the tested database with an error that is smaller than 1 Å RMS. However, for structure prediction, 80% coverage is not sufficient. This raises the questions of how many building blocks are required to cover the entire (say above 98%) span of conformations of short fragments that exist in the current PDB structural database.

The question of the trade off between the number of required building blocks and the accuracy of the representation was recently extensively studied by Kolodny *et al.* [2002]. In this comprehensive study, the polypeptide chain was represented by a sequence of rigid fragments and concatenated without any degrees of freedom. Fragments chosen from a library of representative fragments were fit to the native structure using a greedy build-up method. This gives a one-dimensional representation of native three-dimensional protein structure, whose quality depends on the nature of the library. A library is characterized by two parameters: One is the length (f) of the utilized fragments, which in this study was varied between 4 and 7 residues. The other is the size (s) of the library, i.e. the number of building blocks, which was varied between 10 and 300. These two measures were combined to define the complexity of the library as $s^{1/(f-3)}$. It was found that the accuracy depends on the library complexity and varies from 2.9 Å for a library of complexity 2.7 using fragments of length 7 to 0.76 Å for a library of complexity 15 using fragments of length 5. The representation that offers the best tradeoff between accuracy and size was shown to be a library of 100 fragments of length 5 amino acids.

Another related question to be studied in this context is the accuracy of superposition between overlapping building blocks. Using a library of building blocks for structure prediction is based on concatenating building blocks in an overlapping manner, such that the prefix of one block is fused, by superimposition, to the suffix of the previous building block. The level of "superimposability" (i.e. how well the current prefix matches the previous suffix) provides an indication to the probability of the concatenation. If the superimposition accuracy is too low, it suggests that the two fragments in question do not belong together. On the other hand, if the required accuracy of the superimposition is too high, then the concatenation is done in a "rigid" manner that does not allow the growing chain enough flexibility that is needed to reconstruct the overall conformation.

2.2. USING VARIABLE SIZE BUILDING-BLOCKS

Most of the current procedures aim to produce a set of building blocks with a fixed size: The early work at Unger *et al.* used hexamers, fragments of size six amino-acids. In subsequent studies, other fixed sizes ranging from five [de Brevern *et al.* 2000] to eight were used. In the studies by Baker *et al.* [1998], a combination of fragments of lengths, 3, 9, and 12 was used. However, in all these studies the length of the building blocks used was fixed and was not allowed to be adjusted individually to each building block.

As a radical alternative, we would like to suggest exploring the possibility of using a library of building blocks of variable size. Consider the analogy of building blocks to words in a language. As words in any language have various lengths, so should the building blocks. In fact, the analogy to linguistics will help us to think about a method to detect the "natural" size of each building block.

Assume that you are given a text in an unknown language, where blanks and any punctuation marks have been omitted. Is it possible to detect words in this language? One approach is to test all the one letter extensions of a given prefix and to see if there are non-random preferences. For example, if we examine extension of the prefix "compu", we will notice that there is a preferred extension letter which is "t"; thus we can assume that most likely "compu" is not generally a stand alone word, but rather, that it is commonly part of a larger unit "comput". If we continue the process, we will notice that the most frequent extension to "comput" is "compute", and then the next extension is "r" to form "computer". When we analyze all the letters that appear after the unit "computer" we will see a wide distribution, probably similar to that of the distributions of letters in English. At that point we can conclude that indeed "computer" is a stand-alone word.

This approach was used by Brendel at al [1986] to search for sequential motifs in DNA. Similar ideas can be used to "fish-out" structural "words" i.e. motifs in the language of protein structure. One can start with a set of short fixed-size building blocks, for example a set of pentamers, and try to extend each one to a building block of six amino-acids. If, for a set of the occurrences of a given pentamer, the relative three dimensional position of the sixth residue is similar (or can be clustered to a very small number of possibilities) then the situation is similar to the that of the word "compu", i.e. most likely we are dealing with a pentamer that is not a stand-alone unit, but rather represents the beginning of a longer unit. If, on the other hand, it is found that , the position of the sixth residue has a wide distribution, then one can conclude that the given pentamer is a structural unit by itself. In turn, the building blocks that are shown to be prefixes will continue to grow until they will lose their structural integrity. By this technique it should be possible to construct a library of "natural" building blocks, with sizes ranging from 3–4 to 10–12 amino acids, whose sequence to structure signal may be cleaner than the currently available libraries.

2.3. CONSTRUCTING A LIBRARY OF LONGER STRUCTURAL MOTIFS

Most studies have focused on constructing libraries of short structural motifs. Fewer studies have dealt with cataloging longer motifs ranging in size from about 10 to 25 amino acids [e.g. Rooman *et al.*, 1990, Wintjens *et al.*, 1998, Sun and Jiang, 1996]. Most of the studies concentrated on the properties of well known super secondary structure motifs such as helix-turn-helix. In the building block framework, the purpose should be different, that is to produce a catalog of **all** possible conformations of longer fragments of sizes of about 10 to 25 amino-acids, and not only of the well defined "super secondary structure motifs". The structural representation by these motifs is significantly less precise than that of the shorter motifs, (say 2–3 Å compared to 1 Å); still there is a lot of information, largely untapped, that can be gained by their analysis. The main usage for these longer motifs is in controlling the quality of the concatenation process of the short building blocks. When fragments of the chain grow to a size of 15–20 amino acids, it should be possible to compare them to the library of longer structural motifs, in order to check whether the build-up combination is reasonable.

2.4. ENHANCING THE SEQUENCE SIGNAL OF BUILDING BLOCKS

Improving the ability to select a small set of appropriate building blocks that are compatible with the sequence of a given fragment is a major component of the prediction scheme. The naïve approach [Unger *et al.*, 1989] is to produce for each building block a frequency matrix of the amino acids of the fragments that are represented by this building block. For example, in the case of hexamers, this yields a matrix of 6 by 20 of the observed frequency of each amino acid at each position of the fragment. These matrices are later used as profiles that are used to select for a given sequence fragment its building block match. One of the problems with this approach is that it tends to "clamp" together sequences that might come from different sequence motifs that happened to be represented by the same structural building block. This will significantly blur the sequence specificity of each matrix. As an extreme example imagine that SSSSSS and LLLLLL tend to assume a similar three-dimensional shapes and thus are mapped into the same building block. In this case, the sequence matrix associated with this building block will be a combination of S and L, and will match a sequence fragment like SLSLSL which might have a different structural preference. Another problem is that it uses only information from proteins whose structure has been solved and ignores the wealth of information that is available in the much larger sequence databases.

We are aware of two recent studies that attempted to overcome, these problems using two different approaches. In [de Brevern *et al.*, 2000], the problem of "clamping" together different sequence families was treated by a two-stage process. First, fragments were divided into structural clusters according to structure similarity, then the members of each cluster were further divided into sequence sub-families using a clustering algorithm that minimizes the sequence entropy in each sub-family. In [Bonneau *et al.*, 2001], multiple sequence information was utilized, in an indirect way, by repeating the prediction for a set of sequences that are homologous to the target sequence and then selecting predictions that are consistent across the set. Both these studies resulted in improvements of performance compared to the standard methods.

Another more radical approach is to integrate the ability to retrieve the clusters by sequence signal into the clustering procedure itself. Fragments are admitted to the same cluster if they maintain both similar structural and sequence features. There are two situations that reduce the sequence signal: structures that can accommodate several sequence motifs and sequences that are compatible with several structural motifs. Those cases should be identified by the clustering algorithms and considered as separate classes, thus achieving a stronger signal for the other fragments. Multiple sequence data can be included by considering the sequence variation of each fragment derived from searches in the sequence database.

2.5. ALGORITHMS AND SCORE FUNCTIONS TO CONCATENATE BUILDING BLOCKS INTO A COMPLETE STRUCTURE

Two main algorithmic tools can be used to concatenate the building blocks into a complete structure using the obtained data set. One approach is based on graph algorithms and the other on optimization algorithms such as Monte Carlo and Genetic Algorithm. The graph algorithm approach considers the building blocks as nodes in a graph, and the overlap

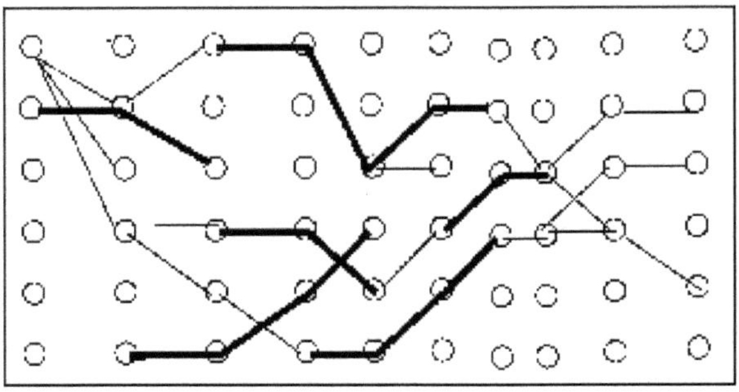

Figure 2. Schematic view of the layer graph that represents the concatenation process of building blocks into a complete structure. Each node represents a building block, and alternative building blocks for each sequence fragment are drawn on the same column. An edge between two nodes means that these two building blocks are fused together in an overlapping manner. Multi-edges (thick lines) represent longer predefined structural motifs. Thus, a path through the graph represents a conformation of the structure. Each node and edge might have a different probability.

between consecutive overlapping building blocks is represented as short edges in a "layer graph" where each position in the sequence represents one layer of the graph. A path through the graph that includes one node from each layer determines a conformation. Probabilities will be assigned to nodes and edges such that the probability of the nodes reflects the ranking of each building block based on the sequence signal. The probability of the edges reflects the quality of the superposition between the two building blocks it connects. Information about longer units, like secondary structure prediction, or similarity to known super secondary structure motifs can be encoded into the graph as predefined segments of the path with their own probability. The predefined segments are presented as thick "multi-edges" in Figure 2, which shows a schematic description of such a graph. To simplify the diagram, only a very small set of nodes, edges, and path segments (in thick lines) are presented. Thus, the probabilities are not shown on the figure, and the figure actually shows the elements with the highest probabilities in the graph. By giving low weights to elements with high probability, shortest paths algorithm can be used to produce a large number of different paths that have overall high multiplicative probability.

Note that this small graph, although it contains some dead-end edges, can still produce a large number of conformations. Note also that the size of the full graph would still be reasonable, on the order of few thousands nodes (on the order of a couple of dozens nodes for each position (i.e. each layer) times the length of the protein). Because of the locality of the edges, and even of the "multi-edges", their number is on the same order of magnitude as the nodes.

The second approach that we suggest is to use Genetic Algorithms to concatenate building blocks: Genetic Algorithm is a parallel computation paradigm based on the idea that repeated mutations, crossovers and selections, will efficiently evolve solutions to difficult problems. Several problems related to protein structure have been addressed by genetic algorithms [for example Unger and Moult, 1993, Pedersen and Moult, 97a, 97b, Yadgari et al, 1998]. To implement a genetic algorithm for a specific problem, one needs to represent the solutions as strings, to define the genetic operators of mutation and crossover on these strings, and to define a fitness function to score each solution.

For the problem of concatenating building blocks, these definitions are easily derived. Solutions, i.e. conformations, can be represented by a string of numbers of the length of the proteins, where the ith number refers to the identity of the building block used in the ith position of the structure. A mutation operation entails replacing one building block in the structure, and a crossover represents the fusion of two conformations by mixing and matching their parts. The fitness function scores a conformation for its overall structure (e.g. avoiding collisions and maintaining a globular structure) as well as the degree to which it is consistent with the other data such as secondary structure prediction, existence of super secondary structure motifs, etc. The mechanism of the genetic algorithm guarantees that the final population will be enriched with conformations that obey the constraints that are imposed. Note that by their nature, genetic algorithms generate large populations, which can be used as the ensemble for the next stage of screening.

Recent studies [Bonneau *et al.*, 2001a, Kihara *et al.*, 2001, Glick and Goldblum, 2000] demonstrated that looking for commonalities within a large population of structures often leads to good results in structure optimization searches. Thus, the first step in selecting the optimal structure should be locating common features that are frequent in the ensemble. Such features could include sets of intra-molecular distances, secondary structure distribution, common topology, etc. These features can be used as filters to select a smaller sub-population of conformations, each containing a large number of these common features.

In subsequent steps, a backbone structure is built for each conformation and the energy function described by Samudrala and Moult [1998], which is suitable for a simplified chain model, can be used to rank the structures. Next, a full atom model is constructed for those structures with good scores from the previous step. Then the full atom models can be scored by full atom energy functions [for example Avbelj, 1992; Avbelj and Moult, 1995]. The best conformations are chosen to represent the structure prediction for the protein in question.

2.6. COMBINING EXPERIMENTAL DATA AND COMPUTATIONAL PREDICTIONS

The idea of combining experimental data and computational predictions was suggested by several authors. We mention three important studies that have taken this approach. In [Dundekar and Argos, 1997] it was demonstrated that various structural data such as the existence of S–S bonds, protein side-chain ligands to iron–sulfur cages, crosslinks between side-chains, and conserved hydrophobic and catalytic residues can be used by genetic algorithms to improve the quality of protein structure prediction. The improvement was significant, usually improving the prediction by more than 2 Å. However, even with the improvement, the overall prediction quality was poor, usually differing by more than 5 or 6 Å from the target structure. This was probably due to the small number and the diverse nature of the experimentally obtained structural data.

Another, more focused, study was presented in [Kolinski and Skolnick, 1998]. In this study the authors simulate the use of long-range distance restraints, in addition to inclusion of complete secondary structure assignment information, in a Monte-Carlo algorithm to predict the structure of several proteins. Using a few dozen distance constraints (in general, the authors used N/7 distance constraints where N is the length of the protein) in a detailed Monte-Carlo scheme, good predictions were achieved. For most proteins tested, the distance of the lowest energy structure from the native structure was below 4 Å.

In recent studies from Baker's group, it was demonstrated that a building block approach combined with NMR data leads to a good "de novo determination" of protein structure. Bowers *et al.* [2000] showed that distance constraints derived from NMR experiments, can be used to improve the performance of a building block approach. In [Rohl and Baker, 2002] it was shown that Residual Dipolar Coupling (RDC) NMR measurements can be used to achieve structure determination. The results of both studies are promising, often achieving predictions with RMS to the correct structure of less than 3 Å. However, a large number of experimental constraints are needed, on the order of hundreds, thus rendering this method more of a tool to facilitate NMR structure determination than a rapid method for de novo structure prediction.

These studies clearly demonstrate that a combination of experimental data and computational predictions is an effective way to significantly improve the quality of protein structure prediction. Currently, there is a strong correlation between the amounts of experimental data included in the computation and the quality of the prediction. The challenge is to come up with an experimental method that will yield, with reasonable effort, sufficient data to enable reliable predictions.

3. Conclusions

The road to protein structure prediction is paved with building blocks. The building block approach simplifies the protein folding problem on many levels. In essence, the building block approach changes the protein folding problem from a continuous to a discrete problem. Practically, the problem becomes a one dimensional problem instead of a three dimensional one. Furthermore, the problem does not require predicting a new structure for a given sequence, but rather selecting the correct modules to be used in building the correct conformation.

However, just moving to the realm of building blocks does not automatically solve the protein folding problem. The problems lay in the two aspects of the process: How to select a small number of candidate building blocks for each fragment, and how to combine these candidates into the correct structure. While the statistical significance of the sequence preference of the building blocks has been demonstrated, using this preference to accurately select the correct building block has proven to be a difficult task. Similarly, while ensembles of conformations can be produced that may contain the correct structure, it is often difficult to choose the correct conformation from the many decoys.

It seems reasonable to suggest that the building block approach will be most valuable if it can be used in conjunction with additional sources of data to further narrow down the possibilities in selecting the building blocks and in choosing the right conformation from the created ensembles. These data should be achieved for a large number of proteins with much less experimental effort than the current structural determination methods of X-ray crystallography and NMR. Further advances in proteomic research might be able to supply such data, for example by techniques including epitope mapping, and intra-molecular distance measurements based on florescence, RDC NMR, and related techniques. Thus, a realistic near term goal would be to develop a structure prediction method based both on experimental data and on the computational building block approach.

4. References

Avbelj, F. (1992) Use of a potential of mean force to analyze free energy contributions in protein folding. Biochemistry. **31**: 6290–6297.

Avbelj, F. and Moult, J. (1995) Determination of the conformation of folding initiation sites in proteins by computer simulation. Proteins. **23**: 129–141.

Baker, D. and Sali A. (2001) Protein structure prediction and structural genomics. Science. **294**: 93–96.

Bonneau, R., Strauss, C.E. and Baker, D. (2001a) Improving the performance of Rosetta using multiple sequence alignment information and global measures of hydrophobic core formation. Proteins. **43**: 1–11.

Bonneau, R., Tsai, J., Ruczinski, I., Chivian, D., Rohl, C., Strauss, C.E. and Baker, D. (2001b) Rosetta in CASP4: progress in ab initio protein structure prediction. Proteins. Suppl **5**: 119–126.

Bowers, P.M., Strauss, C.E. and Baker, D. (2000) De novo protein structure determination using sparse NMR data. J Biomol NMR **18**: 311–318.

Brendel, V., Beckmann, J.S. and Trifonov, E.N. (1986) Linguistics of nucleotide sequences: morphology and comparison of vocabularies. J Biomol Struct Dyn. **4**: 11–21.

Bystroff, C. and Baker, D. (1998) Prediction of local structure in proteins using a library of sequence-structure motifs. J Mol Biol. **281**: 565–577.

Chothia, C. (1992) One thousand families for the molecular biologist. Nature. **357**: 543–544.

Claessens, M., Van Cutsem, E., Lasters, I. and Wodak, S. (1989) Modelling the polypeptide backbone with 'spare parts' from known protein structures. Protein Eng. **2**: **335**–345.

Dandekar, T. and Argos, P. (1997) Applying experimental data to protein fold prediction with the genetic algorithm. Protein Eng **10**: 877–893.

de-Brevern, A.G., Etchebest, C. and Hazout, S. (2000) Bayesian probabilistic approach for predicting backbone structures in terms of protein blocks. Proteins. **41**: 271–287.

Glick, M. and Goldblum, A. (2000) A novel energy-based stochastic method for positioning polar protons in protein structures from X-rays. Proteins. **38**: 273–287.

Jones, T.A. and Thirup, S. (1986) Using known substructures in protein model building and crystallography. EMBO J. **5**: 819–822.

Kabsch, W. and Sander, C. (1983) Dictionary of protein secondary structure: pattern recognition of hydrogen-bonded and geometrical features. Biopolymers. **22**: 2577–2637.

Karplus, M. and Weaver, D.L. (1994) Protein folding dynamics: the diffusion-collision model and experimental data. Protein Sci **3**: 650–668.

Kihara, D., Lu, H., Kolinski, A. and Skolnick, J. (2001) TOUCHSTONE: an ab initio protein structure prediction method that uses threading-based tertiary restraints. Proc Natl Acad Sci USA. **98**: 10125–10130.

Kolinski, A. and Skolnick, J., (1998) Assembly of Protein Structure From Sparse Experimental Data: An Efficient Monte Carlo Model. Proteins **32**: 475–494.

Kolodny, R., Koehl, P., Guibas, L. and Levitt, M. (2002) Small libraries of protein fragments model native protein structures accurately. J Mol Biol. **323**: 297–307.

Lessel, U. and Schomburg, D. (1997) Creation and characterization of a new, non-redundant fragment data bank. Protein Eng. **10**: 659–664.

Micheletti, C., Seno, F. and Maritan, A (2000) Recurrent oligomers in proteins: an optimal scheme reconciling accurate and concise backbone representations in automated folding and design studies. Proteins. **40**: 662–674.

Moult, J. (1999) Predicting protein three-dimensional structure. Curr Opin Biotechnol. **10**: 583–588.

Moult, J., Pedersen, J.T., Judson, R. and Fidelis, K. (1995) A large-scale experiment to assess protein structure prediction methods. Proteins. **23**: ii–v.

Moult, J. and Unger, R. (1991) An analysis of protein folding pathways. Biochemistry. **30**: 3816–3824.

Panchenko, A.R., Luthey-Schulten, Z., Cole, R. and Wolynes, P.G. (1997) The foldon universe: a survey of structural similarity and self-recognition of independently folding units. J Mol Biol **272**: 95–105.

Pedersen, J.T. and Moult, J. (1997a) Ab initio protein folding simulations with genetic algorithms: simulations on the complete sequence of small proteins. Proteins. Suppl **1**: 179–184.

Pedersen, J.T. and Moult, J. (1997b) Protein folding simulations with genetic algorithms and a detailed molecular description. J Mol Biol. **269**: 240–259.

Ponder, J.W. and Richards, F.M. (1987) Tertiary templates for proteins. Use of packing criteria in the enumeration of allowed sequences for different structural classes. J Mol Biol. **193**: 775–791.

Ramakrishnan, C. and Ramachandran, G.N. (1965) Streochemical criteria for polypeptide and protein chain conformation. Biophys J **5**: 909–933.

Rohl, C.A. and Baker, D. (2002) De Novo Determination of Protein Backbone Structure from Residual Dipolar Couplings Using Rosetta. J. Am. Chem. Soc., **124**: 2723–2729.

Rooman, M.J., Rodriguez, J. and Wodak, S.J. (1990) Automatic definition of recurrent local structure motifs in proteins. J Mol Biol. **213**: 327–336.

Rubin, G.M. (2001) The draft sequences. Comparing species. Nature. **409**: 820–821.

Samudrala, R. and Moult, J. (1998) A graph-theoretic algorithm for comparative modeling of protein structure. J Mol Biol. **279**: 287–302.

Simons, K.T., Bonneau, R., Ruczinski, I. and Baker, D. (1999) Ab initio protein structure prediction of CASP III targets using ROSETTA. Proteins. **37(S3)**: 171–176.

Sun, Z. and Jiang B. (1996) Patterns and conformations of commonly occurring supersecondary structures (basic motifs) in protein data bank. J Protein Chem. **15**: 675–90.

Todd, A.E., Orengo, C.A. and Thornton, J.M.. (2001) Evolution of function in protein superfamilies, from a structural perspective. J Mol Biol **307**: 1113–1143.

Tsai, C.J. and Nussinov, R. (2001) The building block folding model and the kinetics of protein folding. Protein Eng **14**: 723–733.

Unger, R., Harel, D., Wherland, S. and Sussman, J.L. (1989) A 3D building blocks approach to analyzing and predicting structure of proteins. Proteins. **5**: 355–373.

Unger, R. and Moult, J. (1993) Genetic algorithms for protein folding simulations. J Mol Biol. **231**: 75–81.

Unger, R. and Moult, J. (1996) Local interactions dominate folding in a simple protein model. J Mol Biol. **259**: 988–994.

Wintjens, R., Wodak, S.J. and Rooman, M. (1998) Typical interaction patterns in alpha-beta and beta-alpha turn motifs. Protein Eng. **11**: 505–522.

Yadgari, J., Amir, A. and Unger, R. (1998) Genetic algorithms for protein threading. Proc Int Conf Intell Syst Mol Biol. **6**: 193–202.

Biodata of **Leszek Rychlewski**, author (with coauthors J. M. Bujnicki and D. Fischer) of the chapter ***"Protein Fold—Recognition and Experimental Structure Determination."***

Dr. Leszek Rychlewski obtained his MD degree in 1998 at the Charite, Humboldt University zu Berlin. Between 1996 and 1998 he was appointed as Research Fellow at the Scripps Research Institute (TSRI) La Jolla, CA. USA. Between 1998 and 1999 he was employed as Postdoctoral Researcher at the University of California at San Diego (UCSD), USA. In 2001 he founded the dependent Institute BioInfoBank Institute in Poznan (Poland). The main research goal of this new laboratory was Th e development a bioinformatics technological platform for the analysis of genes, proteins and the design of new drugs and some other functions.

E-mail: **leszek@bioinfo.pl**

Biodata of Janusz Bujnicki

Dr. Janusz M. Bujnicki obtained his PhD degree in 2001 at the University of Warsaw. He is currently a contract Professor, head of the Laboratory of Bioinformatics and Protein Engineering at the Institute of Molecular and Cell Biology in Warsaw (Poland). His scientific interests are in the areas of protein structure prediction by bioinformatics and protein fold determination by low-resolution methods, molecular evolution, and engineering of proteins, in particular enzymes with new functions. Dr. Bujnicki is an awardee of the EMBO/Howard Hughes Medical Institute Young Investigator Programme and a holder of a Fellowship for Young Scientists from the Foundation for Polish Science.

E-mail: **iamb@genesilico.pl**

Biodata of Daniel Fischer

Dr. Fischer is a Senior Lecturer at the Ben-Gurion University of the Negev, Beer-Sheva, Israel. He obtained his PhD in 1994 at the Tel-Aviv University in computer sciences. He was employed by Intel, at the city of Haifa and spent his postdoctoral at NIH, USA. Then he served as an assistant researcher at the Molecular Institute UCLA, in California and from 1998 to present he is at the Ben-Gurion University.

E-mail: **dfischer@cs.bgu.ac.il**

Leszek

Janusz Bujnicki

Daniel Fischer

J. Seckbach and E. Rubin (eds.), The New Avenues in Bioinformatics, 189–199.
© 2004 *Kluwer Academic Publishers. Printed in the Netherlands.*

PROTEIN FOLD-RECOGNITION AND EXPERIMENTAL STRUCTURE DETERMINATION

LESZEK RYCHLEWSKI[1], JANUSZ M. BUJNICKI[2] and DANIEL FISCHER[3]

[1]*BioInfoBank Institute, ul. Limanowskiego 24A, 60-744 Poznan, Poland,* [2]*Bioinformatics Laboratory, International Institute of Molecular and Cell Biology, ks. Trojdena 4 02-109 Warsaw, Poland, and* [3]*Bioinforma-tics, Dept. Computer Science, Ben Gurion University Beer-Sheva 84015, Israel. Corresponding author*

1. Introduction

The value of a protein's three-dimensional (3D) structure in connection with its function is enormous. Protein 3D structure provides a solid framework for planning experiments and for the interpretation of their results, represents obligatory information for docking simulations or for rational structure-guided drug design, and allows evolutionary classification when the sequence information is insufficient. The growing gap between the number of protein sequences awaiting interpretation, and the number of known protein structures, has prompted the birth of structural genomics projects. These projects aim at determining the 3D structures of a carefully chosen set of representative proteins from different families so that it will be possible to model the 3D-structures of most family members using computational techniques (Chance et al., 2002). Thus, computational structure modeling has become an important research area, and a cornerstone of structural genomics (Fischer et al., 2001a).

Here we review the state of the art of computational structure prediction, with focus on the sub-area named fold-recognition (FR), also known as "threading". We describe how the use of computational protein structure predictions can in some cases be of help during the experimental protein structure determination process. We illustrate the value and applicability of this approach with two examples of recently determined protein structures for which the computational predictions suggested that the experimental structures may be erroneous. Subsequent confirmation by the authors of the structures demonstrated that our predictions were correct. We suggest that computational prediction can become a useful tool in structure determination and refinement.

[1]*leszek@bioinfo.pl*; [2]*iamb@genesilico.pl*; [3]*dfischer@cs.bgu.ac.il*

2. Fold Recognition

FR is a special class of bioinformatics methods that are aimed at predicting the 3D fold of a target protein from its amino acid sequence through the detection of template proteins of known structure. As opposed to Homology Modeling, FR is aimed at target sequences that share no significant sequence similarity to any protein of known structure, and is often able to generate relatively accurate 3D models for many target proteins that have no close relatives of known 3D structure. FR methods rely on the observation that the number of different 3D folds in Nature is relatively small, and that evolutionarily related proteins usually have similar structures, even in the absence of significant sequence similarity. FR generates sequence-structure alignments which are influenced by the combination of the evolutionary relationships between target and template and by the sequence-structure compatibility evaluation (Fischer et al., 1996).

2.1. EVALUATING THE ACCURACY OF FR METHODS

There exist a number of popular evaluation experiments conducted by the protein structure prediction community, which provide an assessment of the capabilities and limitations of current methods. These experiments use soon-to-be-determined or newly determined structures as prediction targets. The biannual CASP meeting (Moult et al., 2001) brings together over a hundred bioinformatics groups that assess their prediction capabilities by comparing the blindly predicted models to the experimental structures. The related CAFASP experiment (Fischer et al., 2001b) uses the same targets as CASP, but evaluates only fully automated procedures, available through web-based servers. Fully automated benchmarking procedures like LiveBench (Bujnicki et al., 2001a) or EVA (Rost and Eyrich, 2001) complement this effort by conducting weekly evaluation runs on newly deposited PDB structures, resulting in a more robust, statistically significant assessment because the sample of test targets is much larger.

2.2. THE SUM IS BETTER THAN THE PARTS

A major finding from the latest prediction experiments is that better structure predictions can be obtained by combining the results produced using different methods. In particular, the last CASP4 experiment showed that the group named CAFASP-CONSENSUS, which filed predictions extracted from a number automated servers, performed considerably better than any individual server, and better than all but six human predictors. Due to this success, automated "meta-predictors" (http://bioinfo.pl, http://bioserv.infobiosud.univ-montp1.fr and http://cubic.bioc.columbia.edu) and consensus-generating servers have been deployed (Bujnicki et al., 2001b; Fischer, 2003). LiveBench results have demonstrated that these new meta-predictors have significantly improved the sensitivity and specificity of FR in the last year.

2.3. FR SUCCEEDS FOR ABOUT HALF OF THE TARGETS

To estimate the applicability of FR, consider the targets used in LiveBench. These correspond to a set of newly released structures that share no clear sequence similarity to previously determined structures. The LiveBench results show that current methods are able to generate useful models for about half of them. As the methods improve, and as our structural knowledge expands, this fraction is likely to grow. Thus, within a few years, it is likely that more or less accurate computational models will be at hand well before the experimental structures of the vast majority of proteins is determined.

2.4. IDENTIFYING HIGHLY RELIABLE PREDICTIONS

Another important result of large-scale evaluation experiments such as LiveBench is that a better estimate of the specificity of each server can be obtained. Because FR methods assign a score to each prediction, similar to the scores produced by standard sequence comparison tools such as BLAST, it is possible to know whether a particular score corresponds to a confident prediction or not. Knowing the specificity of any prediction method is essential for its wider applicability. The main outcome is that today it is easier for a biologist to determine when a particular prediction can be trusted. Further confirmation of the correctness of a prediction can often be obtained if the predictions of a number of independent servers agree with each other and if the meta-predictors produce above-threshold scores (threshold scores can be estimated for each server using the LiveBench data). Other factors affecting the reliability of a given prediction are: the completeness of the sequence-structure alignment, the number and the size of gaps, and the patterns of conserved residues in homologs of the target and homologs of the template (for example, are important residues, such as those in the active site, well conserved in the protein families of the target and of the structure template?). Further confidence can be obtained if the template(s) used to build the FR model correspond to highly refined structures and if structure verification tools assess the FR model as "correct" (see below).

2.5. HIGHLY RELIABLE FR MODELS ARE USEFUL

Here we focus on highly reliable FR models fulfilling the above criteria. LiveBench has demonstrated that such highly reliable FR predictions are excellent, rough structural models, sometimes comparable to the accuracy of medium-resolution X-ray or nuclear magnetic resonance (NMR) models. Admittedly, such highly reliable FR models are not often obtained, nor do the FR models reach the level of accuracy of medium to high-resolution experimental structures. However, as the gaps in our structural knowledge are being rapidly filled, the number of highly reliable computational models is also growing fast.

3. Fold Recognition Detects Errors in Newly Determined Structures

In the course of the last LiveBench-4 experiment, several cases of targets have been observed, where highly reliable FR models showed significant differences from certain protein structures experimentally solved at medium to low resolution (2.3-3.1 A). Inconsistencies between the FR-modeled and experimentally-solved structures ranged from residue index shifts in the tracing of the chain into the electron-density-map, to tracing in the opposite direction, to completely different folds; after communicating these inconsistencies to the authors of the structures, the crystallographic models were confirmed to be dubious. In one case, our findings prompted the experimentalists to revise their data. In another, the experimental structure was withdrawn.

3.1. *DEINOCOCCUS RADIODURANS* LARGE RIBOSOMAL SUBUNIT STRUCTURE (LSU)

The first example that we present illustrates the use of FR models as an aid in the interpretation of the recently released low-resolution structure of bacterial LSU (D50S; Protein Data Bank entry 1kc9 (Harms et al., 2001)). With the LiveBench Meta-Server, FR models were generated for a number of D50S proteins, using only their amino-acid sequences and the high-resolution 3D templates of distantly related, previously solved structures (including the high-resolution structure of the remotely related archaeal LSU; 1jj2 in PDB (Klein et al., 2001)). The FR models obtained were highly reliable (see Section 2), but when compared to the 1kc9 experimental structure, a number of inconsistencies were revealed. After a detailed analysis of the latter, it became clear that a number of ribosomal proteins had been incorrectly traced into the electron-density map of D50S. Based on the FR sequence-structure assignment and the original 1kc9 backbone, alternative models of D50S proteins were generated and deposited in the PDB as 1gs2 (Bujnicki et al., 2002a). The authors of the D50S structure were immediately notified and provided with the names of the potentially mistraced chains. They confirmed that the 1kc9 entry contained a number of shifts, and deposited a replacement entry (1kpj), in which several proteins had been retraced without explicitly utilizing information from our models. Comparison of the 1kpj entry with the FR models suggested further corrections (Bujnicki et al., 2002b), most of which have been partially introduced into the third version of the D50S crystallographic model (1lnr). Figures 1 and 2 compare our proposed model for the ribosomal protein L24

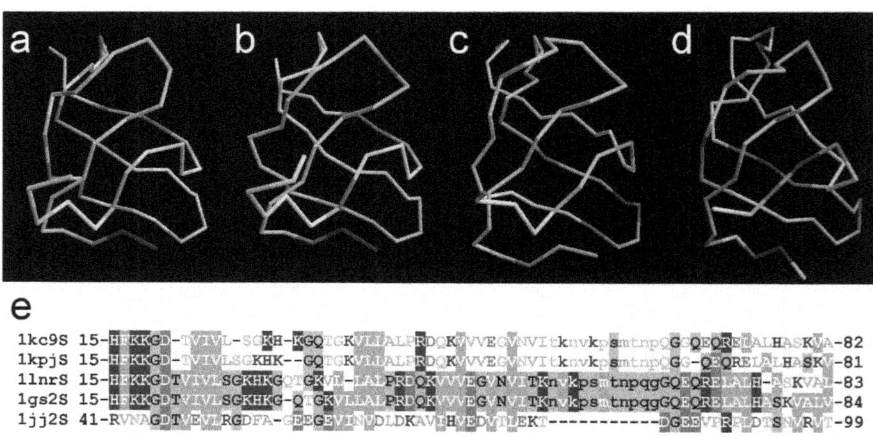

Figure 1. Progressive refinement of the *D.radiodurans* L24 protein structure prompted by inconsistency of the initial model with the experimental results. a-d) differences between the FR-based amino acid sequence assignment (1gs2) and the experimental models mapped onto the respective C-alpha coordinates of the globular core (aa 15–55, 68–66) and expressed as an RMSD spectrum (deep blue—identical position, red—spatial shift). (a) initial model 1kc9S; (b) model 1kpjS; (c) model 1lnrS re-revised; (d) template structure 1jj2S; {e) structure-based amino acid sequence alignment showing progressive refinement of the bacterial L24 protein structure (1kc9 to1kpj to 1lnr) prompted by the FR model (1gs2) based on the archaeal L24 structure (1jj2). Lowercase letters indicate the non-globular, "looped-out" insertion of *D.radiodurans* L24 (residues 56–67), not present in the archaeal 1jj2. Note that the final crystallographic model (1lnr) is very similar to the FR model, which was generated when only the 1kc9 and 1jj2 structures were available.

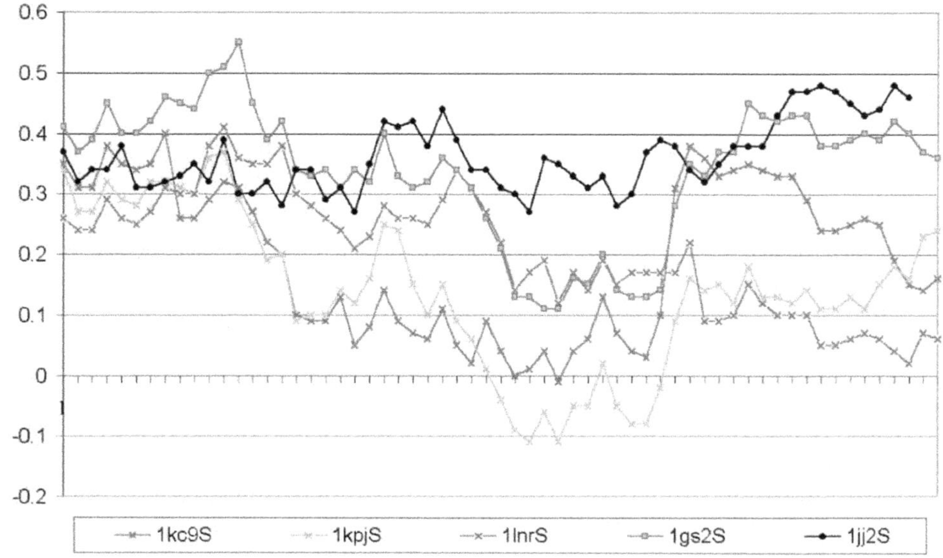

Figure 2. VERIFY3D (Luthy et al., 1992) plots of the five models shown in Figure 1. The x-axis corresponds to the residue numbers, and the y-axis corresponds to the VERIFY3D score of each residue. Only the globular core (aa 15–55, 68–86) was analyzed, since VERIFY3D was not designed to evaluate non-globular regions. Scores close-to or below-zero, indicate possible problems in the models. The plot shows that 1kc9S and 1kpjS have regions with such low scores. No negative regions appear in the theoretical model 1gs2S nor in the latest correction 1lnrS. The VERIFY3D plot of 1jj2S is shown for comparison. Verify3D is a method for assessment of the compatibility of a protein sequence with its structure, using three-dimensional (3D) profiles. The structural environments are described by: the area of the residue buried in the protein and inaccessible to solvent, the fraction of side-chain area that is covered by polar atoms, and the local secondary structure. The quality of the sequence-structure fit is analyzed using a moving-window scan. This method has been shown to correctly identify misfolded models as well as incorrectly modeled segment in an otherwise correct structures.

with its corresponding versions in 1kc9, 1kpj, 1lnr, and 1jj2. This example demonstrates that FR methods are capable of detecting register shifts in low-resolution structures and are able also to overtly suggest how to correct them, by providing evolutionary links to previously solved, high-resolution structures.

3.2. MarR TRANSCRIPTION FACTOR

The second example of the use of FR to detect potential errors is obtained by the FR analysis of the MarR transcription factor (Alekshun *et al.*, 2001; PDB code 1jgs, deposited on June 2001), which was also used as a LiveBench target. FR methods suggested that MarR is a member of the "winged helix" (wH) superfamily (multiple related structures are known) being closely related to the SarA transcription factor. There are two forms of SarA in PDB, free (1fzn) and in a complex with DNA (1fzp) (Schumacher et al., 2001a). 1fzn and 1fzp differ due to the apparent conformational change in the C-terminus (beta-helix transition), but they share a major portion of the common core. Curiously, comparison of the MarR structure (1jgs) and all its potential templates revealed that MarR is indeed highly similar to the wH proteins structures, while the fold of both forms of SarA is completely different. Observation of different folds for closely related proteins suggests that one of these structures may be wrong. The results of the analysis of the SarA sequence using the FR meta-server suggested that this protein is also a member of the wH superfamily, despite its unusual fold. This suggested that either (i) MarR and all other wH structures are wrong (nearly impossible), (ii) SarA changed its fold in the evolution (unlikely, but cannot be excluded) or (iii) the SarA structure is wrong. It is noteworthy that the 1fzn entry has been withdrawn from the PDB on January 2002 (Schumacher et al., 2001b). As in the D50S case, this study demonstrated again that the initial use of FR models could have helped the crystallographers identify alternative solutions and suggest further refinement before submission of a dubious structural model.

4. Difficulties In The Interpretation Of Protein Structure Determination

4.1. EXPERIMENTAL STRUCTURES ARE INTERPRETATIONS

X-ray or NMR structures are often considered by non-experts as flawless by definition. However, in both these techniques the protein structure is seldom calculated directly from the measurements, but usually represents an interpretation—fitting of the model to NMR restraints or to the X-ray diffraction data. For instance, only the intensities of the diffracted X-ray waves can be measured and not their phases, hence the traditional structure solution process starts with guessing how the structure may look like, followed by a calculation of how the model would diffract X-rays and a comparison with the observed data. Thus, solving a protein structure can become a complex and challenging task, especially if the experimental data is of low resolution or difficult to interpret. Although the vast majority of the solved structures correspond to correct interpretations of the data, occasionally, errors can be introduced. The most severe errors in protein structure determination, like assigning a completely wrong fold or tracing the polypeptide chain in an opposite direction happen very rarely. More common are errors where the secondary structure elements are recognized but connected incorrectly or when the sequence gets out of register with the electron-density during the tracing in areas of low-resolution (Kleywegt and Jones, 1995). Some representative examples of such errors, which hampered the identification of functionally important residues, have been reviewed by Branden and Jones (1990).

4.2. MODEL VERIFICATION TOOLS

Consequently, it is imperative that 3D structures be extensively verified in order to avoid public release of incorrect models. There are several approaches to measure the quality of a protein structure. The most common measures involve the comparison of theoretical parameters calculated from the model with the directly observed experimental data, for instance, the R-factor and R-free in crystallography and the comparison of experimental and theoretical restraints in NMR. Methods have also been developed (Kleywegt, 1997) to assess C-alpha-only models (as can be the case for low-resolution, not fully-refined models). Other methods analyze features of the model using statistics obtained from previously deposited, high-resolution (and undoubtedly correct) structures (e.g. VERIFY3D, Luthy et al., 1992; PROSAII, Sippl, 1993; PROCHECK, Laskowski et al., 1993), or use empirical effective energy functions to estimate the stability of protein folds (Guerois et al., 2002). These methods can also be applied to verify the quality of models not derived from experimental data, such as FR or homology models. However, none of the above methods are able to directly suggest alternative models.

After years of efforts in developing verification tools, many experimentalists already apply these methods routinely, and the vast majority of published structures are essentially error-free. However, we continue to occasionally observe cases of publication of apparently wrong protein structure models that are subsequently revised. This may partially be attributed to the fact that the application of these tools requires some level of expertise by the user and that the iterative refinement process can sometimes be time-consuming. Thus, with the increasing pressure to publish fast, the verification process is not or can not always be applied thoroughly.

4.3. NEW FR-BASED VERIFICATION METHOD

We propose here that in some cases of low-resolution, difficult to interpret data, FR models can be used as another, independent and easy-to-use verification tool (available at http://bioinfo.pl/), which is able also to suggest an alternative model. This procedure can be used to help fit the amino-acid sequence into a low-resolution electron-density map, and can be applied in at least two scenarios (see below). In both of them, a prerequisite is the generation of a reliable FR model using the sequence of the target protein. Hence, FR can be confidently used as an aid in structure determination/validation/refinement only if an undoubtedly correct, preferably high-resolution structure of a protein of the same fold exists in the database, and if a robust evolutionary link can be identified between that protein and the target protein. In what follows, we briefly describe two of the suggested scenarios where, in addition to a pletory of structure validation tools, FR models may be of help to the crystallographer.

4.3.1. Initial Tracing

If a preliminary, incomplete C-alpha trace can be deduced from the electron-density map of the target protein, superimpose the FR model onto the C-alpha trace. From the FR model, the directionality and the connectivity of the fragments of the polypeptide chain, as well as the amino acid identities can be assigned to the preliminary C-alpha coordinates. This procedure can provide a jump-start for further refinement and has a great potential to be incorporated into the automated structure prediction pipelines of structural genomics.

4.3.2. Verification Of Evolutionary Consistency Of The Solved Structure and Other Remotely Related Proteins

First superimpose the coordinates of the target protein and the FR model. If any sequence shifts are observed in the core regions, analyze the multiple sequence alignments and perform thorough structure evaluation checks to identify potential sources of problems in the target as well as in the template structures. This procedure may help identify errors in previously published structures or avoid publication of a new, erroneous structure.

The application of the above procedures may have allowed the early identification of errors in the two examples presented above.

5. Conclusions

In the two cases described in Section 2, only comparison of new, presently unavailable, high-resolution structures of the bacterial LSU and the SarA transcription factor with the existing counterparts (archaeal LSU and wH proteins, respectively) and their thorough analysis with structure assessment tools will allow to conclude whether the target or the template structures exhibit errors or whether an unusual change of fold has occurred since the corresponding protein pairs diverged from their common ancestors. However, even before the new data become available, the FR models available for the D50S proteins and the SarR protein could be compared with the existing data and used as starting points to obtain potentially better interpretations.

Current FR models do not reach the level of accuracy and detail as do high quality experimental structures, and are usually biased towards the template(s) of known structure that were used to build them. Thus, FR models cannot accurately model the subtle details of the structural differences that exist between the target and the template(s). Consequently, computational models cannot yet replace the need of high-resolution experimental structures. However, highly reliable FR models are sufficiently accurate to guide or verify the initial tracing of the chain or to identify other errors in the experimental structure. This is so because of the rich information embedded in a FR model. This information includes (i) evolutionary information between the target and the template(s) and (ii) accurate structural information from the high-resolution template structure(s). Because the FR model is not biased by the experimental data, it provides an independent, additional source of valuable information that can readily be exploited during the structure determination process. Nevertheless, it must be emphasized that the quality of the FR model critically depends on the quality of the available templates. Thus, FR is not intended to replace experimental structure determination, but in this particular context only facilitates detection of evolutionary links between remotely related structures.

The FR-based verification method proposed here may be particularly useful in low-resolution, hard-to-interpret experimental data where some level of subjective, interactive interpretation by the crystallographer is necessary. Another possible application of FR in structure determination is the use of FR models as phasing models for Molecular Replacement (MR). In macromolecular crystallography, the initial determination of phases by molecular replacement (MR) is often attempted if the structure of a similar or homologous macromolecule is known (Blow & Rossman, 1961). MR involves the placement (i.e. rotation and translation) of that similar structure ("search model") in the unit cell of the target crystal in order to obtain the best agreement between calculated and observed diffraction data. The optimally placed search model is used to obtain initial phases for structure building and refinement. While MR has traditionally been applied only if a closely related template exists, here we propose that FR models, using distantly related templates, may also be of aid. It is noteworthy to point out that D. Jones has recently discussed the potential of using FR models as candidate phasing models for MR (Jones, 2001).

Further tests of the applicability of these various uses of FR during the structure determination process are currently being applied as part of the ongoing CAFASP3 experiment (MR-CAFASP; http://www.cs.bgu.ac.il/~dfischer/CAFASP3). The interplay of theory and experiment will undoubtedly have significant implications for Structural Genomics.

6. References

Alekshun, M.N., Levy, S.B., Mealy, T.R., Seaton, B.A., Head, J.F. (2001). The crystal structure of MarR, a regulator of multiple antibiotic resistance, at 2.3 A resolution. Nat Struct Biol. 8, 710–714.

Blow, D. M. & Rossmann, M. G. (1961). The single isomorphous replacement method. Acta Cryst. 14, 1195–1202.

Branden, C.I. and Jones, T.A. (1990). Between objectivity and subjectivity. Nature 343, 687–689.

Bujnicki, J.M., Elofsson, A., Fischer, D., and Rychlewski, L. (2001a). LiveBench-2: Large-scale automated evaluation of protein structure prediction servers. Proteins 45, 184–191.

Bujnicki, J.M., Elofsson, A., Fischer, D., and Rychlewski, L. (2001b). Structure prediction Meta Server. Bioinformatics 17 750–751.

Bujnicki, J.M., Rychlewski, L., and Fischer, D. (2002a). Fold Recognition Detects an Error in the PDB. Bioinformatics, 18, 1391–1395, 2002.

Bujnicki, J.M., Feder, M., Rychlewski, L., and Fischer, D. (2002b). Errors in the D. radiodurans large ribosomal subunit structure detected by fold recognition and structure evaluation tools. FEBS Lett. 525, 174–175, 2002.

Chance, M.R., Bresnick, A.R., Burley, S.K., Jiang, J.S., Lima, C.D., Sali, A., Almo, S.C., Bonanno, J.B., Buglino, J.A., Boulton, S., Chen, H., Eswar, N., He, G., Huang, R., Ilyin, V., McMahan, L., Pieper, U., Ray, S., Vidal, M., and Wang, L.K. (2002). Structural genomics: a pipeline for providing structures for the biologist. Protein Sci. 11, 723–738.

Fischer, D., Rice, D., Bowie, J.U., and Eisenberg, D. (1996). Assigning amino acid sequences to 3-dimensional protein folds. FASEB J. 10, 126–136.

Fischer, D., Baker, D., and Moult, J. (2001a). We need both computer models and experiments. Nature 409, 558.

Fischer, D., Elofsson, A., Rychlewski, L., Pazos, F., Valencia, A., Rost, B., Ortiz, A.R., and Dunbrack, R.L., Jr. (2001b). CAFASP2: The second critical assessment of fully automated structure prediction methods. Proteins 45 Supp.l 5, 171–183.

Fischer, D. (2003). 3D-SHOTGUN: A Novel, Cooperative, Fold-Recognition Meta-Predictor. Proteins In the press.

Guerois, R, Nielsen, J.E. and Serrano, L (2002). Predicting changes in the stability of proteins and protein complexes: A study of more than 1000 mutations. J Mol. Biol. 320, 369–387.

Harms, J., Schluenzen, F., Zarivach, R., Bashan, A., Gat, S., Agmon, I., Bartels, H., Franceschi, F. and Yonath, A. (2001). High-resolution structure of the large ribosomal subunit from a mesophilic eubacterium. Cell 107, 679–688.

Jones, D.T. (2001). Evaluating the potential of using fold-recognition models for molecular replacement. Acta Crystallogr. D. Biol Crystallogr. 57, 1428–1434.

Klein, D.J., Schmeing, T.M., Moore, P.B., and Steitz, T.A. (2001). The kink-turn: a new RNA secondary structure motif. EMBO J. 20, 4214–4221.

Kleywegt, G.J. (1997). Validation of protein models from C-alpha coordinates alone. J Mol. Biol. 273, 371–376.

Kleywegt, G.J. and Jones, T.A. (1995). Where freedom is given, liberties are taken. Structure 3, 535–540.

Laskowski, R.A., MacArthur, M.W., Moss, D.S., and Thornton, J.M. (1993). PROCHECK: a program to check the stereochemical quality of protein structures. J. Appl. Crystallogr. 26, 283–291.

Luthy, R., Bowie, J.U., and Eisenberg, D. (1992). Assessment of protein models with three-dimensional profiles. Nature 356, 83–85.

Moult, J., Fidelis, K., Zemla, A., and Hubbard, T. (2001). Critical assessment of methods of protein structure prediction (CASP): Round IV. Proteins 45 Suppl. 5, 2–7.

Rost, B. and Eyrich, V.A. (2001). EVA: large-scale analysis of secondary structure prediction. Proteins Suppl. 5, 192–199.

Schumacher, M.A., Hurlburt, B.K., and Brennan, R.G. (2001a). Crystal structures of SarA, a pleiotropic regulator of virulence genes in S. aureus. Nature 409, 215–219.

Schumacher, M.A., Hurlburt, B.K., and Brennan, R.G. (2001b). Correction: Crystal structures of SarA, a pleiotropic regulator of virulence genes in S. aureus. Nature 414, 85.

Sippl, M.J. (1993). Recognition of errors in three-dimensional structures of proteins. Proteins 17, 355–362.

Biodata of **Ori Sasson** author (with coauthor Professor **Michal Linial**) of the chapter ***"Protein Clustering and Classification."***

Mr. Ori Sasson is currently completing his Ph.D. dissertation in Computer Science in the Hebrew University of Jerusalem. He holds a M.Sc. and B.Sc. in Computer Science from the Hebrew University. He is the author of several technical books.

E-mail: **Ori@osasson.com**

Professor Michal Linial is a member of the Center for Molecular and Cellular Neurobiology in Jerusalem. In this scope her laboratory study the dynamic processes of nerve terminals and the molecular aspects of synapse functioning, plasticity and maturation. She obtained her Ph.D. from the The Hebrew University of Jerusalem in 1986. Dr. M. Linial served as the head of teaching division for Structural and Molecular Biochemistry in the Hebrew University. from 1999 and she is the head of a joint program for undergraduate and master level studies in Life Science and Computer Science. In recent years M.L. is part of the Structural Genomics Informatics initiatives. Jointly with her collegues Michal Linial studying the protein metric space and its application towards automatic methods for protein annotation and functional predictions.

E-mail: **michall@cc.huji.ac.il**

Ori Sasson

Michal Linial

J. Seckbach and E. Rubin (eds.), The New Avenues in Bioinformatics, 201–226.
© 2004 *Kluwer Academic Publishers. Printed in the Netherlands.*

PROTEIN CLUSTERING AND CLASSIFICATION

ORI SASSON[1] AND MICHAL LINIAL[2]
[1]*The School of Computer Science and Engeeniring and* [2]*The Life Science Institute, The Hebrew University of Jerusalem, Israel*

1. Introduction

Proteins are the building blocks of all organisms. The name *protein* is derived from the Greek word 'protos' which means first or primal. Indeed proteins are the most fundamental substance of life, as they are the key component of the protoplasm of all cells. In addition to their role as the building blocks of cells and tissue, proteins also play a role in executing and regulating most biological processes. Enzymes, hormones, transcription factors, pumps and antibodies are examples for the diverse functions fulfilled by proteins in a living organism.

Proteins are macromolecules, and consist of combinations of amino acids in peptide linkages, that contain carbon, hydrogen, oxygen, nitrogen, and sulfur atoms. There are only 20 different types of amino acids, and they can be combined to generate an infinite number of sequences. In reality, only a small subset of all possible sequences appears in nature.

In the study of proteins, there are three important attributes of proteins: sequence, structure, and function. The *sequence* is essentially the string of amino acids which comprises the protein. The *structure* of the protein is the way the protein is outlaid in the three dimensional space. Perhaps most important, yet most elusive, is the protein *function*. The protein function is its actual role in the specific organism in which it exists. Understanding the protein function is critical for most applications, such as drug design, genetic engineering, or pure biological research.

The advent of advanced techniques for sequencing proteins in the last two decades, spur the explosive growth witnessed today in protein databases. Due to this rate of growth, the biological function of a large fraction (between one third and one half, depending on the organism) of sequenced proteins remains unknown. Furthermore, the functional annotation available is often partial or inaccurate.

The difficulty of assigning a certain function to a particular protein stems from the inherent complexity of the definition of protein function. For example, the function of a protein is often defined by its interaction with other proteins. Another difficulty arises from the fact that a protein is a dynamic entity, and it may be subjected to large number of modifications that may affect its function and fate. In this survey we will not address any of the dynamic properties of the proteins and will only address the protein as a fixed string of amino-acids.

A common way to tackle the complexity of protein function prediction uses database searches to find proteins similar to a new protein, thus inferring the protein function. This method is generalized by protein *clustering* or *classification*, where databases of proteins are organized into groups or families in a manner that attempt to capture protein similarity.

We survey the field of protein clustering and classification systems. Such systems use the protein sequence, and at times structure, to classify proteins into families. The classification may be leveraged towards function inference.

In order to provide a foundation for our survey, the next section describes the most commonly used algorithms for sequence similarity, and the way they can be used directly for protein classification. The following sections describe classification systems based on the methodology used: motif-based classifications, full-sequence analysis classifications, phylogenetic classification, and structure based classifications, aggregated classifications making use of the results of other classification systems. The last section provides a summary.

It is important to note that vast amounts of research were made in the field of proteomics which are related to the topic of this survey. Herein, we focus primarily on publicly available software systems for protein classification. We chose to describe a selected subset of systems and methods available, which represents the full spectrum of methods and directions.

2. Sequence Similarity

One of the most common approaches towards classifying proteins is using sequence similarity. Sequence similarity is a well studied subject, and numerous software packages suited for biological sequences are available. Such packages (e.g. BLAST) are probably the most widely used software in the fields of biology and bioinformatics. A comprehension of sequence similarity algorithms is beyond the scope of this survey; the interested reader may refer to Gusfield (2001).

Sequence similarity algorithms (and software) take as input two sequences and provide a measure of distance or similarity between them. Note that these quantities are related in the sense that the higher the distance the lower the similarity, and vice versa.

The notion of distance between sequences has been formalized by Levenshtein (1965), who has introduced a dynamic programming algorithm for determining this distance, referred to as 'edit distance'. The 'edit distance' between two strings is defined as the number of insertions, deletions, and replacements of characters from the first string required to obtain the second string. Some variants of the edit distance allow for reversals of sub-strings.

The edit distance problem is strongly related to the problem of string alignment. The problems are essentially equivalent, as the alignment can be easily produced from a set of insertions and deletions of characters. In the context of biological sequences, similarity and alignment were first studied by Needleman and Wunsch (1970). The Needleman-Wunsch sequence similarity and sequence alignment are usually referred to as *global* sequence alignment. In other words, this is alignment of full length sequences. In practice, only extremely similar sequences can be globally aligned in a meaningful way. However, many proteins exhibit strong *local* similarity, for example when they share a domain or motif. The local alignment problem was studied by Smith and Waterman (1981).

Algorithms for calculating either local or global similarity for protein sequences do not give equal weight to all amino-acids. Instead, scoring matrices are used to achieve an alphabet-weighted similarity. An alphabet-weighted edit distance can also be defined using such scoring matrices, giving different weight to replacement of different characters. The most commonly used scoring matrices are BLOSUM (Henikoff and Henikoff, 1992), and the earlier PAM (Dayhoff *et al.*, 1978).

The most popular algorithm and software package used for global and local similarity calculation (both for proteins and nucleic acid sequences) is BLAST, along with a variant of it called PSI-BLAST. Other algorithms are FASTA and Smith-Waterman's algorithm.

2.1. SMITH-WATERMAN

The Smith-Waterman method (Smith and Waterman, 1981) searches for local alignment. In other words, instead of looking at the entire length of each sequence, it compares substrings of all possible lengths. The Smith-Waterman algorithm is based on *dynamic programming*. This is an algorithmic technique where a problem is solved by caching the solutions of sub-problems, and these are used in later stages of the computation.

In the case of sequence alignment, the idea is to calculate the best local alignment score at a certain location along the string based on the best scores in the previous locations. This process is typically described in a table, where the rows correspond to one sequence and the columns to another sequence (see Figure 1). An important property associated with dynamic programming is that it improves computation time without losing any accuracy, so in our case it is guaranteed that the best local alignment is obtained.

The alignment scores are based on the notion of 'edit distance', counting the number of transformation one sequence is required to obtain the second sequence. Transformations include substituting one character for another, insertion of a string of characters, or deletion of string of characters. The actual score is calculated using score matrices which associate a weight with each pair of characters. The algorithm uses two penalties, one for opening gaps and another for extending them. The formula shown in Figure 1 combines these penalties

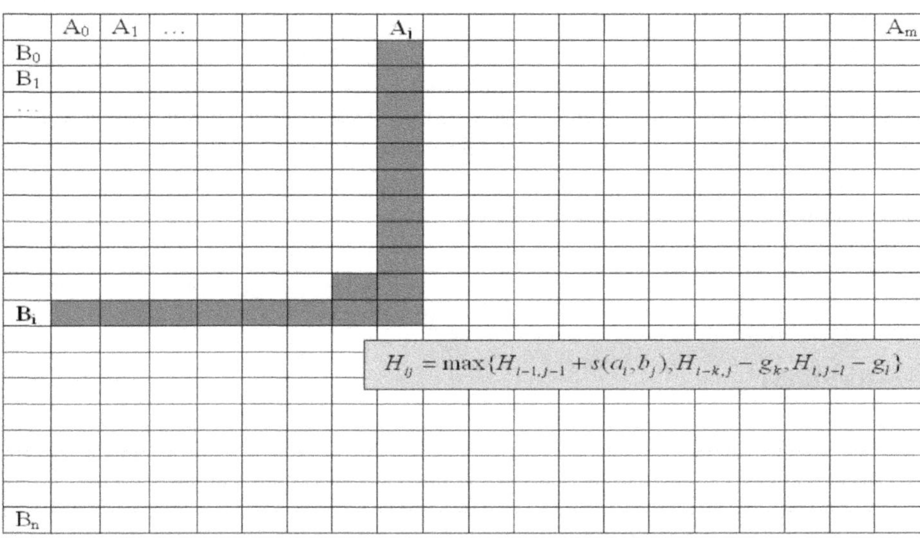

Figure 1. Smith-Waterman Algorithm Dynamic Table.

into one penalty g_k which is the penalty for opening a gap of length k. In the same figure, *s(a, b)* denotes the score matrix entry associated with the amino acids *a* and *b*.

2.2. FASTA

The Smith-Waterman algorithm provides a good measure of local alignments, but at a cost. In a naïve implementation, its running time is cubic in the lengths of the compares sequences. As sequence database grew, the need for a more efficient comparison method emerged.

FASTA (Pearson, 1990) is a heuristic method which provides an approximation of the local alignment score. It is based on the following observation: good local alignments typically stem from identities in sequences. The FASTA algorithm constructs a lookup table which stores all instances *k*-tuples of amino-acids appearing in the sequence. The typical value of *k* used is k = 2, which means the lookup table stores all instances of pairs of amino-acids. Using this lookup table, the best regions with the highest density of identities are identified. These identities, viewed on full dynamic programming table, can be considered as k-length diagonals. With these diagonals at hand, FASTA searches for the best 'diagonal runs', which are sequences of consecutive identities on a single diagonal. Finally, a dynamic programming algorithm is used for these areas only (using the Needleman-Wunch algorithm).

2.3. BLAST

BLAST (Altschul *et al.*, 1997) is another heuristic method for local alignment. Instead of looking for identical k-tuples in sequences, it looks for *similar* k-tuples. BLAST looks for k-tuples) in one sequence that score at least T when aligned with the other sequence, again using a scoring matrix. The typical value of *k* used is k = 3. Such local similarities are the extended in both directions in an attempt to find locally optimal un-gapped (i.e. continuous) alignments, with a score of at least S. These alignments are called high scoring pairs (HSPs), and are used to provide the best local alignment of the two strings. The neighborhood threshold T and the score threshold S are tunable parameters.

BLAST is typically faster than FASTA, as it uses a restricted form dynamic programming. It is considered to be as sensitive (and thus as accurate) as FASTA for most biological queries, and for this reason it is currently the most popular sequence search and comparison tool, both for amino acid and nucleic acid sequences.

BLAST outputs the similarity score, sequence alignments, and the statistical significance of the similarity, referred to as E-score. The latter can be used to estimate the probability of having similarity of this quality with a random string.

2.4. PSI-BLAST

PSI-BLAST (Altschul *et al.*, 1997) is a variant of BLAST which performs databases searches in an iterative fashion. Given a query sequence and a database of sequences, BLAST is used to find the sequences in the database that are most similar to the query sequence. The sequences found are aligned to the query sequence, and based on this alignment, a profile is generated. The profile can be represented as a position specific scoring matrix which holds the probability of having each amino-acid in each on of the positions in the sequence. Such a matrix can be used for searching for new local alignments in the protein database, seeking for local alignments, using an algorithm similar to BLAST (recall that BLAST, FASTA, and the Smith-Waterman algorithm all use a scoring matrix to define the similarity between a pair of amino-acids). Such a search outputs a possibly new set of sequences in the database, which may be used to construct a new profile. This process can be repeated for this set and so on, until convergence occurs (i.e. the BLAST result is identical to the set of sequences from which the profile was constructed) or for a fixed number of iterations.

PSI-BLAST is considered the program of choice for detecting remote homologues. While being a very effective tool at that, it has the distinct disadvantage of generating significant amounts of noise in the form of false hits, especially when conducting numerous iterations.

2.5. CLASSIFICATION WITH SEQUENCE SIMILARITY

Several inherent challenges exist in classifications proteins that are based on their sequences similarities. The current protein databases combine proteins with different evolutionary history. For example, sequences of proteins that have evolved from a common ancestor were already diverged beyond detection by any of the search programs described above. Other proteins may still exhibit extremely high degree of conservation in their sequences despite very long evolutionary distances. Thus, it is evident that varying the distance matrix selected, defining the penalty for opening and extending gaps in the alignments and choosing the preferred statistical and computational parameters are fundamental for any sequence-based classification. In other words, different proteins evolve at different rates, and for each degree of divergence different search parameters are required.

The Sequence similarity measures such as BLAST and Smith-Waterman are used as the building block for some classification systems, detailed in Section 4. Such measures can be used directly for classification, using the well-known "Nearest-Neighbor" paradigm for supervised learning. In other words, a new protein is associated with the protein nearest to it in the database of proteins with known function. The BLAST software package even provides a clustering program, called blastclust.

Similarity based classification is typical done by searching for all proteins similar to a given protein up to a specified threshold. One caveat of such classification is the choice of threshold. If the threshold is chosen to be too restrictive (e.g. too low in terms of E-value), it is possible no matches are found. Alternatively, if the threshold is chosen to be too permissive, it is likely that a lot of non-related sequences, or 'false positives' shall be retrieved. This problem frequently surfaces when classifying highly divergent sequences, and thus the similarity is difficult to detect without having a large fraction of false positives.

3. Classification Based on Domain and Motif Analysis

While sequence comparison is the most widely used tool for classifying proteins, it is not sufficient in all cases, as explained above. An alternative approach is based on the notion that perhaps domains form the building block of proteins and not single amino acids. Based on this notion, proteins with similar domains are associated with each other.

A variety of classification systems were developed which are based on domain and motif analysis. Such systems use multiple sequence alignments to detect conserved regions in sequences, and differ in underlying computational representation of a motif or domain. One of the main drawbacks of classification based on domain and motif analysis is the difficulty in providing full coverage of the protein space.

The problem of detecting domains and motifs in an automatic manner is a difficult problem. Part of the difficulty lies in defining what exactly is a domain or motif. Even with a clear-cut definition, problems such as overlapping domains (or motifs), or slight variations of the same pattern make it very difficult to correctly identify all domains and motifs.

3.1. PROSITE

PROSITE (http://www.expasy.ch/prosite, Falquet *et al.*, 2002) was the first motif-based classification of proteins. The goal set out by PROSITE is to identify and represent all biologically significant signatures (or 'fingerprints'). Signatures are described either as *regular expressions* or as PROSITE profiles (similar to the profiles described above for PSI-BLAST). Release 17.31 dated December 15th 2002 contains 1, 156 families described by 1585 different patterns, rules, and profiles.

A regular expression is a concise way to describe a family of strings. Regular expressions in PROSITE are described using the following rules:

1. Standard one-letter codes for amino-acids are used
2. The symbol 'x' is used as a wildcard, in a position where any amino acid is accepted.
3. At positions where one of several amino acids may be used, square parentheses are used. For example [LK] stands for 'L' or 'K'.
4. At positions where most amino acids can be used, curly brackets are used. For example {LK} stands for any amino-acid other than 'L' or 'K'.
5. Elements in a pattern are separated by a hyphen ('-').
6. Repetition of an element is indicated by a number in parenthesis. For example x(3) means x-x-x and x(2, 3) means x-x or x-x-x.
7. '<' and '>' indicate when a pattern is restricted to either the N- or C- terminal of a sequence respectively.

For example, the PROSITE entry PS00029 describes the *Leucine zipper* pattern as L-x(6)-L-x(6)-L-x(6)-L. This reads as 'L' followed by any 6 amino acids, followed by another L with 6 more amino acids, follows by an L with 6 more amino acids, and a final L.

PROSITE is essentially a manually maintained database. This allows each entry to be associated with all relevant literature references. In fact, many of the entries are generated via published multiple-alignments. In addition, cross references to PDB (Berman *et al.*, 2000) are provided when applicable. PDB is a database of proteins for which three-dimensional structures are known (Figure 2).

Regular expressions are not suitable for classifying families whose members are highly diverged. To this end, PROSITE was extended to include profiles, thus extending its coverage.

NiceSite View of PROSITE: PS01359

General infromation about the entry

Entry name	ZF_PHD_1
Accession number	PS01359
Entry type	PATTERN
Date	APR-2002 (CREATED); JAN-2003 (DATA UPDATE); FEB-2004 (INFO UPDATE).
PROSITE documentation	PDOC50016
Name and characterization of the entry	
Description	Zinc finger PHD-type signature.
Pattern	C-x(1,2)-C-x(5,45)-[VMFLWIE]-x-C-x(1,4)-C-x(1,4)-[WYFVQHLT]-H-x(2)-C- x(5,45)-[WFLYI]-x-C-x(2)-C.

Numerical Results

- Swiss-Prot release number: **42.11**, total number of sequence entries in that release: **145587**.
- Total number of hits in Swiss-Prot: **148 hits in 106 different sequences**
- Number of hits on proteins that are known to belong to the set under consideration: **146 hits in 104 different sequences**
- Number of hits on proteins that could potentially belong to the set under consideration: **0 hits in 0 different sequences**
- Number of false hits (on unrelated proteins): **2 hits in 2 different sequences**
- Number of known missed hits: **21**
- Number of partial sequences which belong to the set under consideration, but which are not hit by the pattern or profile because they are partial (fragment) sequences: **0**
- Precision (true hits / (true hits + false positives)): **98.65 %**

Figure 2. The information shown in a typical PROSITE entry, this includes the pattern, known sequences and literature references. The numerical result that describe the precision and recall for this signature (PS01359—Zinc finger PHD-type) are presented.

While PROSITE provides a dictionary of motifs and domains it has several drawbacks as a classification system. One problem is that of missing patterns. As patterns are detected mostly manually, not all patterns existing in real-world proteins have been detected. Another problem is that of low-information patterns. Since pattern lengths can vary from a few amino-acids to hundreds of amino-acids, some patterns (the shorter ones) have little value in classifying a new protein which is not already classified with PROSITE.

3.2. BLOCKS

BLOCKS (http://www.blocks.fhcrc.org, Henikiff *et al.*, 2000) is a database of highly conserved protein regions. A 'block' is defined as a contiguous segment corresponding to the most conserved regions of proteins. Such blocks are automatically detected. In contrast to PROSITE, blocks are not associated with function or with known literature, as the process is completely automated. Blocks are derived by performing multiple alignments of protein families as defined in InterPro, and searching for segments with a high number of identities.

Version 13.0 of the Blocks database (dated August 2001) consists of 8656 blocks, representing 2101 groups documented in InterPro 3.1. These blocks are keyed to SwissProt 39.17.

It is important to note that the PROSITE pattern is not used to build to Blocks database, and a Block entry may or may not contain the PROSITE pattern corresponding to the InterPro group from which the entry was derived.

An important application of BLOCKS was the construction of amino acid substitution matrices used for the sequence similarity algorithms mentioned above (Smith-Waterman, FASTA, and BLAST). The BLOSUM (Henikoff and Henikoff, 1992) matrices were derived from the BLOCKS database, and are currently the substitution matrices most widely used for sequence comparison.

3.3. PRINTS

PRINTS (http://www.bioinf.man.ac.uk/dbbrowser/PRINTS, Attwood *et al.*, 2002) is a database of domain-family fingerprints. A fingerprint is defined as a group of conserved motifs used to characterize a protein family. Fingerprints are more powerful constructs than single motifs, in the sense that they can narrow down families more effectively.

The fingerprinting method used in PRINTS is based on an iterative process of multiple sequence alignments. The process starts with only a small number of proteins, for which a set of conserved regions forming a fingerprint is identified. The full database is then scanned to find matching proteins. These two steps are repeated until convergence (i.e. until the set of matching proteins in the second step is identical to the input set of proteins in the first step).

PRINTS shares the main drawback of other domain-based systems, which is the lack of full coverage of the protein space. Another problem is that fingerprints are at times too restrictive (Silverstein *et al.*, 2001). Thus it is possible that a new protein will only match a portion (several motifs) in a fingerprint. In such cases, the protein might belong to a sub-family, but there is no clear-cut classification for it.

Figure 3 shows an example of the fingerprint given by PRINTS for *Prokaryotic zinc-dependent phospholipase* C.

PRINTS release 35.0 dated July 2002 contains 1, 750 database entries, relating to 10, 626 motifs.

3.4. PFAM

Pfam (http://www.sanger.ac.uk/Software/Pfam, Bateman *et al.*, 2000) is a database of protein alignments and HMMs. HMM stands for ‘Hidden Markov Model’. Version 7.8 dated November 2002 contains 5049 families.

A hidden Markov model is an abstraction used to statistically describe the consensus sequence for a protein. Such a model consists of a sequence of node, with a designated begin state and end state. Each node in an HMM has three ‘invisible’ states associated with it (hence the name ‘hidden’). These are a match state (M), insert state (I), and delete state (D). Each has a transition probability associated with it. This probability is node-specific, and hence is position specific in terms of the sequence. The match state has a probability of matching a particular amino-acid. This probability is referred to as ‘emitting’ probability. Similarly, in the insert state there is a probability associated with each amino-acid. The probability of no amino-acid associated with a node (i.e. a ‘gap’ occurring) is captured by the probability of transitioning into the delete state.

RPHPHLPASEC *Prokaryotic zinc-dependent phospholipase C signature*

RPHPHLPASEC is a 5-element fingerprint that provides a signature for prokaryotic zinc-dependent phospholipase C. The fingerprint was derived from an initial alignment of 5 sequences: the motifs were drawn from conserved regions largely spanning the central portion of the alignment - motif 3 includes the region encoded by PROSITE pattern PROKAR_ZN_DEPEND_PLPC (PS00384), the 2 conserved histidines and the conserved aspartate of which provide 3 of the zinc ligands; and motifs 1 and 4 contain other zinc-binding residues. Two iterations on OWL27.0 were required to reach convergence, at which point a true set comprising 11 sequences was identified. A single partial match (LIU190352) was also found, a lecithinase fragment.

An update on SPTR37_9f identified a true set of 18 sequences.

SUMMARY INFORMATION

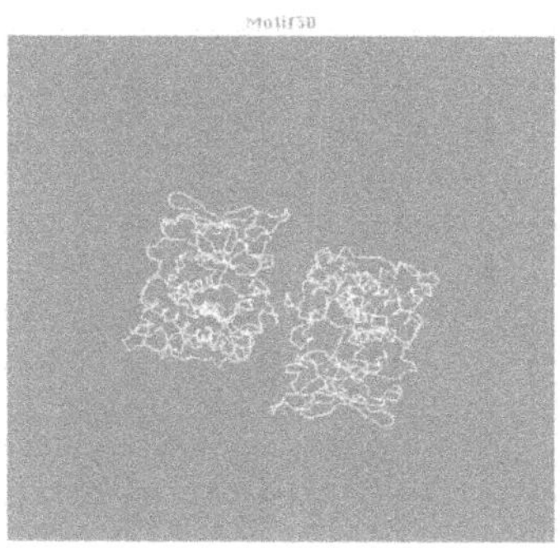

```
18 codes involving  5 elements
 0 codes involving  4 elements
 0 codes involving  3 elements
 0 codes involving  2 elements
```

COMPOSITE FINGERPRINT INDEX

```
5| 18  18  18  18  18
4|  0   0   0   0   0
3|  0   0   0   0   0
2|  0   0   0   0   0
--+-------------------------
 |  1   2   3   4   5
```

TRUE POSITIVES

PHLC_CLOPE	Q60121	Q59305	Q59304
Q59303	Q57317	Q46281	Q46280
Q46279	Q46246	Q46282	P94658
Q46150	PHLC_CLOBI	PHLD_BACCE	O52864
PHLC_BACCE	PHLC_LISMO		

Figure 3. PRINTS fingerprints for Prokaryotic zinc-dependent phospholipase C signature. A view of the structure representative is available from the PDB and via Motif3D tool.

HMMs can be automatically generated from multiple alignments. The software package used in Pfam, called HMMER (Sonnhammer, *et al.,* 1998), is one of the most popular software packages for generating HMMs.

Given a new sequence, it is possible to calculate the probability of that sequence belonging to the family modeled by certain HMM. A similarity score is associated with a new sequence based on the most probable path through the HMM which generates the input sequence.

Pfam comes in two flavors: Pfam-A is a manually curated version of Pfam, and Pfam-B is an automating clustering of the remaining proteins in SwissProt and TrEMBL. Pfam-A covers roughly 65% of the SwissProt database.

The Pfam-A database is generated in a semi-automated process, starting from a seed multiple alignments taken either from the literature or from other databases such as Prosite or ProDom. After manual inspection an HMM profile is built, and used to search the database. Members are added to the seed alignment and the process is repeated. Pfam-A HMMs do not overlap.

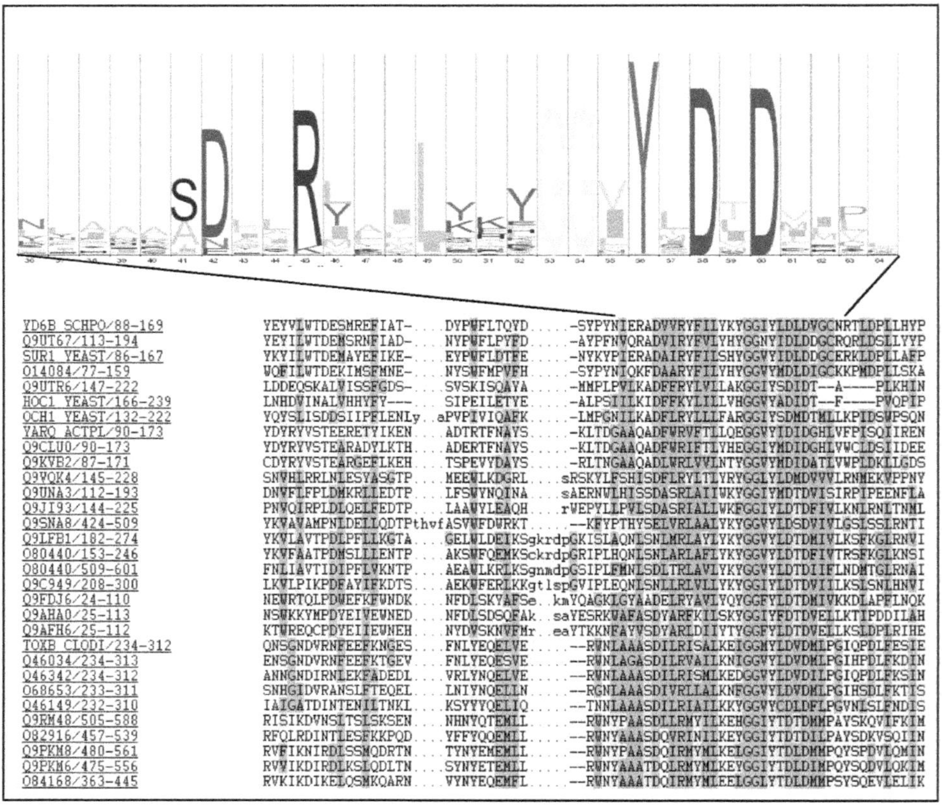

Figure 4. Pfam Multiple Sequence Alignment and a partial of the consensus presented by the family HMM Logo.

Pfam offers a comprehensive Web-based interface which provides links to InterPro (see below) and other systems. Each entry includes the HMM itself, and structural information if available. Pfam also offers a graphical representation of families.

Figure 4 shows a multiple alignment for the same family. Other visualization methods include a phylogenetic tree and domain organization by species.

The advantage of HMMs is their being a powerful modeling tool, thus facilitating more accurate modeling of families, with less chance similarities occurring. The disadvantage of HMMs (compared to regular expressions for example) is that they are typically quite large and thus difficult to understand. Another caveat is that HMMs require a large quantity of inputs in order to identify distant homologues, and that they are unable to efficiently identify long-distance correlations within a sequence (Eddy 1998).

3.5. PRODOM

ProDom (http://protein.toulouse.inra.fr/prodom, Corpet *et al.*, 2000) is a database of automatically generated protein domains.

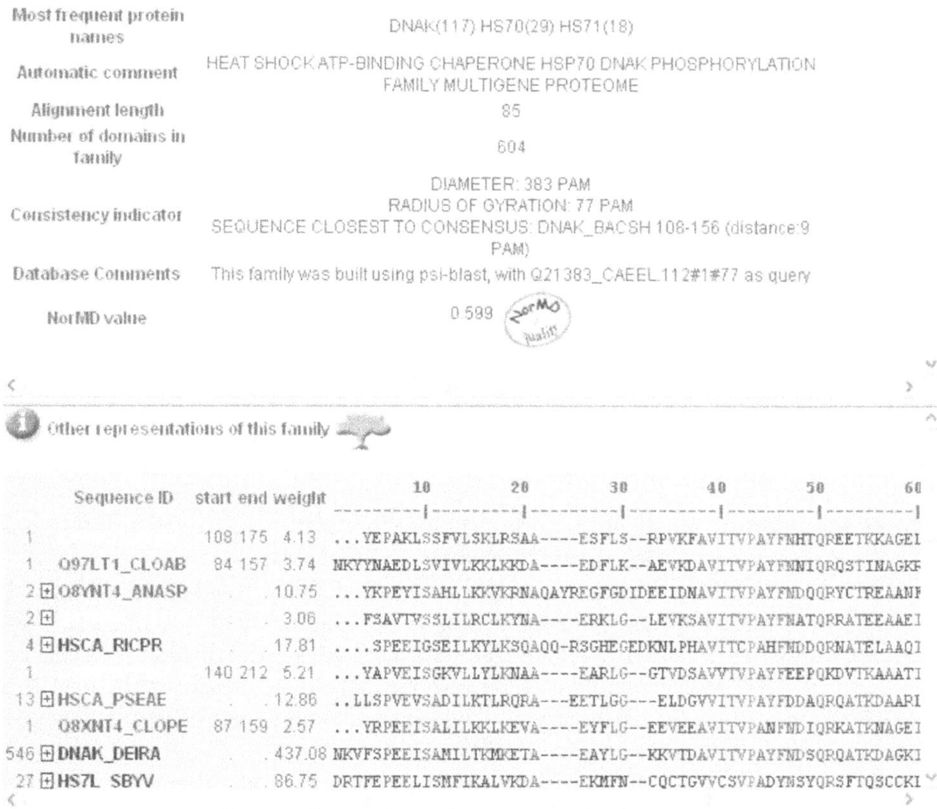

Figure 5. ProDom Entry PD000089, Heat Shock ATP-Binding chaperone.

ProDom identifies domains using BLAST search. Originally it used plain BLAST to identify domains, and now it used PSI-BLAST. ProDom groups all SwissProt and TREMBL sequences into domains, and generated 365,172 families in version dated May 2002.

ProDom clustering is generated under the assumption that the shortest full-length sequence is a single-domain protein (while in theory this might be questionable, this heuristic appears to work in practice). PSI-BLAST is used to find homologous domains, which are then clustered together with that sequence. The process is repeated for the remaining sequences.

The ProDom Web-site provides various search facilities, and the output provided for a ProDom entry is by default a multiple alignment. For example, Figure 5 shows the multiple alignments for the ProDom entry PD000089, *Heat Shock ATP-Binding chaperone*.

3.6. DOMO

DOMO (http://www.infobiogen.fr/services/domo, Gracy and Argos, 1998a) is a protein domain database generated in a fully automated fashion. The current version of DOMO, dated December 2002, includes 8877 multiple sequence alignments and 99058 protein domains. The database was generated from 83054 non-redundant protein sequences taken from SwissProt and PIR.

DOMO clustering is based on iterative sequence similarity search followed by multiple sequence alignments. Global similarities are detected from the pairwise comparison of amino acid and dipeptide compositions of each protein.

One representative sequence is chosen from each of the generated clusters. For these representatives, a suffix tree is constructed. This suffix tree is then used to detect local sequence similarities. Finally, wherever local similarities are found, sequences are multiply aligned (using a specially designed algorithm described in Gracy and Argos, 1998b), and the alignments are analyzed to detect domain boundaries.

A comparative study (Gracy and Argos, 1998b) indicated that the performance of DOMO (in terms of correctness of classification) is significantly better of that exhibited by ProDom, and falls only slightly short of PROSITE (which is manually maintained).

3.7. SMART

SMART (http://smart.embl-heidelberg.de, Schultz *et al.,* 2000) is a database of domains. SMART version 3.4 dated December 2002 contains 654 HMMs.

SMART builds HMMs based on multiple sequence alignments of hand-picked family members. HMMs are built using a combination of several homology searching software packages. Furthermore candidate homologue sequences are subject to additional analysis (e.g. using BLAST) to estimate the statistical significance of sequence similarities). The SMART database is constructed by carefully choosing and calibration a variety of thresholds used in the process of constructing the HMMs.

SMART provides a facility for searching for domains in a query proteins. Matching proteins are displayed graphically.

3.8. S-BASE

S-BASE (http://www.icgeb.org/sbase, Vlahovicek *et al.,* 2002) is a protein domain library. S-BASE provides clustering of functional and structural domains.

S-BASE uses a fairly simple strategy to construct domains. A manually annotated database of subsequences is derived from the literature and from other databases. This database is an attempt to create a comprehensive dictionary of domains and motifs. Classification is based BLAST searches against this database.

S-BASE provides the facility for BLAST of PSI-BLAST searches against the domains database.

3.9. TIGRFAMS

TIGRFAMs (http://www.tigr.org/TIGRFAMs, Haft *et al.,* 2001) is a database of protein families based on HMMs. TIGRFAMs is built using manually collected multiple sequence alignment, along with associated functional information. TIGRFAMs places an emphasis on protein function, and is biased towards microbial genomes.

The TIGRFAMs Web-based interface shows both the name and literature reference for each HMM generated. The TIGRFAMs database also includes Pfam HMMs, and provides reference to both Pfam and InterPro families.

4. Classification Based On Full Protein Sequence

Domain-based classifications are somewhat limited in the sense that many proteins have several domain appearances, and that some proteins do not have any domain (or at least not a recorded domain). In addition, for small families (e.g. with 2 members) it is not possible to define domains.

A different approach to classification is offered by several systems which classify proteins based on the full sequence. This is typically achieved by sequence comparison.

The basic tenet of most full-sequence based classifications is that of homology transitivity. The idea is that homology (the relation between two proteins which have evolved from the same protein) is a transitive relation (see Yona *et al.,* 2000a for a detailed discussion of this concept). The main caveat of all such classifications is chance similarities, which result in misclassification, and multi-domain proteins which may be related to several families.

4.1. PROTOMAP

ProtoMap (http://protomap.cs.cornell.edu, Yona *et al.,* 2000a) provides a fully automated hierarchical clustering of the protein space. ProtoMap takes a graph-based approach, where the sequence space is represented by a directed graph where vertices are protein sequences and edges represent similarity between proteins. The weight associated with an edge measures the similarity, or the significance of the relationship. ProtoMap uses a combination of Smith-Waterman, FASTA, and BLAST to determine similarity.

The ProtoMap algorithm constructs a partitioning of the protein database at different levels of granularity. At first only the most significant relationships are considered. The subgraph induced by all edges with high similarity (e.g. E-value of 1e-100 or less) is used, and each connected components is considered a single cluster. These clusters are then iteratively merged in agglomerative hierarchical clustering, using an average-link where the average is geometric mean of sequence similarity scores.

Hierarchical clustering, as the name suggests, is a clustering techniques which generates a hierarchy of clusters, where at each level clusters are generated by merging clusters of a lower level, and at the bottom level each cluster is a single entity (e.g. a single protein in our case). There are two fundamental paradigms for hierarchical clustering: agglomerative (bottom-up) and divisive (top-down). The former starts from the bottom, from single entities, and repeatedly cluster pairs of clusters into larger clusters. At each step the pair of clusters with the highest similarity is merged. In the divisive approach, the whole set of data items is considered, and is recursively split into smaller clusters.

In practice, when using hierarchical clustering for proteins, usually agglomerative clustering is used. ProtoMap uses average-link clustering, in an iterative manner. Once the initial hierarchical clustering procedure is for 1e-100 completed, the threshold used is increased. The same process is repeated for 1e-95, 1e-90, etc, up to 1e0. Each time the threshold is increased, strongly connected clusters are merged and hierarchical clustering applied again.

ProtoMap offers a Web-based interface which allows browsing the clusters both textually and graphically. ProtoMap also maintains SwissProt annotation and keywords, as well as links into BioSpace (see 6.3 below). The cluster page details each member protein, along with its SwissProt keywords. ProtoMap provides individual protein pages linking each protein to all clusters containing it.

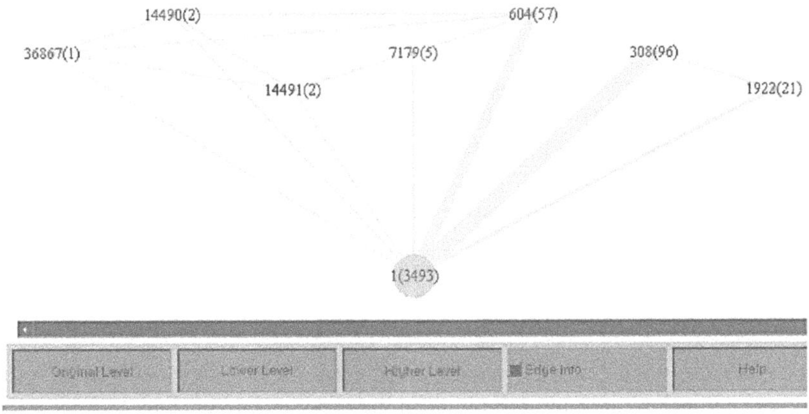

Figure 6. ProtoMap Cluster Relationships Diagram. Each circle represents a cluster at the selected level (1e-5 here), and is denoted by its cluster number, followed by cluster size in parenthesis. The edges indicate new links between clusters which were revealed upon lowering the threshold. The width of an edge between two clusters is proportionate to the number of connections between them.

Each cluster also has a summary page with some additional information pertaining to the sequences contained in it, such as PROSITE families and taxonomy. ProtoMap provides a graphical display of cluster relationships (see Figure 6), as well as a tree-like representation.

Release 3.0 of ProtoMap covers 365,174 proteins taken from SwissProt/TREMBL.

ProtoMap was used as a platform for target selection for mapping 3D structures of proteins (Portugaly *et al.*, 2002).

4.2. PROTONET

ProtoNet (http://www.protonet.cs.huji.ac.il, Sasson *et al.*, 2002) provides a fully automated hierarchical clustering of the SwissProt proteins. ProtoNet implements an average-link hierarchical agglomerative clustering algorithm. The novelty of ProtoNet is the use of several averaging methods. The averaging methods provided are arithmetic mean, geometric mean, and harmonic mean. Each one of these averaging methods generates a different clustering hierarchy (or dendogram).

ProtoNet offers a variety of methods to traverse through clusters and analyze their contents. The hierarchical clustering is fully traversable, thus providing users with varying granularity, based on individual requirements.

The ProtoNet Web-based interface provides detailed information in the cluster level, as shown in Figure 7. The top portion of the cluster details window shows a graphical representation of the sequence of merges taking place for the creation of this cluster and subsequent clusters.

Each protein recorded in the ProtoNet database is associated with detailed taxonomical representation as well as a variety of keywords (including InterPro and PROSITE entries). Protonet also provides motif and domain information taken from PFAM, Prints, ProDom, Prosite, and SMART, when such information is available.

ProtoNet provides a unique feature by which users may classify their own proteins. In contrast to other systems, where users may classify only a single sequence at a time,

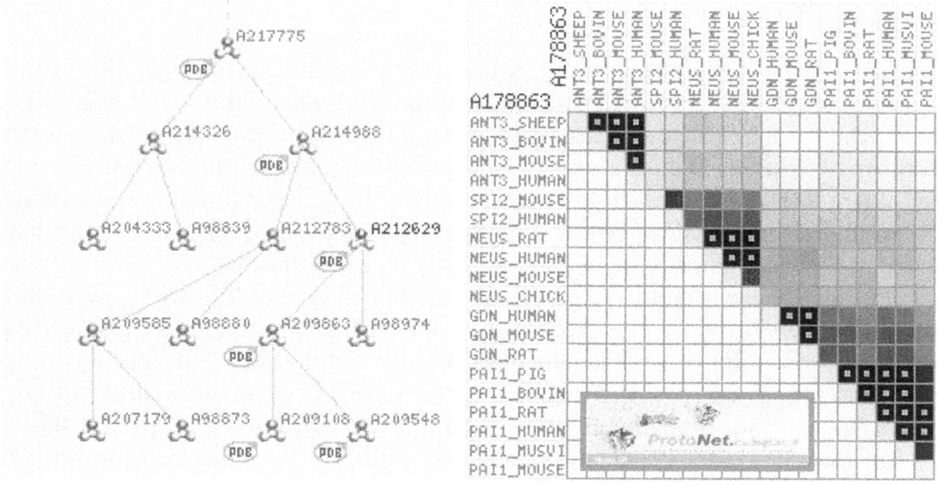

Figure 7. ProtoNet Cluster Card. The cluster card shows the member proteins as well as the chain of mergers. Proteins that were structurally solved are marked by the PDB symbol (left). Activating the visualization tools allows navigating in the cluster tree. A matrix reflecting the Blast-based scores for all proteins in a cluster is shown (right). The significance of the scores is depicted by a color gradient.

ProtoNet stores sequences classified by users, and specifically, multiple sequences may be classified concurrently.

ProtoNet provide statistical measures for the purity and the sensitivity for each merging in the process as compared with other type of classifications.

ProtoNet version 2.1 dated December 2002 covers 114,000 proteins from SwissProt release 40.28. The hierarchical nature of ProtoNet allows analyzing the protein space at different level of granularity. At a level of 10,000 clusters (of which ~2500 are singletons), very good correspondence can be observed between ProtoNet clusters and the InterPro entries. Version 3.0 supports over one million sequences from UniProt (Swissprot and TrEMBL) and is assiated with additional external databases such as SCOP, InterPro, GO and more.

4.3. PIR-ALN

PIR-ALN (http://www-nrbf.georgetown.edu/pirwww/dbinfo/piraln.html, Srinivasarao *et al., 1999)* provides a simple classification of protein sequences based on sequence alignment.

Classification is made into three categories. Families include sequences which are at least 45% identical. Superfamilies are multiple alignments containing sequences from different families. Homology domain alignments contain homologous subsequences (segments) from different proteins. Homology searching and multiple alignment are done using FASTA and with tailor-made software. The PIR-ALN version release in December 2000 contains 3508 alignments includes 994 superfamilies and 386 homology domain alignments.

4.4. SYSTERS

Systers (http://systers.molgen.mpg.de, Krause *et al.,* 2002) is classification of the protein space based on single linkage agglomerative clustering. Single-linkage clustering (in contrast to the average-linkage clustering used in ProtoMap and ProtoNet) defined the similarity between two clusters as the highest similarity between pairs from the two clusters.

Systers clustering is based on Smith-Waterman all-against-all sequence comparison. To avoid problems resulting from asymmetric alignment scores, a symmetric E-value is computed for each pair with significant similarity (in either direction).

The agglomerative clustering generated a single linkage tree. This tree is the result of successive merging of any two clusters which are associated with at least one significant similarity. From this tree, superfamilies are derived, and this is done automatically without user intervention (and specifically without the need to define an arbitrary cut-off value), in an attempt to choose the optimal level of detail. Once superfamilies are identified, they are broken down into 'family' clusters. Families are defined as connected components in the graph generated by omitting all edges below a certain threshold from the similarity graph of proteins within the superfamily (in other words, two proteins are in the same family if they have a similarity above the threshold, or have a 'chain' of similarities above the threshold which connects them).

Systers clusters are annotated with Pfam domains for the contained protein sequences, and links to PROSITE and PDB where applicable.

Release 3 of Systers includes 290,811 non-redundant sequences (as well as annotations for 583,448 redundant sequences) taken from SwissProt, TREMBL, PIR, FlyBase, Wormpep 20, and MIPS Yeast protein translations. These proteins are sorted into 82,450 disjoint clusters. Of these 55,182 are singleton (i.e. single sequence) clusters, and the remaining 235,629 sequences are contained in 27,268 clusters ranging in size from 2 to 8,821 non-redundant sequences per cluster.

4.5. PROCLUST

ProClust (http://promoter.mi.uni-koeln.de/~proclust, Pipenbacher *et al.,* 2002) is a protein clustering derived using a graph-based approach. A graph of proteins is constructed where edge-weights are based on Smith-Waterman similarity scores scaled with respect to self-similarity. After removing all edges with insignificant similarities (based on some arbitrary threshold), clustering proceeds by seeking maximal strongly connected components in the graph. In other words, this means that every two proteins in the cluster are connected in the graph in both ways (the graph here is directed, and the requirement is for connectivity in both directions).

The ProClust software is available for download, and the test cases used to validate it include SwissProt proteins in release 39 (June 2000).

4.6. CLUSTR

CluSTr (http://www.ebi.ac.uk/clustr, Kriventseva *et al., 2001)* provides a classification of SwissProt/TREMBL proteins. The clustering is based on single-linkage hierarchical clustering (similar to Systers), where the underlying scores are based on the Smith-Waterman algorithm. The scores are processed to generate a 'Z-score' representing the statistical significance of the similarity. The 'Z-score' for two sequences is calculated by averaging the Smith-Waterman similarity score for the first sequence with a set of shuffled instances of the second sequence. Unlike E-values generated by BLAST, this Z-score is independent of the sequence database and thus facilitates easier updating of the database.

4.7. PICASSO

Picasso (http://www.ebi.ac.uk/picasso, Heger and Holm, 2001) is a global classification of protein sequences.

Picasso uses a clustering algorithm similar to hierarchical clustering. Clusters are merged based on a threshold E-value obtained from BLAST, and the threshold is gradually increased. The clustering process involves sequences or parts of sequences detected by BLAST multiple alignment. The outcome of the process is 10,000 unified domain families (not counting singletons).

4.8. TRIBE-MCL

TribeMCL (http://www.ebi.ac.uk/research/cgg/tribe, Enright *et al.*, 2002) is a protein clustering software package. TribeMCL takes as input the results of a BLAST calculation, and output a clustering of proteins into families.

TribeMCL uses a novel clustering method called Markov Clustering. The authors claim that clustering algorithm addresses the difficulties in protein clustering such as multi-domain proteins, protein fragments, and promiscuous domains. These are addressed via the mechanics of their algorithm.

The TribeMCL software is available for download, but there is no Web-based classification of common databases into families.

5. Phylogenetic Classification

COGS (http://www.ncbi.nlm.nih.gov/COG, Tatusov *et al.*, 2001) provides a clustering of proteins derived from 43 complete genomes. COGs applies single linkage hierarchical agglomerative clustering to get clustering of orthologous proteins or orthologous sets of paralogs. The former refers to genes from different special which have evolved from the same protein, and the later to genes from the same genome which are related by duplication.

Each cluster consists of at least three species. The clustering process starts by forming a minimal COG with three elements, and then proceeds by merging COGs sharing an edge. The COGs clustering algorithm has the capability of splitting COGs which were incorrectly merged.

6. Classification Based on Protein Structure

Full-sequence analysis helps in classifying proteins with known sequences. However, it is thought that protein function stems from structure (see e.g. Brenner 2000). This gives rise to the idea of classifying proteins based on structure. The drawback of structure-based clustering is its incompleteness. In other words, the number of sequences for which the structure is available (or the structure is 'solved') is relatively small, due to the complexity of obtaining high-resolution structural information.

Structure based clustering takes into account similarity in the three-dimensional structure proteins. Several different algorithms and software packages are available for measuring the similarity between two protein structures. Examples are CE (Shindyalov and Bourne, 1998), Dali (Holm and Sander, 1998), PrISM (Yang and Honig, 2000), VAST (Gibrat *et al.*, 1996) and STRUCTAL (Orengo *et al.*, 1999). A full description of these methods is beyond the scope of this survey.

6.1. SCOP

SCOP (http://scop.berkeley.edu, Lo Conte *et al.*, 2002) is a structural classification of proteins into a four-level hierarchy, based on manual analysis by experts.

- **Family**—Proteins with significant sequence similarity (typically 30% or greater identity), and with clear evolutionary relationship.
- **Superfamily**—Proteins with low sequence similarity, but with structural and functional features suggesting a common evolutionary origin.
- **Fold**—Superfamilies with major structural similarity.
- **Class**—High level classification (e.g. All-alpha, All-beta, Alpha/Beta, Alpha+Beta, Membrane proteins, etc.)

SCOP is organized as a tree, and on top of all classes there is the SCOP "root".

6.2. CATH

CATH (http://www.biochem.ucl.ac.uk/dbbrowser/cath, Orengo *et al.,* 1999) is a structural classification into a different 4-level hierarchy:

- Homologous superfamily (H level)—Sequences with high similarity. Several conditions are defined combining sequence and structure (e.g. at least 35% sequence identity and at least 60% structural similarity).
- Topology (T level)—Structure comparison is used to group together protein structures into fold families.
- Architecture (A level)—Structures are grouped based on the overall shape of the domain structures.
- Class—Structures are determined according to secondary structure composition, similar to SCOP classes.

The name of the system is based on the level names (Class, Architecture, Topology, Homology). Assignment of structures to families and topologies is automatic, based on similarity. At the architecture and class levels, manual considerations are included into the classification. Figure 8 shows a sample page (for domain 1cuk03) in CATH.

Figure 8. CATH Domain Page for E. coli Ruva protein, PDB 1cuk03.

6.3. BIOSPACE

BioSpace (http://biospace.cornell.edu, Yona *et al.*, 2000b) is a protein classification system combining sequence and structure information.

BioSpace uses a novel clustering algorithm which gives preference to structural similarity over sequence similarity. This is based on the common perception that structural similarity implies strong functional similarity. BioSpace uses SCOP as a starting point, to cluster sequences based on the ProtoMap algorithm. It then uses PSI-BLAST to cluster sequences outside of SCOP (note that SCOP sequences represent a small fraction of protein sequences considered in BioSpace). Finally, new clusters are created using structural comparison (using Gerstein and Levitt, 1998), and profile comparisons.

The BioSpace Web interface allows browsing the various clusters, providing multiple sequences alignment and 3D models where available.

BioSpace revision 1 (dated January, 2000) covers most protein sequences and structural models publicly at the time of release, including SwissProt, TrEMBL, PIR, SCOP, PDB, and 22 complete genomes.

6.4. FSSP

FSSP (http://www.ebi.ac.uk/dali/fssp, Holm and Sander, 1996) provides fold classification using structure-structure alignment of proteins. The classification used hierarchical clustering based on an exhaustive all-against-all structure comparison using Dali (Holm and Sander, 1998) structure comparison.

FSSP provides a Web-based interface which facilitates 3D superimposition and multiple alignments of structures.

The recent version of FSSP (dated June 2002) contains 3242 sequence families representing 30,624 protein structures (taken from PDB).

6.5. SUPERFAMILY

SUPERFAMILY (http://supfam.mrc-lmb.cam.ac.uk/SUPERFAMILY, Gough and Chothia, 2002) is an HMM library which models structure. It focuses on the so called 'superfamily' level. At this level, all proteins with any structural evidence for a common evolutionary ancestor are grouped together.

SUPERFAMILY uses a library of 1073 SCOP families (taken from SCOP 1.59), each represented with a group of HMMs. HMMs are generated using the SAM (Karchin and Highey, 1998) software.

SUPERFAMILY provides sequence searching, allowing users to search their own sequences against the HMM library. In addition it provides a view of alignments of a superfamily, and assignments of structural domains to genome sequences.

7. Aggregated Systems

Each classification system has its own strengths as well as its own weaknesses. In order to leverage on the strength of several systems, it is possible to combine several systems to generate one 'aggregated' classification. Such a classification is appealing in the sense that it might be more 'valid'. Several attempts were made to combine multiple classifications.

7.1. INTERPRO

InterPro (http://www.ebi.ac.uk/interpro, Apweiler *et al.*, 2001) provides a unified domain database, based on Pfam, PRINTS, PROSITE, ProDom, SMART, and TIGRFAMs.

InterPro version 5.3, dated November 2002, includes 6725 entries, representing 1,453 domains and 5121 families. The InterPro Web-site provides for each InterPro entry (corresponding to a domain or family) the list of member proteins, literature references, references to the underlying databases, and an abstract of the biological context for the entry. In addition, graphical representation of domains is available.

InterPro is considered the 'state of the art' in terms of protein classification. The classification provided by IntePro is based on a carefully selected set of rules and considerations, which balance the different underlying classifications.

InterPro versions are released in conjunction with SwissProt and TrEMBL releases on a routine basis, and thus it is continuously updated. This stands in contrast to other domain-based systems which are only infrequently updated.

7.2. METAFAM

MetaFam (http://metafam.ahc.umn.edu, Silverstein *et al.*, 2001) is a protein clustering system obtained by seeking maximal agreement between several clustering systems, including BLOCKS, DOMO, Pfam, PIR-ALN, PRINTS, ProDom, PROSITE, Protomap, Systers, and SBASE.

Metafam automatically creates supersets of overlapping families. It covers a non-redundant protein set from SwissProt and PIR. The main difference between InterPro and METFAM is that the latter is not restricted to domain-based classifications.

7.3. IPROCLASS

iProClass (http://pir.georgetown.edu/iproclass, Wu *et al.*, 2001) is an integrated database linking PIR-ALN, Prosite, Pfam, and BLOCKS. It contains a set of non-redundant SwissProt and PIR proteins, classified into 28,000 families. iProClass is an extension of an older system called ProClass which provided classification based on PROSITE patterns and PIR superfamlies. The iProClass classification is based on Pfam domains, PROSITE motifs, PIR-ALN classifications, and PIR superfamilies.

iProClass provides a Web-based interface to a powerful database implemented using Oracle8i.

Release 2.27, dated July 2003 contains 1,094,370 entries. The database consists of non-redundant PIR and SwissProt/TREMBL sequences organized into more than 36,200 PIR superfamilies, 5720 domains, and 1300 motifs.

The advantages of iProClass over other system include powerful database search capability, as well as including a large number of protein sequences.

7.4. CDD

Conserved Domain Database (Marchler-Bauer*et al.*, 2002) is a database of domains consolidating SMART, Pfam and some NCBI contributions (e.g. from COGs). CDD is available at http://www.ncbi.nlm.nih.gov/Structure/cdd/cdd.shtml.

CDD uses a set of specially constructed score matrices prepared for each conserved domain (using a multiple alignment). The score matrices are derived from the domain definitions given by the source databases (e.g. SMART). CDD search and relies on a variant BLAST called reverse-position-specific BLAST (RPS-BLAST).

8. Summary

The field of protein classification provides a varied landscape. Diversely different approaches are used to generate results which are quite impressive considering the complexity of the problem. We have excluded from this survey numerous protein classification systems, mostly those which were applied only to a single family or superfamily of proteins. Detailed accounts of such systems are available in each on Nucleic Acid Research database issues (published every January).

Despite the work invested by numerous different research groups, the 'holy grail' of protein classification is yet to be found. In practice, the best approach to protein classification is combining the use of several systems, preferably trying to leverage the advantages of each system. An important point to keep track of is the frequency in which classifications are updated.

The primary objective of protein classification efforts is finding a mechanized shortcut in the laborious task of functional annotation and functional prediction. However, a very important byproduct of those efforts is global analyses of the protein space, both sequence and structure. The classification of protein families has become an essential building block in genomic comparative studies, in tracing evolutionary processes and in systematic methods for prediction.

The current challenges in the field of protein classification are providing firm indications of the validity of classifications (possibly via cross-validation with manually maintained classifications), as well as consolidating the different considerations (e.g. structure and sequence). On a more technical note, the issue of correctly dealing with multi-domain proteins will become more crucial as additional complete genomes of higher organisms become available (as multiple domains are more frequent is such organisms). Another technical issue is tacking the very different rate of evolution in different protein families. Current method usually use different score matrices for different purposes (e.g. BLOSUM62 for homology searches, and BLOSUM45 for 'remote homologue' searches).

We witness two distinct directions in which this field is currently progressing. On one hand, there are systems and algorithms specializing in unique genomes, systems, and protein families. Such systems are typically limited in their scope of application, but are highly reliable in the context for which they were designed. On the other hand, there is a trend of combining non-sequence based information sources into the sequence and structure data. Examples of such 'external' information sources are transcript profiling, protein-protein interaction, and phylogenetic profile. The most ambitious effort undertaken to map all protein-related information is the Gene Ontology (GO) database (Ashburner *et al.*, 2000). The objective of the GO consortium is to provide an all-encompassing dynamic controlled vocabulary of all known biological processes, molecular functions, and cellular components. GO holds tremendous potential both for cross-validating classification systems and for improving classification systems.

Similar to other fields in bioinformatics, the field of protein interaction is still in its infancy. Only time will tell how the limitations of classification based on sequence and structure can be overcome in the quest for automatically identifying protein function and understanding protein machines in the cell.

9. References

Altschul, S. F., Madden, T. L., Schäffer, A. A., Zhang, J., Zhang, Z., Miller, W. and Lipman, D. J. (1997). Gapped BLAST and PSI-BLAST: a new generation of protein database search programs. Nucleic Acids Res. *25*, 3389–3402.

Apweiler, R., Attwood, T. K., Bairoch, A., Bateman, A., Birney, E., Biswas, M., Bucher, P., Cerutti, L., Corpet, F., Croning, M. D., *et al.* (2001). The InterPro database, an integrated documentation resource for protein families, domains and functional sites. Nucleic Acids Res. *29*, 37–40.

Ashburner, M., Ball, C. A., Blake, J. A., Botstein, D., Butler, H., Cherry, J. M., Davis, A. P., Dolinski, K., Dwight, S. S., Eppig, J. T., *et al.* (2000). Gene ontology: tool for the unification of biology. The Gene Ontology Consortium. Nat. Genet. *25*, 25–29.

Attwood, T. K., Blythe, M. J., Flower, D. R., Gaulton, A., Mabey, J. E., Maudling, N., McGregor, L., Mitchell, A. L., Moulton, G., Paine, K., and Scordis, P. (2002). PRINTS and PRINTS-S shed light on protein ancestry. Nucleic Acids Res *30*, 239–241.

Bateman, A., Birney, E., Durbin, R., Eddy, S. R., Howe, K. L., and Sonnhammer, E. L. (2000). The Pfam protein families database. Nucleic Acids Res *28*, 263–266.

Berman, H. M., Bhat, T. N., Bourne, P. E., Feng, Z., Gilliland, G., Weissig, H., and Westbrook, J. (2000). The Protein Data Bank and the challenge of structural genomics. Nat. Struct. Biol. *7 Suppl*, 957–959.

Brenner, S. E. (2000). Target selection for structural genomics. Natu. Struct. Boil., *7 Suppl*, 967–969.

Corpet, F., Servant, F., Gouzy, J., and Kahn, D. (2000). ProDom and ProDom-CG: tools for protein domain analysis and whole genome comparisons. Nucleic Acids Res *28*, 267–269.

Dayhoff, M.O., Schwartz, R. M., and Orcutt, B.C. (1978). A model of evolutionary change in proteins. Atlas of Protein Sequence and Structure *5*, 345–352.

Eddy, S.R. (1998) Profile hidden Markov Models. Bioinformatics *14*, 755–763.

Enright, A. J., Van Dongen, S., and Ouzounis, C. A. (2002). An efficient algorithm for large-scale detection of protein families. Nucleic Acids Res *30*, 1575–1584.

Falquet, L., Pagni, M., Bucher, P., Hulo, N., Sigrist, C. J., Hofmann, K., and Bairoch A. (2002). The PROSITE database, its status in 2002. Nucleic Acids Res. *30*, 235–238.

Gerstein, M., and Levitt, M. (1998). Comprehensive assessment of automatic structural alignment against a manual standard, the SCOP classification of proteins. Protein Science *7*, 445–456.

Gibrat, J-F., Madej, T., Bryant, S.H. (1996) Surprising similarities in structure comparison. Current Opinion in Structural Biology *6*: 377–385.

Gough, J., and Chothia, C. (2002). SUPERFAMILY: HMMs representing all proteins of known structure. SCOP sequence searches, alignments and genome assignments. Nucleic Acids Res. *30*, 268–272.

Gracy, J., and Argos, P. (1998a). DOMO: a new database of aligned protein domains. Trends Biochem Sci *23*, 495–497.

Gracy, J., and Argos, P. (1998b). Automated protein sequence database classification. II. Delineation of domain boundaries from sequence similarities. Bioinformatics *14*, 174–187.

Gusfield, D. (1997). Algorithms on Strings, Trees and Sequences: Computer Science and Computational Biology. *Cambridge University Press.*

Haft, D. H., Loftus, B. J., Richardson, D. L., Yang, F., Eisen, J. A., Paulsen, I. T., and White, O. (2001). TIGRFAMs: a protein family resource for the functional identification of proteins. Nucleic Acids Res *29*, 41–43.

Henikoff, S. and Henikoff, J.G. (1992). Amino acid substitution matrices from protein blocks. Proc. Natl. Academy Science *89*, 915–919.

Henikoff, J. G., Greene, E. A., Pietrokovski, S., and Henikoff, S. (2000). Increased coverage of protein families with the blocks database servers. Nucleic Acids Res. *28*, 228–230.

Holm, L., and Sander, C. (1996). The FSSP database: fold classification based on structure-structure alignment of proteins. Nucleic Acids Res *24*, 206–209.

Holm, L., and Sander, C. (1998). Touring protein fold space with Dali/FSSP. Nucleic Acids Res 26, 316–319.

Huang, J. Y., and Brutlag D. L. (2001). The EMOTIF database. Nucleair Acids Res *29*, 202–204.

Heger, A., and Holm, L. (2001). Picasso: generating a covering set of protein family profiles. Bioinformatics *17*, 272–279.

Karchin, R., and Hughey, R. (1998). Weighting hidden Markov models for maximum discrimination. Bioinformatics *14*, 772–782.

Krause, A., Haas, S. A., Coward, E., and Vingron, M. (2002). SYSTERS, GeneNest, SpliceNest: exploring sequence space from genome to protein. Nucleic Acids Res *30*, 299–300.

Kriventseva, E. V., Fleischmann, W., Zdobnov, E. M., and Apweiler, R. (2001). CluSTr: a database of clusters of SWISS-PROT+TrEMBL proteins. Nucleic Acids Res *29*, 33–36.

Levenshtein, V. I. (1965). Binary codes capable of correcting deletions, insertions and reversals. Doklady Akademii Nauk SSSR *163*, 845–848.

Linial, M. and Yona, G. (2000). Methodologies for target selection in structural genomics. Progress in Biophysical and Molecular Biology *73*, 297–320

Lo Conte, L., Brenner, S. E., Hubbard, T. J., Chothia, C., and Murzin, A. G. (2002). SCOP database in 2002: refinements accommodate structural genomics. Nucleic Acids Res. *30*, 264–267.

Marchler-Bauer, A., Panchenko, A. R., Shoemaker, B. A., Thiessen, P. A., Geer, L. Y., and Bryant, S. H. (2002). CDD: a database of conserved domain alignments with links to domain three-dimensional structure. Nucleic Acids Res *30*, 281–283.

Needleman S. B. and Wunsch, C. D. (1970). A general method applicable to the search for *similarities in the amino acid sequence of two proteins*. J. Mol. Biol. **48**, 443–453.

Orengo, C. A., Pearl, F. M., Bray, J. E., Todd, A. E., Martin, A. C., Lo Conte, L., and Thornton, J. M. (1999). The CATH Database provides insights into protein structure/function relationships. Nucleic Acids Res. *27*, 275–279.

Pearson, W.R. (1990). Rapid and Sentive Sequence Comparison with PASTP and FASTA. Methods Enzymol. *183*, 63–98.

Pipenbacher, P., Schliep, A., Schneckener, S., Schonhuth, A., Schomburg, D., and Schrader, R. (2002). ProClust: improved clustering of protein sequences with an extended graph-based approach. Bioinformatics *18 Suppl 2*, S182–191.

Portugaly, E., Kifer, I., and Linial, M. (2002). Selecting targets for structural determination by navigating in a graph of protein families. Bioinformatics *18*, 899–907.

Sasson, O., Linial, N., and Linial, M. (2002). The metric space of proteins-comparative study of clustering algorithms. Bioinformatics *18 Suppl 1*, 14–21.

Schultz, J., Copley, R. R., Doerks, T., Ponting, C. P., and Bork, P. (2000). SMART: a web-based tool for the study of genetically mobile domains. Nucleic Acids Res *28*, 231–234.

Shindyalov, I.N., Bourne, P. E. (1998) Protein structure alignment by incremental combinatorial extension (CE) of the optimal path. Protein Engineering *11* (9), 739–747.

Silverstein, K. A., Shoop, E., Johnson, J. E., and Retzel, E. F. (2001). MetaFam: a unified classification of protein families. I. Overview and statistics. Bioinformatics *17*, 249–261.

Smith, T.F. and Waterman, M.S. (1981). Identification of common molecular subsequences, J. Mol. Biol. *147*, 195–197.

Sonnhammer, E. L., Eddy, S. R., Birney, E., Bateman, A., and Durbin, R. (1998). Pfam: multiple sequence alignments and HMM-profiles of protein domains. Nucleic Acids Res. *26*, 320–322.

Tatusov, R. L., Natale, D. A., Garkavtsev, I. V., Tatusova, T. A., Shankavaram, U. T., Rao, B. S., Kiryutin, B., Galperin, M. Y., Fedorova, N. D., and Koonin, E. V. (2001). The COG database: new developments in phylogenetic classification of proteins from complete genomes. Nucleic Acids Res *29*, 22–28.

Srinivasarao, G. Y., Yeh, L. S., Marzec, C. R., Orcutt, B. C., and Barker, W. C. (1999). PIR-ALN: a database of protein sequence alignments. Bioinformatics *15*, 382–390.

Vlahovicek, K., Murvai, J., Barta, E., and Pongor, S. (2002). The SBASE protein domain library, release 9.0: an online resource for protein domain identification. Nucleic Acids Res *30*, 273–275.

Wu, C. H., Xiao, C., Hou, Z., Huang, H., and Barker, W. C. (2001). iProClass: an integrated, comprehensive and annotated protein classification database. Nucleic Acids Res. *29*, 52–54.

Yang, A. S. & Honig, B. (2000). An integrated approach to the analysis and modeling of protein sequences and structures. I. Protein structure alignment and quantitative measure for protein structural distance. J Mol Biol. *301*(3), 665–678.

Yona, G., Linial, N., and Linial, M. (2000a). ProtoMap: automatic classification of protein sequences and hierarchy of protein families. Nucleic Acids Res. *28*, 49–55.

Yona, G. and Levitt, M. (2000b) A unified sequence-structure classification of protein sequences: combining sequence and structure in a map of protein space. The proceedings of RECOMB, 308–317.

VI. Education & Legal Aspects

Biodata of **Racheli Kreisberg-Zakarin** author of ***"Training at the Bioinformatics Unit of Tel-Aviv University".***

Dr. Racheli Kreisberg-Zakarin is the Head of the Bioinformatics Unit at the G.S. Wise Faculty of Life Sciences, Tel-Aviv University, Israel. She received her Ph.D. from the Department of Molecular Microbiology and Biotechnology at the Tel-Aviv University in 1993. In addition she is studying towards her M.B.A. degree from Tel-Aviv University. She is involved in the establishments of numerous bioinformatics and cheminformatics training programs.

E-mail: **racheli@post.tau.ac.il**

J. Seckbach and E. Rubin (eds.), The New Avenues in Bioinformatics, 229–235.
© 2004 *Kluwer Academic Publishers. Printed in the Netherlands.*

TRAINING AT THE BIOINFORMATICS UNIT OF TEL-AVIV UNIVERSITY

RACHELI KREISBERG-ZAKARIN
Head of the Bioinformatics Unit,
George S. Wise Faculty of Life Sciences,
Tel Aviv University, Israel

1. Introduction to Bioinformatics

Bioinformatics is the application of information technology (IT) in Biology (1). As such, it deals with the acquisition, storage, management and analysis of biological data. During the 1990s, enormous amounts of biological sequence data were acquired due to the revolution in sequencing technology and molecular biology. One of the most famous projects that employed both sequencing and molecular biology has been the Human Genome Project, whose goal was to sequence, store, analyze and annotate the entire human genome (Baltimore, D., 2001).

The entrance of computer scientists into the field of biology led to the development of methodologies, algorithms and software tools that enabled analysis and integration of the large amounts of accumulating biological data. Advanced computer hardware satisfied the computational requirements of software tools and the storage capacity requirements of databases.

Bioinformatics is a rapidly changing field driven by developments in biology, as well as improvements in computer hardware and software. This has led to an increase in the degree of complexity of Bioinformatics. In its infancy, researchers and bioinformatics software tools focused on the analysis of single biological entities, such as individual nucleotide sequences, protein sequences or three-dimensional protein structures. Today, new fields are emerging, such as proteomics (Burbaum, J. and Tobal, G.M., 2002) and functional genomics (Lennon, G.G., 2000), which focus on large sets of proteins and expressed genes, respectively, under varying conditions.

Despite the availability of vast amounts of valuable biological data, and the existence of cutting-edge software and state-of-the-art hardware, effective usage of the data and the tools requires proper training in the field of bioinformatics. Very often, the interdisciplinary nature of bioinformatics and its rapid development serve as a barrier for professional researchers, as many of them are trained in one discipline, such as biology or computer science.

The aim of this chapter is to describe and to analyze the organization of bioinformatics training programs, and to provide general guidelines for setting up other inter-disciplinary programs. The experience accumulated over the years in the Bioinformatics Unit of the George S. Wise Life Science Faculty of Tel Aviv University will be shared with our readers.

2. Need for Training in Bioinformatics

The need for extensive training in bioinformatics arises, as there is a great variety of data that exists in different databases, as well as a growing selection of bioinformatics tools. There is a continuing undersupply of bioinformaticians, due to the fact that bioinformatics requires interdisciplinary expertise that includes knowledge of biology, as well as advanced computing skills. On one hand, an expert molecular biologist is a user of bioinformatics tools, but usually lacks a proper background in computer science and lacks the know-how regarding usage of computers, software and databases. On the other hand, skilled computer scientists miss the profound knowledge of biology for the proper interpretation of the analyses carried out using bioinformatics tools, which could even be programmed by the computer scientists themselves.

3. Bioinformatics Training

Bioinformatics training has many facets, including training subjects, modes of instruction, training personnel, the academic background of the participants, training facilities, methodologies, and administration. These aspects will be described in detail using the experience gathered at the Bioinformatics Unit of the George S. Wise Life Science Faculty of Tel Aviv University (TAU) (5) as a case study. The Bioinformatics Unit was established in the mid 1990s. From the beginning of its existence, training was one of the major activities of the Unit.

3.1. BIOINFORMATICS TRAINING SUBJECTS

The selection of training subjects at the Tel Aviv University's Bioinformatics Unit can be divided into several chronological stages. In the mid 1990s, there was a need for basic bioinformatics training that focused on the sequence analysis of nucleotides and proteins using the GCG software package, internet-based tools and freeware. Due to increased interest in the three-dimensional structure of proteins, structure visualization and analysis techniques, using freeware, were taught as well.

In the late 1990s, commercial software for protein structure analysis, InsightII of Accelrys Inc., was introduced. As the expertise of the Unit's staff members lies mostly in the field of protein analysis, guest lecturers with experience in nucleotide analysis were invited in order to provide training to researchers interested in genome-related projects. The computational nature of bioinformatics led to the introduction of programming tools for bioinformaticians, such as PERL, BioPERL and Javascript.

Nowadays, new fields of interest are emerging, such as functional genomics and proteomics. These subjects require integration between training at the experimental level, and the analysis of the vast quantities of results.

3.2. MODES OF INSTRUCTION IN BIOINFORMATICS

Bioinformatics can be taught through individual consultations, workshops, and academic and commercial courses. Consultation requires of the trainer in depth knowledge of a number of techniques, out of which the consultant for a specific consulting project is using several. Workshops and courses demand that the trainer be acquainted with a majority of the techniques. This requires very intense preparation before the workshop or course takes place. In addition, the lecturer has to develop the ability to tackle new and unexpected issues on short notice in the course of the lesson.

Training activities started in the form of personal consultation. Due to the large community of bioinformatics users, it was decided to provide bioinformatics training through workshops and to hold personal consultation sessions for specific and advanced subjects only. The participants in the workshops include undergraduate students, graduate students, researchers and technicians. Graduate students' demands soon led the way to the conversion of some of the workshops into academic courses for credit. The courses require students to be present at a minimum amount of lectures and to carry out an assignment in the form of a scientific article that focuses on their research.

Industry demand led to the development of commercial training programs. These programs are taught during non-working hours, and include a compilation of bioinformatics subjects, such as sequence analysis, structure analysis, and programming for bioinformaticians.

3.3. BIOINFORMATICS TRAINING PERSONNEL

Bioinformatics training is dependent on a small group of highly qualified bioinformaticians with teaching skills. In general, there is a shortage in qualified training personnel as a result of the need for: 1) an appropriate academic background, preferably post-graduate, in Life Sciences; b) the ability to locate and to use diverse computer programs and databases; c) the ability to operate the programs on both the Windows and Unix operating systems; d) the desire and talent to teach individuals and heterogeneous groups.

The Tel Aviv University Bioinformatics Unit has developed expertise in the fields of nucleotide and protein sequence analysis, protein structure analysis and programming for bioinformaticians. The core of the training team is made of the staff of the Bioinformatics Unit, as well as teaching assistants. The latter are students studying for their M.Sc. and Ph.D. within the Faculty of Life Sciences. In the field of genomics, inviting guest lecturers from academia and industry fills the knowledge gap.

3.4. ACADEMIC BACKGROUND OF PARTICIPANTS

The participants in the bioinformatics training activities come from academia, from governmental institutions, such as hospitals, and from the biotechnology industry. In addition, in the commercial bioinformatics training programs, a large number of the participants are private individuals.

The participants have different research and personal goals based on their affiliation. Their choice of instruction mode reflects the underlying difference between the participants. Participants who attend our workshops and academic courses require knowledge in a specific topic for their research projects, regardless of their affiliation. Research and development in academia and in industry make use of the same cutting-edge tools. Those who attend our commercial programs are interested either in studying multiple bioinformatics subjects or making a career change. Both need to become acquainted with multiple bioinformatics tools in a short period of time because of the pressure of work or due to the need for employment.

3.5. BIOINFORMATICS TRAINING FACILITIES

The training takes place in computer classrooms that are set up to enable both frontal lectures and hands-on experience. Two different types of computer classrooms are used—personal computer (PC) classrooms and classrooms containing Silicon Graphics (SGI) workstations. The choice of the computer classroom is dependent on the subjects.

Training that takes place in PC classrooms focuses on sequence analysis, protein visualization and programming for bioinformaticians. These subjects employ software accessible through web browsers and software installed on personal computers that can be accessed through command-line or graphical user interfaces.

The Silicon Graphics classroom hosts training on protein structure analysis that is both CPU-intensive and requires advanced graphical capabilities. The commercial software and freeware with which protein structure analyses are carried out are installed on a powerful R12000 Silicon Graphics workstation with 2GB of memory and four processors.

The number of participants is restricted by the classroom size. PC classrooms can host a much larger number of people than expensive Silicon Graphics classrooms. Luckily, there is a balance between the demands and the limitations set by the sizes of the classrooms, as the demand for PC-based training is much higher than that based on SGI workstations.

Another important factor is the technical assistance provided by the PC and SGI system administrators. Both classrooms have to be maintained in terms of hardware, software and auxiliary equipment, such as printers and storage facilities.

In addition, a PC system administrator who is continuously present throughout the oral presentation and the hands-on training sessions deals with potential technical problems that might arise during the lessons in the PC classroom. These problems include difficulties operating the various PCs, communications facilities and software applications.

Training activities tie up hardware and software resources that are being used by researchers. This creates conflicts between the service given by the Bioinformatics Unit to already knowledgeable users, and the services provided to potential users that have to be trained. A way to tackle this problem is to provide bioinformatics training during weekday evenings and on non-working days and holidays.

3.6. BIOINFORMATICS TRAINING METHODOLOGIES

Hands-on experience is an important ingredient in bioinformatics training. The theoretical knowledge obtained has to be translated into practice. The barrier to implementing bioinformatics techniques in research and development is mainly overcome by practice, which involves repetition of learned techniques, as well as the introduction of new challenges. In addition, hands-on training aids biologists in overcoming the fear of using computers, and in mastering the computational techniques.

The lecturers prepare their oral presentations on Powerpoint slides. These slides are provided to students in hardcopy format as well as online during and after the training. In addition, hands-on exercises are provided on web pages and in hardcopy format.

Providing the participants with printed matter requires that all presentations and exercises will be ready well in advance of commencement of the program. Very often, however, last-minute changes are made before or even after the lesson. This causes a conflict between the will to provide printed matter in advance and reality, which might lead to frustrations being experienced both by the participants and the lecturer alike. Flexibility and mutual understanding are the solution to this problem. Unlike the printed matter, on-line presentations can be easily modified following the current lesson and before the next lesson.

3.7. ADMINISTRATION

Offering a variety of bioinformatics training alternatives requires an excellent administration system. This system takes care of the planning of the training time table, the inlaying of the training personnel, teaching assistants, and guest lecturers, classroom coordination, promotion of the training activity, subscription through electronic forms, charging of tuition fees, and handling of cancellations.

4. Summary

Education is one of the fundamentals of society. A complex field such as bioinformatics requires training that accounts for its interdisciplinary and multi-faceted nature. The fundamentals of the bioinformatics training system can serve as the basis for the establishment of other interdisciplinary programs, such as computer-aided drug design and chem-informatics (6), which involve the integration of chemistry, biology, and computer science.

5. References

1. Baltimore, D. (2001) Our genome unveiled, Nature **409**, 816–818
2. Burbaum, J. and Tobal, G.M. (2002) Proteomics in drug discovery, Current Opinion in Chemical Biology, **6**(4), 427–433
3. Lennon, G.G. (2000) High-throughput gene expression analysis for drug discovery, Drug Discovery Today **5**(2), 59–66
4. Luscombe, N.M, Greenbaum, D. and Gerstein, M. (2001) What is Bioinformatics? A proposed definition and overview of the field, Method Inform. Med. **40**, 346–358
5. http://www.tau.ac.il/lifesci/bioinfo/
6. http://www.tau.ac.il/lifesci/bioinfo/drug-design-program-2002-English.doc

Biodata of **Mark P. W. Einerhand** the co-author (with J. van Melle) of the chapter ***"Patenting of Inventions in the field of Bioinformatics."***

Dr. Mark Einerhand currently a patent attorney at Vereenigde in The Hague, the Netherlands. He graduated in molecular biology (1987) at the University of Amsterdam, and then he obtained his Ph.D. (1987) in gene therapy. Dr. Einerhand served as a senior scientist and project manager in a fast-growing gene therapy institute in the NL. In 2001, he was registered as a Dutch patent attorney.

E-mail: **m.einerhand@vereenigde.nl**

Biodata of Johannes van Melle author (with Mark Einerhand) of the chapter ***"Patenting of Inventions in the field of Bioinformatics."***

Johannes van Melle is a Dutch Patent Attorney with Vereenigde. He graduated in physics at the University of Utrecht in 1991, then (1993) graduated in Dutch law. Mr. Johannes worked in the business sector for some years, e.g. as product developer of internet applications. He has a wide knowledge in numerous technical fields, including informatics, semiconductor products and optics.

E-mail: **j.vanmelle@vereenigde.nl**

Mark Einerhand

Johannes van Melle

J. Seckbach and E. Rubin (eds.), The New Avenues in Bioinformatics, 237–254.
© 2004 *Kluwer Academic Publishers. Printed in the Netherlands.*

PATENTING OF INVENTIONS IN THE FIELD OF BIOINFORMATICS

MARK P.W. EINERHAND AND JOHANNES VAN MELLE
Vereenigde. Nieuwe Parklaan 97, 2587 BN the Hague, The Netherlands.

1. Introduction

Knowledge is one of the more important assets of a scientist. Knowledge is typically communicated through papers. However, once published it is available to the scientific community in general and no longer the sole property of the creator. Patenting inventions is a way in which scientists may continue to profit from knowledge obtained also after publication thereof.

This chapter intends to provide the reader with some guidance as to what is patentable in bioinformatics and what not. While there is much to be said about patenting inventions in general, this chapter particularly focuses on aspects of patenting that are relevant for inventions made in the field of bioinformatics. This chapter intends to give an overview of the possibilities for obtaining patent protection for both the tools and the results obtained through the use of these tools. General aspects on the patentability of inventions, and the requirements for novelty, inventive step and sufficiency of disclosure of the invention will be discussed first. Subsequently these issues will come back in the context of patenting of tools and results of bioinformatics. As knowledge about the competition is almost as important as one's own inventions, a final section is devoted to Internet access of patent information.

National law ultimately governs patent law. As both authors have their base in Europe, much of the issues referred to in this chapter are influenced by the rules and regulations of the European Patent Office (EPO). This is of course a restriction of this chapter. However, since the principles used in patent law are harmonized between the major markets, many of the statements made herein have wider applicability. Where major differences exist between the system of the EPO and the United States Patent and Trademark Office (USPTO) these have been indicated.

While some of you may already have extensive experience with patents, other readers may not. To appeal to both the novice and the expert we have laced general aspects with more detailed discussions on particular interesting or controversial points.

2. Basic Principles of Patentability in The European Patent Convention

Patents are granted by or on behalf of national governments. With the grant of a patent a government grants the owner a temporary monopoly for the commercialization of his/her invention. To structure this process, a system was created by which a government can decide on the grant in an objective way. The basics of the system as implemented by means of the European Patent Convention (EPC) are listed below. The EPC entered into force for seven countries on 7 October 1977[1]. The EPC has been an overwhelming success. Since the start by the first seven contracting states twenty other states have acceded to the treaty. Over 300.000 patents have been granted through this system. The basic requirements for the patentability of inventions are given in article 52 (EPC). Article 52 contains four paragraphs that are discussed per paragraph below. The first paragraph reads:

A. 52 (1) European patents shall be granted for any inventions which are susceptible of industrial application, which are new and which involve an inventive step.

From this paragraph can be distilled that there are four basic requirements for patentability. (i) There must be an invention, (ii) the invention must be susceptible of industrial application, (iii) the invention must be new, and (iv) the invention must involve an inventive step.

2.1. INVENTIONS

The convention does not positively define what is meant by an invention. Apparently it is not the intention of the legislators to restrict the concept of an invention too much. The legislator has however, considered it prudent to state explicitly what is not regarded as an invention. Since some areas of exclusion relate to the handling, use and presentation of information, this paragraph is of importance for the field of (bio)-informatics. Some of the exclusions are listed in article 52 (2) EPC.

A. 52 (2) The following in particular shall not be regarded as inventions within the meaning of paragraph 1:

(a) discoveries, scientific theories and mathematical methods;

(b) aesthetic creations;

(c) schemes, rules and methods for performing mental acts, playing games or doing business, and programs for computers;

(d) presentations of information.

This paragraph should be read in conjunction with the next paragraph of article 52 (EPC):

A. 52(3) The provisions of paragraph 2 shall exclude patentability of the subject-matter or activities referred to in that provision only to the extent to which a European patent application or European patent relates to such subject-matter or activities as such.

Article 52(3) is of importance because it limits the exclusions of article 52(2) to inventions that only relate to the excluded fields. Items on the list of article 52(2) are not patentable *as such*. Inventions may encompass one or more of the features listed above but there must be at least one other aspect in the invention. For instance, a program for computers is not patentable as such, however, there are circumstances where programs may be claimed, see also in detail below. It is noted that the exclusions on the list are either all

[1] Belgium, Switzerland, the Federal Republic of Germany, France, the United Kingdom, Luxembourg and the Netherlands

abstract (i.e. discoveries, scientific theories etc.) or non-technical (i.e. aesthetic creations or presentation of information). The list at least implies that an invention is patentable when it has a technical character. This view is shared by the boards of appeal of the European Patent Office (EPO) in its recent decisions. The requirement for a technical character of the invention is particularly relevant for the patentability of information and information handling such as is often the case in computer programs, Internet applications and (bio)-informatics in general. This point will therefore be more elaborately discussed below.

Further exclusions of inventions from patentability are directed toward exclusion of subject matter that the patent office might consider to contain inventions but that it does not want to grant patents for. The reasons for the exclusion are either of ethical nature or because of interference with other systems for the grant of monopolies (i.e. breeders rights for plant varieties). The exclusions are detailed in article 52(4), article 53 and parts of rule 23. Many of these exclusions are of importance to what can be protected of the results of bioinformatics. Since protection is of commercial importance for third parties desiring to use the information gathered with bioinformatics, knowledge of these exclusion should also aid the bioinformatics artisans in their work.

Art 52 (4) Methods for treatment of the human or animal body by surgery or therapy and diagnostic methods practiced on the human or animal body shall not be regarded as inventions which are susceptible of industrial application within the meaning of paragraph 1. This provision shall not apply to products, in particular substances or compositions, for use in any of these methods.

The actual exclusion criterion used in this article is "not susceptible to industrial application"; however, it is generally considered that the motives for introducing this exclusion were guided by ethical perceptions. The legislator does not want to have the activities of medical personal interfered with through the patent system. Medical persons should be free to use anything in their power to help the needing. This perception has the consequence that any claim is excluded from patentability if it includes at least one feature defining a physical activity or action that constitutes a method of treatment excluded under article 52(4).

It is commonly known that much of the progress in the treatment of diseases comes from large pharmaceutical companies that invest heavily in the development of novel therapeutics. To allow them to protect and to obtain revenues from these investments, the patent system does allow protection for the products to be used in such medical treatments. This protection allows the proprietor to act against competitors but as mentioned, not against the medical staff using the fruits to help the ill. Of importance to mention here is that the US legislators were not so motivated. In the US it is therefore possible to obtain protection for methods of treatments.

Art. 53(a) is an article that explicitly introduces ethical considerations into the patent system.

A. 53 European patents shall not be granted in respect of:

(a) inventions the publication or exploitation of which would be contrary to "ordre public" or morality, provided that the exploitation shall not be deemed to be so contrary merely because it is prohibited by law or regulation in some or all of the Contracting States;

(b) plant or animal varieties or essentially biological processes for the production of plants or animals; this provision does not apply to microbiological processes or the products thereof.

Ethical considerations relating to biology were first dealt with by the enlarged boards of appeal of the European Patent Office. A noteworthy decision to deal with ethical issues under this article was the so-called "harvard-oncomouse" (decision T19/90). The claims concerned a transgenic non-human mammalian animal whose germ cells and somatic cells contain in the genome an activated oncogene that increases the probability of neoplasm development in the animal. The claims also contained methods for making such an animal. The board considered that a careful weighing up of the suffering of animals and the possible risks for the environment on the one hand, and the usefulness to mankind on the other, must be made before a decision can be given to grant or refuse a patent application. The patent has been granted after this decision but has been subject to opposition. It was upheld in limited form, but maintaining claims directed to a genetically modified non-human animal.

Since this decision, the implementing regulations of the EPC have been amended to include new rule 23(d) EPC which provides that patents shall not be granted in respect of biotechnological inventions which concern a processes for modifying the genetic identity of animals which are likely to cause them suffering without any substantial medical benefit to man or animal, and also animals resulting from such processes. This inclusion makes it less likely that the ethical criteria used to assess the patentability of the harvard-mouse will be modified.

It can be noted that new rule 23(d) is only concerned with genetically modified animals. This can, in our opinion, not be taken as an indication that genetically modified plants will be free from ethical criteria.

The boards defined the concept of "ordre public" as covering the protection of public security and the physical integrity of individuals as part of society. It also encompasses the protection of the environment. Accordingly, inventions of which exploitation is likely to seriously prejudice the environment will be excluded from patentability as being contrary to "ordre public". This criterion will very likely be used to assess inventions relating to genetically modified plants.

Under article 53(b) EPC, plant varieties or animal varieties and essentially biological methods are excluded from patentability. This exclusion was introduced in view of the existence of the system for the protection of plant varieties granted in the UPOV (The International Union for the Protection of New Varieties of Plants), and similar acts in the various contracting states. Plant varieties are defined as any plant grouping within a single botanical taxon of the lowest known rank that can be: (a) defined by the expression of the characteristics that results from a given genotype or combination of genotypes, (b) distinguished from any other plant grouping by the expression of at least one of the said characteristics, and (c) considered as a unit with regard to its suitability for being propagated unchanged. Animal varieties are not defined in the convention. A process for the production of plants or animals is essentially biological if it consists entirely of natural phenomena such as crossing or selection.

As mentioned above, the exclusions relating to the gathering, use and presentation of information are relevant for the patentability of new tools in the field of bioinformatics. The other exclusions mentioned are only relevant insofar as applicability of the results of the information gathered is considered.

2.2. NOVELTY

Of further importance for the assessment of patentability is novelty. The invention must be novel otherwise no patent can be granted. Novelty is at first glance an easy concept but when looked into in more detail actually pretty complicated. The general principle is that one cannot patent what is already in the art. Inventions can be made part of the art in any way conceivable by man. Be it written, or electronically for instance though the Internet, by a product or by means of oral disclosure. For novelty to be destroyed it is critical that the invention was disclosed outside a so-called closed circle of persons that are bound to secrecy explicitly or by an implied tacit secrecy agreement. For instance, members of a research group are typically considered to be part of a closed circle. However, one may argue whether a disclosure to a trusted colleague working down the hall is considered to be a disclosure within the closed circle. In this respect it is better to maintain a rather restricted interpretation of the closed circle and when in doubt, make explicit arrangements.

The time point for determining what art is novelty destroying and what art is not, is the filing date of the application. Any disclosure before this date is considered relevant, any disclosure after this date is considered not relevant for the novelty of the invention. There are two exceptions to this general rule. One is to claim the date of an earlier filed application and the other is the period of Grace for disclosures originating from the inventors. The two exceptions are discussed in more detail below.

It is possible to claim the date of an earlier filed patent application for the invention. In this case the relevant date for assessing novelty shifts to the date of filing of the earlier application. In the patent world this is called claiming priority from the earlier application. A detailed discussion of priority is outside the scope of this chapter. However, some aspects need to be discussed here. An application can serve as a priority document when filed no more than 12 months before the filing of the application in which its priority is claimed. The priority document must have been filed in a Paris Convention country. Most countries, with notable exceptions as Taiwan and some South American states, have acceded to the Paris Convention.

As mentioned above, any disclosure of the invention prior to the filing of the patent application damages the novelty of the invention. It is thus important that the inventors do not talk about or publish the invention prior to the filing of the patent application. This limitation is seen as a serious restriction of the normal exchange of information in the scientific community. Many governments have recognized this problem and have introduced a period of Grace for prior disclosures originating from the inventors. The period of Grace is a period of time before the actual filing of the application within which the patent authority considers a prior disclosure of the invention by the inventors not damaging for the patentability of the invention. The US, Canada and for instance Japan allow for a period of Grace of one year (US and Canada) or six months (Japan) wherein disclosures originating from the inventors are not considered damaging to the novelty or inventive step of the invention. The time point from which such period of grace is calculated is important. Some countries calculate back from the date of priority claimed. Other countries, among which the US, take the date of filing of the national application as the relevant date.

The EPC does not recognize a general period of grace for disclosures originating from the inventors. However, there is one exception. The EPC disregards such disclosures for novelty and inventive step if their disclosure was an evident abuse to the party filing the application. The abused party has a period of 6 months to file the European application. Of note here is that this 6 months period is calculated from the filing date of the European application and not from a priority date.

In any case, the divergence of the date from which Grace periods are calculated in the various countries and the divergence in the amount of time allowed for a previous disclosure, makes it important to decide in advance on where and how to file a patent application in case of disclosures of inventors prior to the filing of the application. This is a topic that should be discussed with the patent attorney handling the filing of the application.

Though the general principle of novelty is easy to understand, it poses some problems for the biotechnology field. For instance, much of the knowledge generated is derived directly or indirectly from nature itself. A gene has been there also before discovery thereof. Is it new? No, because it already existed. Patent systems throughout the world have dealt with this problem in much the same way. It is considered that the isolated or recombinant version is new. This in its turn caused another problem. As soon as the first cellular nucleic acid was isolated and cloned into libraries it fulfilled the feature of at least a recombinant gene. Is a gene identified in the library new when the library is part of the art? Yes, it is considered to be new when not previously identified in the library or in another way. This is important also for bioinformatics. In principle a gene identified in a large database of nucleic acids, as for instance generated by the genome efforts, should be new, when presented as an isolated or recombinant gene. Other problems with respect to novelty relate to the function of a sequence. Is a novel sequence in the absence of information on the function patentable? What if an indication of function is obtained using bioinformatics? These and other related issues will be dealt with in more detail below under the heading “The results”.

2.3. INVENTIVE STEP

An invention is considered to comprise an inventive step if it is not obvious to a person skilled in the art at the time of filing of the application. Although there is much to be said about inventive step in general, we will not discuss these aspects here. A good rule of thumb for patent attorneys is that if the inventor considers it an invention, the invention is likely to comprise an inventive step. The person skilled in the art is a fictitious person. Because one person is not another, patent systems have had deal with the problem of objectively judging inventive step. They have solved this problem by defining the person skilled in the art as someone who has complete knowledge of the art at the time of filing of the application but lacks any inventive skill. It may be clear that this skilled person cannot be found on this planet. It is a common mistake of inventors to equate the level of skill of the person skilled in the art with themselves. This should not be done. A good patent attorney will often be able to argue that a new invention is also inventive.

2.4. INDUSTRIAL APPLICABILITY

This requirement typically regarded as the easiest to meet. The EPC defines this concept in article 57.

Art. 57. An invention shall be considered as susceptible to industrial application if it can be made or used in any kind of industry, including agriculture.

This should be interpreted in the broadest sense possible. Any activity of technical character qualifies. This article does not overrule the mentioned exclusions of article 52(2) but should be seen as a further requirement. In general it is required that the description of the patent application should, where this is not self-evident, indicate the way in which the invention is capable of exploitation in industry. In relation to sequences and partial sequences of genes this general requirement is given specific form in that the industrial application of a sequence or partial sequence of a gene must be disclosed in a patent application. A mere nucleic acid sequence without an indication of a function is currently regarded as insufficient for patentability by the EPC. The US uses the criterion of utility to the same effect but differently worded. A more detailed discussion will be given below.

3. The Tools

Bioinformatics is generally described in the field as the use of computers to handle biological information. In this respect, the investigative work of bioengineers using computers is sometimes referred to as in silico methods, opposing the otherwise traditional concepts of in vitro or in vivo measurements. Hence in silico methods are usually the methods where data recognition, sorting, searching, simulation and computation are the vital ingredients to perform investigative research. Clearly, computing and computational tools form a vital tool in this field.

While large amounts of money are invested in bioinformatics and related research and while the part spent on the information technology is ever increasing, it is traditionally quite difficult to protect the intellectual property rights of these tools. Today's software products are increasingly complicated and difficult to improve, while, by their nature, the tools can be easily copied. Patents traditionally play a dominant role when it comes to protecting the inventive technology, but, in this respect, worldwide, computer-implemented inventions are problematic in terms of patentability.

This problem is primarily caused by the fact that the computer is, in view of patentability, on the one hand just a substitute for a pencil and paper, and hence a vehicle for guiding "just" a mental process. In this view, patentability can not be an issue, as good as any mental process is not patentable, whether it is "pure mental" and only performed "within the mind" or it is assisted by pen and paper, or rather, by the aid of a computer, to better access and support the handling of information and stages wherein information is processed.

On the other hand, the computer may very well form a (vital) link in a complex technical system, where not so much as a mental process is involved but a true technical process is going on. While this vocabulary of "technical process" will be worked on later and a more sophisticated feel will be developed in order to be able to discern what might be patentable and what not, this dichotomy is quite the basic cornerstone on which the issue of patentability is decided.

Returning to the working definition of bioinformatics, the possibilities for patenting your inventions in this field are unfortunately quite limited, since handling of information per se is quite abstract in its form and not primarily related to a down to earth technical process. It is only within the realm of true physical/technical contexts that patentability may come into sight as an option to protect the inventive idea. Hence, it will be expected, and as will be demonstrated later in a few working examples, that generally only derivate issues in the field of bioinformatics may be patentable. Illustrative may the following distinction:

- Genetic information per se is not patentable as abstract concept; neither is it when stored in physical form in a database;
- However, a specific way of accessing said information, solving a technical problem to improve or accommodate access thereof may be patentable.

Once such distinction is clear, the inventor, or quite often the patent attorney assisting the inventor, might look for ways to formulate his invention in terms of such a technical context.

Computer-implemented inventions are the more problematic inventions in terms of patentability since there is no general statement to give what is patentable and what is not. First, there is of course no general statement describing what is patentable, since inventions are by nature unpredictable and may from time to time cause a shift in the opinion on what is patentable and what not. Second, it is recalled that patent law is basically national law and differs quite substantially from country to country. Although far going initiatives have been made to provide uniform standards on patentability, specifically the software-related patentability issues discussed here are solved differently in various countries in the world. In Europe, a lot of emphasis has been put on the presence of a "technical effect". The United States prefers to talk about a "useful, concrete and tangible result". The Japanese Patent Office is concerned with the presence of "a creation of technical ideas utilizing a law of nature". The legal positions in the USA and Japan hence differ substantially from that under the European Patent Convention, all the more since only in Europe patentability of computer programs "as such" is specifically excluded in the European Patent Convention. Since elaborate discussion of these various concepts is outside the scope of this introductive text, in the remainder the discussion will be primarily based on the European approach.

3.1. PATENTABILITY OF COMPUTERS AND PROGRAM THEREFORE

One of the latest leading case law is the Decision of Technical Board of Appeal 3.5.1 dated 1 July 1998 (T1173/97). Herein, it is stated that "a computer program product is not excluded from patentability (...) if, when it is run on a computer, it produces a further technical effect which goes beyond the "normal" physical interactions between program (software) and computer (hardware)".

In this respect, the "normal" interaction of a computer refers to the physics that is needed to operate a processor in a computer, that is: the interplay between for instance memory elements, registers and processor elements, the basic electric components for modifying streams of data. These are, as the boards mentions, a common feature of all those programs for computers which have been made suitable for being run on a computer, and therefore cannot be used to distinguish programs for computers with a technical character from programs for computers as such. The decision elaborates further on what can be considered a "technical effect". This can be the case where a piece of software manages, by means of a computer, an industrial process or the working of a piece of machinery. However, in cases where a program for a computer is the only means, or one of the necessary means, to obtaining such a required technical effect, this would be not sufficient for obtaining a technical effect. For instance, a technical effect of that kind is achieved by the internal functioning of a computer itself under the influence of said program. Thus, as the author sees it, rendering the problem "technical" by merely using a technical means in the form of a computer, cannot render the invention patentable.

Further, the board sees no reason to exclude the computer program itself when it possesses, when run on a computer, such a "further" technical effect. This is quite a remarkable point of view, since it distinguishes herein further from the exclusion of "computer programs as such" which are inhibited by the second paragraph of Article 52 of the European Patent Convention.

From this decision it has become clear that in certain aspects, computer programs may be very well patentable, provided they are not "mere manipulation of data", "mere representation of information", mere "a guiding mental process", terms that have come across in the case law related to this issue. To have a feel for what can be considered a technical effect one can think of more efficient ways to store and/or retrieve data (including data structures); faster access, faster computations, larger information handling possibilities etc. etc. as long as the information itself is not claimed. Further examples may be derived from case law and comprise inter alia: Ensuring optimum exposure with sufficient protection against overloading of the X-ray tube by an X-ray apparatus incorporating a data processing unit; co-ordination and control by software of the internal communication between programs and data files held at different processors in a data processing system; a computer-implemented method for entering a rotation angle value into an interactive draw graphic system allowing the rotation of displayed graphic objects with increased accuracy.

It must be mentioned here that the view of the European Patent Office is not a reflection of a common approach among the European countries. In this respect, recently a proposal has seen the light for a Directive of the European Parliament, regarding the question of the patentability of computer-implemented inventions. In some respect the Directive deviates from the course already set. In this respect, the most remarkable change is that it does not follow the practice of the EPO in permitting claims to computer program products either on their own or on a carrier. Contrary to the earlier discussed decision of the EPO board, the drafters of the Directive are of the opinion that such computer program products are equivalent to the computer programs "as such" that are excluded by the exclusion list of Art 52.2 of the European Patent Convention.

The directive consists only of a few substantive articles, wherein article 4 outlines the basic condition for patentability: (..) A computer-implemented invention is patentable on the condition that it is susceptible of industrial application, is new, and involves an inventive step. The article further stipulates that it is a condition of involving an inventive step that a computer-implemented invention must make a technical contribution.

Interestingly, the discussion of the presence of a technical effect, which is traditionally held regarding patentability per se (which is the domain of art. 52 EPC) has shifted to a discussion regarding inventive step. This is in line with the recent Controlling Pension Benefits System case, in a Decision of Technical Board of Appeal 3.5.1 dated 8 September 2000 (T 931/95). In this decision, it was decided that a computer and a computer program constitute a physical entity or a concrete product, suitable for performing or supporting an economic activity, and are therefore considered to be "invention" within the meaning of Article 52(1) EPC. However, the question of "technical effect" in terms of inventive step has great similarities with the previous discussion. As the memorandum accompanying the Directive describes it, a computer-implemented invention in which the contribution to the prior art does not have a technical character will hence be considered to lack inventive step even if the (non-technical) contribution to the prior art is not obvious.

While it is still unclear what the ultimate legal situation in Europe will be, much has been gained by this express codification, since it incites the national countries to harmonize to a further extent their judicial system on national level. While the European Patent Office is able to evolve in a way a consistent approach towards these patentability issues, the national courts have sometimes diverged from this line based on national law, thereby creating unpredictable outcome throughout the countries of the European Patent Convention. Off course, this is undesirable, and the Directive creates a further incentive to come to a harmonized approach.

3.2. PATENTABILITY OF INFORMATION

While in the preceding, much emphasis has been put on the exclusion of patentability of computer programs as such, the fourth item of the exclusion list in Art. 52.2 is also of importance in the field of Bioinformatics. More precisely the prohibition on patenting information is formulated as an absence of patentability for representations of information, defined solely by the content of information. In the context of Bioinformatics this is quite an important issue, since one of the main objects of this discipline is the storage and retrieval of biological information.

From an example, derived from a report issued by a cooperation of the European patent office, the United States patent office and the Japanese patent office ("Report on comparative study on protein 3-dimensional (3-D) structure related claims"), it becomes clear that a 3D computer model of a protein generated with the atomic coordinates listed in a database is not patentable. Neither the database comprising said information, or otherwise put

"A data array comprising the atomic coordinates of a protein which, when acted upon by a protein modeling algorithm, yields a representation of the 3-D structure of protein P."

This is quite unfortunate, since the proteins and their biological function may themselves be already disclosed and as such not patentable. When only the 3D information regarding their structure is new, this information in itself is not patentable. This applies whether the claim is directed to the information per se (i.e. the above mentioned 3D model) or to the processes and apparatus for presenting the information (i.e. the database defined by the 3D information it comprises). This is only different when the presentation has technical features or generates a technical effect, which criterion is quite the same as discussed above. A specific arrangement or manner of representation could therefore very well form patent eligible subject matter, if a technical effect is present. However, if the model or database comprising the information merely specifies the atomic coordinates of a single protein molecule in space, without any direct technical character; the data merely encode cognitive content in a standard manner. This is not considered patentable.

In this respect, reference is made to another landmark decision of the European Patent Office, wherein it was decided that a record carrier characterized by having functional data recorded thereon is not a presentation of information as such and hence not excluded from patentability. In this Decision of the Technical Board of Appeal dated 15 March 2000

(T 1194/97) it is becomes clear, that a specific form of the data themselves can be patentable, if there is an inherent technical effect present in said form. In the decision, it is explained how the difference between these two types of data can be characterized:

"The significance of the distinction between (patentable—JM) functional data and (non-patentable—JM) cognitive information content in relation to technical effect and character may be illustrated by the fact that in the present context complete loss of the cognitive content resulting in a humanly meaningless picture like "snow" on a television screen has no effect on the technical working of the system, while loss of functional data will impair the technical operation and in the limit bring the system to a complete halt."

From this explanation it becomes clear that functional data will have some functional relationship with a technical system, for example a read/record system for presenting the functional data. Patentability will become an issue when a novel and inventive new functional relationship is present to use this data. In the context of the above-described 3D model it will become clear, that a specific format for describing said model, which allows an improved way of handling the information of said model, might be patentable. This aspect may become interesting when there is a unique way to describe a 3D model of a protein allowing extracting certain relevant information for it for obtaining a technical effect. However, the object of claiming 3D information in itself, without giving further technical aspects, is not patentable.

3.3. WORKING EXAMPLES; COMPARATIVE ANALYSIS

The above-mentioned study of the trilateral project between the European Patent Office (EPO), the Japanese Patent Office (JPO), and the U.S. Patent and Trademark Office (USPTO) has provided a number of very useful working examples and how these examples are regarded by the three offices in terms of patentability. While the study has a much broader scope, and also discusses the patentability of biological matter, such as proteins and parts there of, in the context of this issue we will focus on the patentability issues regarding data (hence data describing the 3D structure of these matter) and computer-implemented methods for researching and describing this matter.

The report distinguishes basically between items regarding patentability of data (information) claims and in silico screening method claims for identifying active biological compounds.

Regarding first mentioned category, this has been primarily discussed in the preceding chapter. All three patent offices conclude that: a computer model of a protein; a data array comprising atomic coordinates of protein; a computer-readable storage medium encoded with atomic coordinates of a protein and a database encoded with data comprising names and structures of compounds are not patent eligible subject matter or statutory inventions. In this respect it is mentioned that such matter are mere presentations of information or abstract ideas which have not been practically applied and therefore not patentable. The European patent office mentions that such matter is not considered to be a patentable invention; it merely presents the atomic coordinates of a single protein molecule in space as such, without any direct technical character (it is non-technical by not solving a technical problem, and it does not have a technical effect in itself). The US patent office states that such claims (..) directed to a computer model, are not tangibly embodied and therefore is nonfunctional descriptive material per se (..). Descriptive material is considered to be an abstract idea and therefore would be rejected. Further, (..) a data array, is a compilation or mere arrangement of data. The 3-D coordinates of a protein constitute nonfunctional descriptive material without physical structure, and therefore are abstract ideas. Finally, the Japanese patent office is of the opinion that such matter (..) is neither directed to a presentation of information with some technical feature in the presentation or the means or method of presentation, nor is it directed to a data structure which concretely realizes information processing using hardware resources. There may be cases where data is a statutory invention if it is featured by data structure (logical structure of data defined by interrelationship among data elements) and information processing by the data structure is concretely realized by using hardware resources. It can be seen that the reasoning of the three patent offices is quite different but that there is a large overlap on the position of patentability of such data claims.

Regarding the second mentioned category of the above, the in silico screening methods, in the report two cases are described of such screening methods. It is remarkable how different the opinions of the three offices are with respect to these cases. In the following we will try to outline the comments and differences expressed by the offices.

The first case mentions a screening method to identify candidates for drug design that can bind to specific 3D structures of the protein P. In this case it was given that the protein was known already, and also the effect (blood lowering pressure) it contains. The 3D coordinates are not described in the prior art and therefore novel. No working examples are given. The method relies on known programs for predicting the binding pocket of the proteins, as well as several screening computer programs that use the predicted binding pocket.

The EPO regards this case in principle as a patentable invention under art. 52 since it refers to a method having a link to a technical contribution. No statement is made with regard to inventive step. However, since no enablement of the disclosure is given by any working examples, it regards the method as non-patentable.

The USPTO contemplates that the claims are novel because the 3-D coordinates are not found in the prior art. However, it is of the opinion, that the claim is obvious since a known algorithm is used, and the difference between the prior art and the invention (..) is limited to descriptive material stored on or employed by a machine. (..) A method of using a known comparator for its known purpose to compare data sets does not become nonobvious merely because new data becomes available for analysis. Nonfunctional descriptive material cannot render non obvious an invention that would have otherwise been obvious.

The JPO is of the opinion that (..) the claimed invention is considered a computer software-related invention with the technical feature of an information processing method by software. The difference between the prior art and the claimed invention as a whole is limited to the 3-D molecular model. Data that does not alter the processing method should be considered as mere contents. The claim is therefore not regarded as patentable subject matter.

Remarkably, here, the EPO appears to be the most lenient towards this invention, although it considers the invention not patentable since it is not disclosed fully. However, if working examples would have been provided, it seems that the invention could have been protected

The second case regards a method comprising a number of specific steps for identifying compounds which can bind to protein P by comparing the 3-D structure of candidate compounds with a specific novel 3-D molecular model of protein P. (The 3-D molecular model presents the positions of heteroatoms in the aminoacids constituting the binding pocket of protein P wherein said heteroatoms can form hydrogen bonds with hydrogen bonding functional groups in a candidate compound.) The specific steps describe a known data processing method in which a) the coordinate data of the 3-D molecular model is input in a data structure such that the interatomic distances between the atoms of protein P are easily retrieved, and b) the distances between hydrogen-bonding heteroatoms of different candidate compounds and the heteroatoms that form the binding pocket in the 3D molecular model are compared thereby allowing the identification of those candidate compounds which would theoretically form the most stable complexes with the 3-D molecular model binding pocket of protein P, based on optimal hydrogen bonding between the two structures. In this example, the binding pockets are identified and working examples are provided.

Here, the EPO is of the opinion that the screening method is considered to be a patentable invention. This activity is not regarded as a presentation of information or as a pure mathematical method, but to the use of the structural data. Since the prior art did not disclose or suggest the 3-D coordinates of protein P, the claimed method applying the use of the coordinates is considered to be new, non obvious and industrial applicable.

Again, as in the previous case, the USPTO finds that the key factor in analyzing the obviousness of these claims over the prior art is the determination of whether the claimed data processing method used to identify compounds that can potentially bind protein P, would have been obvious to one skilled in the art. In this case, the claimed method would have been prima facie obvious over the prior art because use is made of a known data processing method.

Likewise, the JPO considers the invention a computer software-related invention with the technical feature of an information processing method by software. The difference between the prior art and the claimed invention as a whole is limited to the 3-D molecular model. Data that does not alter the processing method should be considered as mere contents. Hence the requirement of novelty/inventive step is not met.

The USPTO and the JPO do not consider the use of the 3D model patentable. The offices use different arguments: basically the USPTO finds this use of new data in a known data processing method is obvious, whereas the JPO considers these data as non technical and therefore not adding to novelty/inventive step. Again, the EPO is of the opinion that use of the 3D model is technical in nature since it solves a "technical" problem of identifying suitable compounds. The novelty/inventive step issue is not extensively addressed, but it appears that there seems a good ground to expect that such method would be indeed patentable.

The preceding discussion and the very diverse ways patentability matters are solved in different countries make it very difficult to obtain a good prognosis for the patentability of such subject matter as computer programs, modeling, algorithms, data etc. etc what can be considered as the technical tools of bioinformatics. While each invention has its own merits and needs to be addressed by a specific approach, a general statement could be that there are far going possibilities in protecting these matters but the chances on success vary and depend heavily on the technical details. When something more is going on than mere data-manipulation, it needs seriously be contemplated whether this "something" may be patentable. This may be al the more the case, when technical problems are solved and concrete, workable results are rendered. Patents matters remain however a very specialist field and professional advice is quite indispensable.

4. The Results

The result of a bioinformatics manipulation is knowledge. This knowledge as such is not patentable, as we have seen. However, it is often thought that the information obtained using bioinformatics provides an argument in favor of the patentability of a novel sequence. A newly identified sequence can be used to screen a sequence database for homologues sequences, similar structural configurations etc. Any homologous or similar sequence may function as a starting point for finding a function of the newly identified sequence. Many domains with specific functions have already been identified. Consensus sequences have been identified for phosphatases, phosphate kinases, zinc finger proteins and many other classes of proteins but also for regulatory sequences regulating for instance gene expression, translation and protein degradation. A search for such sequences can help to identify a function of an unknown gene or sequence.

The strategy mentioned above is now so common place that the question of patentability of sequence provided with a function through bioinformatics has been put to the three major patent offices in the world: The European Patent Office, the Patent and Trade Mark Office of the United States (USPTO) and the Japanese Patent Office (JPO).

The EPO requires that the function performed by a claimed sequence and, if any, the protein encoded by it, should be certain to the degree that a specific utility for the sequence becomes apparent beyond the realm of speculation. If the alleged function of a claimed nucleic acid molecule is not credible beyond mere speculation the EPO will request experimental evidence demonstrating the function in accordance with Rule 27(1)e EPC, which requires that the description of a patent should describe at least one way of carrying out the invention claimed

According to the USPTO, an invention must be supported by a specific, substantial and credible utility. General statements of utility are not acceptable. This criterion was developed after scientist at the NIH attempted to patent thousands of expressed sequence tags (ESTs). Concerns centered on the possibility that if broad claims were allowed for sequences encompassing such ESTs, such claims would dominate claims toward the later discovered gene and thereby reduce interest in the development of new medically important genes. The office further holds that the description should contain sufficient information such that a person skilled in the art can make and use the invention. The USPTO does not have any ab initio requirement for experimental evidence to demonstrate the function or utility of any invention. The burden of proof for unpatentability lies on the examiner.

The JPO requires a function from which we can assume the specific utility (the specific function) or the specific function recognized from common general knowledge as of the filing date of the application.

Ten hypothetical cases were put to the patent offices for examination with respect to inventive step and industrial application. From the examinations it is apparent that the EPO rejected most of the cases put forward for both lack of industrial application and lack of inventive step. The USPTO accepted the inventive step (non-obviousness) for all cases put forward but industrial application (utility) was only credible if bioinformatics resulted in a high homology with a known sequence leading to the assignment of a specific function. Cases where homology found was low but where a specific function could be assigned could not be classified as such as the particulars of the case could influence the assessment of patentability. Each individual application should be assessed. Like the EPO, the JPO rejected most of the cases put forward. However, the criterion by which the patentability was refused varied. Some cases were refused for lack of industrial application, whereas others where refused for lack of inventive step.

From the case-by-case analysis one could draw the conclusion that information gathered through bioinformatics is not sufficient to confer patentability on a novel sequence in the EPO and the JPO, whereas there is a good chance of patentability before the USPTO. However, it should be noted that the subject is extremely controversial and the outcome highly dependent on the particular case put forward. One should therefore not be dissuaded to attempt patent protection, if only, because the field is still in development and the perception of patent offices may change. In addition, bioinformatics is still developing, leading to an ever more precise estimation of function.

5. Access to Patent Information

As patent information is becoming more and more important, it might be interesting to know that much of the patent procedure is public. Patent applications are typically published 18 months after publication. As of late the US did not publish patent applications, however, recently the USPTO has begun publishing US filed patent applications for which the right of foreign filings has been requested. This means that applications that are filed for the US only still remain secret until grant.

Published patent applications are made available through the Internet by the respective offices, which also provide publications of granted patents. The respective sites can be found at: http://www.european-patent-office.org/, http://www.uspto.gov/ and http://www.jpo.go.jp/. The Japanese patent office provides a computer translator for quick translations into the English language.

After publication of the patent application it is possible to also inspect the file of the application at the respective patent offices. Through this system of file inspection a third party is given the opportunity to follow the prosecution of relevant cases. The EPO has facilitated this process and allows on line file inspection via Epoline at the following site: http://www.epoline.org/epoline/Epoline?language = EN & page = home & b = NS. Through this site it is possible to obtain status information regarding pending applications (patent register) but also specific information regarding the prosecution of cases, for instance the examiners communications and the answer of the proprietor (online file inspection) and pending claims. This tool can be very helpful indeed.

Statuses of granted patents can also be requested online via the Epoline site (EPO) or the official USPTO site.

6. Conclusion

The authors have tried to provide the reader with a quick overview of the patentability of tools and results in bioinformatics. The chapter was set up to be informative to scientist with a starting interest in Intellectual Property and to scientist with some experience in this area. The chapter provides considerable detail on some points, while other points were not discussed at all. For instance, important secondary requirements of patentability such as sufficiency of disclosure and enablement were not touched upon. Guidance given by this chapter should therefore not be taken as being complete. Though many patent offices accept the filing of a patent application by any person, it is advised to seek the aid of a qualified patent attorney. The filed application is the basis for determining patentability throughout the procedure. Any omissions or mistakes therein cannot be repaired afterwards.

The field of bioinformatics is characterized by the fact that information is created and manipulated. A large part of this chapter deals with the impact of non-patentability of information as such. Where possible we have attempted to provide the reader with a feeling for what additional steps could be taken to translate information into patentable subject matter. Currently, the major hurdle for the tools of bioinformatics seems to be the provision of a technical effect. For the results of bioinformatics, the utility requirement of a novel sequence seems to provide the major hurdle. However, these criteria should not be taken as absolutes. Specific case law in this area is still scarce and perception of patentability may shift in time.

VII. Additional Links

Biodata of Benjamin Yakir, author of the chapter ***"Associating Comt with Schizophrenia: Statistical Tools in Population-Based Genetic Studies."***

Dr. Benjamin Yakir is senior lecturer at the Department of Statistics, The Hebrew University of Jerusalem, Israel. He received his Ph.D. from The Hebrew University in 1991 at the 8 Department of Statistics. Dr. Yakir has obtained special honors, such as The Rothschild Foundation Fellowshiip and the Wolf Foundation (1994); Golda Meir fellowship (1990), and Pulver Foundation Fellowship (1989).

E-mail: **E-mail: msby@mscc.huji.ac.il**

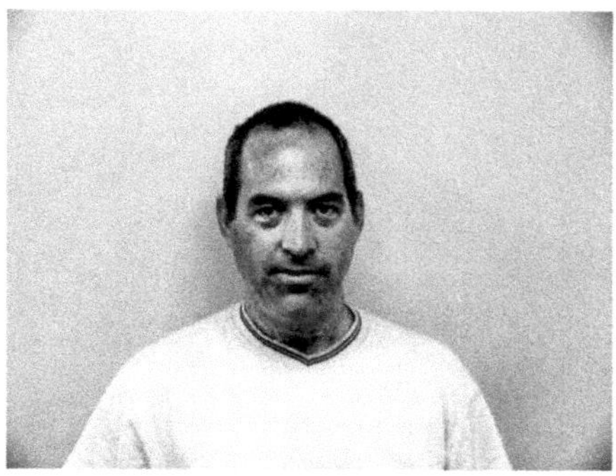

J. Seckbach and E. Rubin (eds.), The New Avenues in Bioinformatics, 257–271.
© 2004 *Kluwer Academic Publishers. Printed in the Netherlands.*

ASSOCIATING COMT WITH SCHIZOPHRENIA
Statistical Tools in Population-Based Genetic Studies

B. YAKIR
Department of Statistics, The Hebrew University, Jerusalem, Israel

1. Introduction

Biological research in human is motivated to a large degree by the promise of discovery of new drugs for the cure of common diseases. This process of developing a new drug is time consuming and expensive. However, the human and economic rewards from putting in the market a new product are immense. It is of no surprise, thus, that extensive resources, both from the academic world and from pharmaceutical companies, are devoted to drug-related research. The process of discovery is typically initiated by the identification of a target: A biological pathway or a protein within a pathway, which is connected to, or even leading to, the development of the given disease. Once such a pathway (or protein) is detected, compounds that affect the biological function may be tested towards becoming the new drug.

An important track for the identification of such targets is via genetic studies. This track has been paved during the last two decades with the emergence of new technologies that allow the "reading" of genetic information from DNA molecules. New vigor was added to this approach in recent years with the completion of the Human Genome Project and the sequencing of the entire human genome. An outgrowth of this project is the mapping of a practically unlimited number of bi-allelic genomic markers, termed Single Nucleotide Polymorphisms or SNPs in short.

Genetic studies in human are typically subdivided into two categories: Family based studies and population based studies. In the former approach, which also comes under the heading "linkage analysis", pedigrees are sampled. The pattern of inheritance of genetic material within families is analyzed and compared to the pattern of inheritance of the investigated phenotype. In the latter approach, also known as "linkage-disequilibrium mapping", unrelated individuals are sampled. Their genomes are compared in conjunction with their phenotype. A typical example of a population-based experimental design is the Case-Control design (CC). According to this design DNA samples are collected from affected cases and unaffected controls, and genotyped. The genetic composition of one group is compared with that of the other group. Loci which show significant divergence in their genetic configuration are presumed to be associated with susceptibility to the disease. A similar design, which is the one we will consider here, is the Case-Random (CR) design. According to this design random controls from the general population are used, instead of assessed disease-free controls. The statistical efficiency for a given sample size is somewhat reduced in this design. However, general DNA resources, like blood banks, can be used in order to obtain control samples. This has a practical importance, since healthy individuals have less of an incentive to participate in a trial, unlike affected individuals. Moreover, when the trait under investigation is not too widely spread in the general population, it is unlikely that the statistical power will be greatly reduced as a result of this less restrictive sampling approach.

This chapter is centered around the concrete example of a study, which investigated the potential association between the catechol-O-methyltransferase (COMT) gene, located on chromosome 22, and schizophrenia (Shifman *et al.*, 2002). The discussion here will focus on some of the statistical considerations, which are related to population-based genetic association studies. The motivation for putting the discussion in the framework of a specific example is to add concreteness. For the same sake, we will also attempt to demonstrate some of the statistical aspects with the aid of computerized simulations. These simulations were conducted in MATLAB. The code for the simulations, as well as the mathematical details of a theorem presented in the sequel, can be accessed on the web via the URL http://pluto.huji.ac.il/~msby.

In the next section we provide some background material on schizophrenia and the COMT gene. We also present the sequential genotyping approach that was used in the trial we consider. This approach was used in order to minimize genotyping costs. In the section that follows we discuss the issue of testing for genetic association in Case-Random trials, and relate the statistical power to the r^2 parameter of linkage disequilibrium. A stage in the sequential approach involves the genotyping of a relatively small number of individuals over a collection of markers, in order to assess the linkage disequilibrium state among those markers. In this context we will introduce the Expectation-Maximization (EM) algorithm for the estimation of parameters of linkage disequilibrium, and illustrate its statistical properties via a simulation. In the concluding section we present the outcomes of the trial and make final comments.

2. Is the COMT Gene Associated with Schizophrenia?

Schizophrenia (SZP) is a serious mental illness that affects about 1% of the general population. The disease is usually diagnosed on the basis of both "positive" symptoms such as delusions or hallucinations, and "negative" symptoms such as social withdrawal, poor motivation, and apathy. The initial stages of the disease are associated mostly with "positive" symptoms, whereas the negative symptoms become more dominant as the disease progresses (Sawa and Snyder, 2002). Twins and adoption studies suggest that SZP has an important genetic factor. However, like most common diseases, it is unlikely that this factor results from the abnormal form of a single gene. It is presumed that an array of genes may be associated with an increased risk to be affected.

One promising candidate for such gene is the catechol-O-methyltransferase (COMT) gene, located on the short arm of chromosome 22. COMT is one of the two principal enzymes that degrade catecholamines such as dopamine, which is part of brain activity. Another reason to suspect association between COMT and schizophrenia is the fact that this gene (as well as several other genes) is located within a micro-deletion at 22q11. This deletion is associated with a syndrome (VCFS—Velo-Cardio-Facial Syndrome) which includes schizophrenia-like symptoms. About 2% of schizophrenia patients harbour this rare micro-deletion (Karayiorgou *et al.*, 1995, Murphy *et al.*, 1999, Usiskin *et al.*, 1999).

About one hundred SNPs were discovered within the COMT gene. Most of these SNPs are located in introns. Some, however, are located in exons, as well as in promoters and other regulatory sites. Special attention was drawn to the SNP rs165688. This is a non-synonymous SNP, which results in a Val/Met substitution. It was demonstrated that this substitution has an effect on the functionality of the translated protein (Lachman *et al.*, 1996). It was also suggested that this substitution may be associated with some behavioral changes (Egan *et al.*, 2001).

The trial we describe was set to explore association between a series of genes, within and around the micro-deletion on chromosome 22, and SZP. The COMT gene was one of the examined genes. Specifically for the COMT, this involved the genotyping of a panel of 12 SNPs (including rs165688) over a sample of about 720 assessed affected individuals, and a sample of about 3,000 random controls, all of Ashkenazi origin.

Currently available technologies of genotyping still require non-negligible expense per reaction. The price-tag involved in conducting individual genotyping over numerous samples is substantial, over $2,000 per SNP in our case. In order to limit expenses, the Pyrosequencing™ genotyping technology was applied for the initial SNP panel on DNA pools. The Pyrosequencing™ technology allows for quantitative genotyping of pools. This quantitative genotyping provides a direct measurement of the allele frequency in the pool. Comparing the difference in measured allele frequency between the case-pool and the control-pool provides an indication of their dissimilarity level. (The drawback of this approach of genotyping in pools, on the other hand, is an increased genotyping error rate, and a loss of information—only the gross allele frequency is measured, no information is available on the frequency of homo- and heterozygotes in the samples. Furthermore, nothing can be said about haplotypes—the joint distribution of several SNPs put together.)

Quantitative genotyping in pools is an efficient method for the initial screen. SNPs that showed in this screen suggestive association with the trait were then genotyped individually. Individual genotyping itself was also conducted in two steps. In the first step all suggestive SNPs were genotyped over a small sample in order to assess the linkage disequilibrium structure among them. Based on this assessment, and based on the outcome of the quantitative pool genotyping, a smaller collection of SNPs was selected for individual genotyping over the full samples. In the next two sections we go into more theoretical details associated with the issue of the statistical inference in association studies, starting with statistical inference based on pools, and then moving to the subject of assessing levels of linkage disequilibrium in small samples. In the last section we will return to the COMT example and describe its outcomes.

3. Statistical Models

The initial screen, which is based on quantitative pool genotyping, suggests the use of the difference in allele frequency between cases and controls as the primary statistic. We open this section with a description of a simple genetic model for which a chi-square test based on this statistic is natural. The statistical power of the test—the chance of detecting true association—is determined by the non-centrality parameter of this chi-square statistic. The form of this non-centrality parameter in the case where the marker is perfectly linked to the functional polymorphism is given and compared to the more likely event where the marker is only partly linked. In the later case the non-centrality parameter, thus the statistical power, is reduced. We show that the ratio of reduction is equal to r^2, the square of the statistical correlation between the marker and the functional polymorphism.

3.1. THE NON-CENTRALITY PARAMETER AT THE FUNCTIONAL LOCUS

We assume a simple genetic model of (at most) one susceptibility functional locus within the region under investigation. This polymorphism, provided it is present, is assumed to consist of two alleles: a susceptibility allele D_1, and wild-type allele D_0. The model does not exclude the possibility of other, unlinked, susceptibility genes, as well as environmental and other effects, all influencing the chance of being affected by the disease. However, all such influencing contributions are independent of the contribution of the given locus. Specifically, the genetic model is built out of three components: A population genetic model, which describes the distribution of the alleles in the locus in the general population; A model for the trait, which describes the marginal contribution of the genotype at the susceptibility locus to the penetrance of the disease; A sampling model, which describes the method of acquiring samples. A combination of the three models gives rise to the distribution of the alleles in the sample of affected and in the sample of controls.

According to the population model, the two haploid genomes within a random individual are selected independently of each other from a general population pool of genomes. This is the assumption of the Hardy-Weinberg equilibrium. An outgrowth of this assumption is the binomial distribution of the number of D_1 alleles within a random genotype:

$$\mathrm{P(Genotype \mid Random\ control)} \sim \mathrm{Binomial}(p_r,2),$$

where p_r is the frequency of allele D_1, and the genotype is specified by the number of such alleles (0, 1, or 2). Adding the assumption of random sampling, gives rise to the observation that the distribution of the alleles in the sample of random controls is binomial. This binomial distribution counts the number of D_1 in a sequence of $2n_r$ independent alleles, where n_r is the sample size of the random controls.

The model for the trait subdivides the population according to the genotype at locus D. The penetrance probability is the probability of being affected, given the genotype at the locus. Denote these probabilities by:

$$\begin{aligned} f_0 &= \mathrm{P(Affected}|D_0, D_0) \\ f_1 &= \mathrm{P(Affected}|D_1, D_0) \\ f_2 &= \mathrm{P(Affected}|D_1, D_1) \end{aligned}$$

according to the number of D_1 alleles in the genotype. The multiplicative model relates the penetrance probabilities to each other via the relation: $f_2/f_0 = (f_1/f_0)^2$. In other words, the genetic relative risk of a D_1 homozygote is the power of the genetic relative risk of a heterozygote. Application of Bayes formula:

$$\mathrm{P(Genotype|Affected)} = \frac{\mathrm{P(Affected|Genotype)P(Genotype)}}{\sum_{\text{All Genotypes}} \mathrm{P(Affected|Genotype)P(Genotype)}},$$

together with the multiplicative and the Hardy-Weinberg assumptions, lead to a binomial distribution for the genome of an affected individual:

$$\text{P(Genotype} \mid \text{Affected)} \sim \text{Binomial}(p_a, 2),$$

where $p_a = [p_r f_1]/[p_r f_1 + (1 - p_r) f_0]$. Therefore, the Hardy-Weinberg equilibrium among random controls, together with the multiplicative model, give rise to the Hardy-Weinberg equilibrium among cases as well, albeit with a different sampling probability of allele D_1 (p_a, instead of p_r). Adding to it the assumption of random sampling of cases results in a binomial distribution of the number of D_1 alleles in the sample of affected. Here $2n_a$ is the length of the sequence of independent alleles, where n_a is the sample size of affected.

In terms of a statistical problem, testing the difference between the distribution at locus D between cases and controls, under the above assumptions, summarizes to testing the equality of two binomial samples. The usual test statistic is based on the standardized difference between the relative frequency of the allele D_1 in the sample of affected and the relative frequency in the sample of random controls:

$$Z^2_{(D)} = \frac{(\hat{p}_a - \hat{p}_r)^2}{\hat{p}(1 - \hat{p})[1/(2n_a) + 1/(2n_r)]}.$$

Here $\hat{p}_a$ is the relative frequency in the sample of affected, $\hat{p}_r$ is the relative frequency in the sample of random controls, and $\hat{p}$ is the relative frequency in the joined samples: $\hat{p} = (n_a\hat{p}_a + n_r\hat{p}_r)\big/(n_a + n_r)$.

When the sample sizes are large, one can use approximate normality to obtain the chi-square (on one degree of freedom) as asymptotic null distribution of this statistic. Under the alternative, the chi-square distribution is distorted by the parameter of non-centrality. This parameter can be computed by substituting the sample values by their population counterparts:

$$\mu^2_{(D)} = \frac{(p_a - p_r)^2}{p_r(1 - p_r)[1/(2n_a) + 1/(2n_r)]}.$$

The non-centrality parameter is increasing in the squared difference between the frequency of alleles and of the sample sizes. The larger the parameter of non-centrality is the better is the chance of detecting association.

Note that the allele test of association is based solely on the relative frequency of the alleles in the samples. Recall that by quantitative genotyping one can measure only this relative frequency in the pooled samples. It can be concluded that, under the assumptions we made, statistical efficiency is not compromised by the use of quantitative genotyping in pools for scanning.

3.2. THE EFFECT OF LINKAGE DISEQUILIBRIUM

In the previous subsection we implicitly analyzed the situation where genotypic information is available for the tentative functional polymorphism. In reality, this scenario is unlikely. Even in the better case, where a candidate gene (such as COMT) is targeted, a functional polymorphism may be any of the dozens, or even hundreds, of polymorphic sites scattered

TABLE 1. Joint distribution of alleles in a population

	D_1	D_0	Total
M_1	p_{11}	p_{10}	$p_{1.}$
M_0	p_{01}	p_{00}	$p_{0.}$
Total	$p_{.1}$	$p_{.0}$	1

about the gene. Consequently, it is safer to assume that the genetic markers actually genotyped are neutral in terms of their biological impact on the trait. Nonetheless, the chances of observing a significant difference between cases and control can still be materialized due to the reflection of the effect of a functional polymorphism on a genotyped marker. The distortion of the distribution of the sample frequencies at the tentative (unobserved) functional polymorphism is transferred, via linkage disequilibrium, and creates a distortion of the distribution of a test statistic computed at a neutral marker.

Linkage disequilibrium is the term geneticists use in order to describe correlation between genomic loci at the population level. Several parameters were proposed in order to measure linkage disequilibrium between a pair of loci. We will consider two: The r^2 parameter of correlation, which is discussed in this section and the D' parameter of association, which will be introduced in the next section.

Consider Table 1, describing the two-by-two joint distribution of alleles of random gametes. We take M_1 and M_0 to be the two alleles of a neutral marker M, and take D_1 and D_0 to be the two alleles of a functional polymorphism. The entries p_{ij} are the relative frequencies of gametes with the allele M_i at locus M and the allele D_j at locus D. We use the dot notation for marginal frequencies. For example, $p_{1.} = p_{11} + p_{10}$ is the marginal probability of sampling the allele M_1. (Note that $p_{.1}$ is the marginal probability of the allele D_1, which we previously denoted simply by p.)

The correlation parameter of r^2 of linkage disequilibrium between the two loci is defined by:

$$r^2 = \frac{(p_{11}p_{00} - p_{10}p_{01})^2}{p_{1.}p_{0.}p_{.1}p_{.0}}.$$

This parameter is the square of the usual Pearson correlation coefficient, which is obtained by assigning numerical values to the alleles. Note, in particular, that the parameter r^2 can take values between zero and one. A value of one is indicative of perfect linkage disequilibrium. In such a case, the marker is equivalent to the functional locus from a statistical point of view. At the other extreme, a value of zero corresponds to perfect linkage equilibrium (or independence, in statistical language). In this case the marker reflects none of the information projected from the functional site.

The following theorem provides a direct relation between the parameter of non-centrality of the chi-square statistic computed at the marker M and the correlation coefficient between this marker and the functional polymorphism D. This relation depends on the notion of statistical sufficiency and on the notion of Hardy-Weinberg equilibrium. We discuss these two notions after the presentation of the theorem:

Theorem: *Assume that the locus D is statistically sufficient for the disease effect, with respect to the marker M. Assume further that Hardy-Weinberg equilibrium holds for both cases and controls. Then*

$$\mu^2_{(M)} = r^2 \times \mu^2_{(D)},$$

where $\mu^2_{(M)}$ *is the parameter of non-centrality computed at the marker M,* $\mu^2_{(D)}$ *is the value of the parameter at D, and* r^2 *is the (squared) correlation coefficient between M and D (evaluated for the general population). In other words, the parameter* r^2 *is the ratio of the disease effect reflected at the marker.*

The intuitive meaning of statistical sufficiency is that D is the only source of the relevant genetic information in the targeted region with respect to the difference between affected and controls. The genetic composition at locus M does not modify the chances of developing the disease. Formally this means that the conditional distribution of the alleles of M, among the subgroup of those that share the same D allele, is the same for affected and controls. Thus, conditional on the status at locus D, the marker M is not associated with the trait. Association is introduced only due to the fact that the analysis is based on the marginal distribution of M. This marginal distribution is computed by taking a linear combination of the conditional distributions, with the frequencies of the alleles of D as linear coefficients. The distribution of D is associated with the trait, which corresponds to the use of different linear coefficients for cases and controls. Consequently, one gets a different marginal distribution of M for cases and for controls, i.e. association between M and the trait.

We also required in the conditions of the theorem that the Hardy-Weinberg equilibrium be valid both for cases and for controls. This requirement may not hold for all models of penetrance. However, as we previously saw, the condition among case follows from Hardy-Weinberg equilibrium among the random controls, together with the multiplicative model of penetrance.

It is worthwhile to note that the parameter r^2 is computed for the random population, and not for affected. It is a property of the population under study and its genetic origin and evolution, and is not related to the disease. Theorem 1 can be interpreted as a diconvolution of the effect at the marker into its two sources: The effect of the gene on the penetrance, and the linkage disequilibrium structure among polymorphic loci within the gene at the population level. One may also identify the parallel roles sample sizes and the parameter r^2 play. Both are proportional multipliers of the parameter of non-centrality. However, they work in opposite directions. Consequently, one may interpret a given level of r^2 as the proportion of reduction in the effective sample sizes due to the genotyping of anonymous markers, instead of the functional polymorphism itself. For example, samples sizes of 720 cases and 3,000 controls translate to an effective sample size of 360 and 1,500 respectively, at a marker with squared correlation $r^2 = 0.5$ to the functional polymorphism.

4. Estimating Linkage Disequilibrium

We turn to the second topic of this chapter: The estimation of linkage disequilibrium parameters from genotypic data. Again, the COMT example is a useful demonstration. The initial screen involved 12 markers. Four of these markers turned out to be non-polymorphic, and one failed on a technical ground. The remaining 7 SNPs were quantitatively genotyped in pools, as well as individually genotyped in a sub-sample of size 70. Quantitative genotyping provided information regarding the association of the SNP and the phenotype. Using the tools that we will present below, information regarding the linkage disequilibrium structure between the SNPs was inferred from the individual genotyping. Combination of these two sources of data provided the basis for an informed selection of SNPs for genotyping over the complete samples of cases and controls.

4.1. HAPLOTYPES

Parameters of linkage disequilibrium are summaries of the distribution of haplotypes. We initiate our discussion with a definition of haplotypes and with a method for the estimation of their distribution in a population, namely the Expectation-Maximization (EM) algorithm. Based on this estimation of distribution, one can construct an estimate of the appropriate parameter of linkage disequilibrium. Such estimation is subject to sampling error, as well as other types of error. We will try to assess these errors in a simple example. The assessment will be conducted via simulations.

Consider a collection of SNPs in a given region. We define a haplotype of these SNPs to be a specification of the alleles of the SNPs over a given gamete. The fact that all the alleles are on the same gamete is denoted by them having the same phase. Note that the total number of different haplotypes that can be constructed from m SNPs is 2^m. The actual distribution of haplotypes in a population can be estimated from a sample. Unfortunately, haplotypes are not measured directly. Instead, one gets to observe, for each sample, only the genotypes at the SNPs, leaving in some cases the phases of the alleles ambiguous. This ambiguity may be sorted out using familial information. However, frequently no such information is available. An alternative approach for overcoming this difficulty relies on statistical estimation. According to this approach, the unknown phases are treated as hidden variables in the process of estimating the unknown haplotype frequencies. A statistical tool for the estimation of these frequencies, in the presence of hidden variables, is applied. This tool is the EM algorithm.

The EM algorithm is based on the principle of maximum likelihood for estimation of unknown parameters. The likelihood of the data, as a function of the parameters, is constructed. Estimates are obtained by the maximization of this likelihood. In order to obtain the maximum, the EM algorithm initiates with a starting value for the parameters. Each iteration updates these values and results in an increase in the likelihood function. The algorithm converges to its final estimates when repeated updates reproduce the same values for the parameters. The EM algorithm uses the Hardy-Weinberg assumption of gamete independence in the construction of the likelihood. The original likelihood function is expanded in order to include the frequency of the different phases as auxiliary parameters. This expansion produces a simple rule for updating the values of the parameters.

In order to demonstrate the EM algorithm we apply it first to the sample of genotypes over two bi-allelic markers. Consider the distribution of the full sample of random controls over the combination of genotypes for the SNPs rs737865 and rs165688 (a total of 2680 individuals who had genotype values for both markers), which is given in Table 2. We can clearly determine from this data that the haplotype C-A appears in the sample at least

TABLE 2. Distribution of genotypes for a pair of markers in the COMT gene

rs737865	rs165688	Frequency	Haplotype 1	Haplotype 2
C/C	A/A	1	C-A	C-A
C/C	G/A	54	C-A	C-G
C/C	G/G	364	C-G	C-G
C/T	A/A	75	C-A	T-A
C/T	G/A	978	C-A and T-G or C-G and T-A?	
C/T	G/G	274	C-G	T-G
T/T	A/A	508	T-A	T-A
T/T	G/A	367	T-A	T-G
T/T	G/G	59	T-G	T-G

$2 \times 1 + 54 + 75 = 131$ times. This follows from the fact that the haplotype is present in two copies in the double-homozygous individual in the first row of the table, and is presented in one copy in the 54 individuals in the second row and in the 75 individuals in the fourth row. However this haplotype may also be present in one copy among some of the individuals in the fifth row of double-heterozygotes. What we do not known is the percentage of such individuals in the fifth row. The EM algorithm treats this percentage as an extra parameter, and tries to estimate it from the data.

Denote by $0 \leq \theta \leq 1$ the ratio of those who have the haplotype C-A (and also T-G) among the double-heterozygous individuals in the fifth row. It follows, that the frequency of the haplotype C-A is $131 + \theta \times 978$. In a similar fashion, one can carry out the computation of the relative frequency of the other 3 haplotypes, and obtain the results presented in Table 3:

Having the frequencies of Table 3, we can turn back and reevaluate the expected percentage θ of (C-A,T-G) double-heterozygotes among all the double-heterozygotes. Assuming the Hardy-Weinberg equilibrium, it results that the probability of obtaining a double-heterozygote is:

$$\frac{131 + \theta \cdot 978}{5360} \times \frac{759 + \theta \cdot 978}{5360} + \frac{1056 + (1 - \theta) \cdot 978}{5360} \times \frac{1458 + (1 - \theta) \cdot 978}{5360}.$$

The probability of obtaining a (C-A,T-G) double-heterozygote, on the other hand, is:

$$\frac{131 + \theta \cdot 978}{5360} \times \frac{759 + \theta \cdot 978}{5360}.$$

TABLE 3. The frequency of the four haplotypes in the sample, given the percentage of (C-A,T-G) double-heterozygotes, and the frequency given its estimated value

Haplotype	Theoretical frequency	Frequency	Probability
C-A	$131 + \theta \times 978.$	154.9	0.0289
C-G	$1056 + (1 - \theta) \times 978.$	2,010.1	0.3750
T-A	$1458 + (1 - \theta) \times 978.$	2,412.1	0.4500
T-G	$759 + \theta \times 978.$	782.9	0.1461
Total	5,360.0	5,360.0	1

Consequently, it turns out that the expected relative frequency of (C-A,T-G) double-heterozygote among the set of all double-heterozygotes, is equal to:

$$\frac{(131 + \theta \cdot 978) \times (759 + \theta \cdot 978)}{(131 + \theta \cdot 978) \times (759 + \theta \cdot 978) + (1056 + (1 - \theta) \cdot 978) \times (1458 + (1 - \theta) \cdot 978)}.$$

Replacing the current value of θ by this update completes an iteration. Reiterating this procedure several times leads to the values of θ converging to a stable value. Plugging this value into the second column of Table 3 provides the EM estimate of the haplotype sample frequencies. These estimates are given in the third column of Table 3. The last column contains the relative frequencies.

4.2. PARAMETERS OF LINKAGE DISEQUILIBRIUM

In the previous section we introduced the squared-correlation parameter r^2 as a measure of linkage disequilibrium between a pair of SNPs. Before going into the issue of estimating this parameter from genotypic information, we would like to introduce another measure of linkage disequilibrium—the coefficient of association, which is denoted by D′. This measure is very popular among population geneticists. It intends at describing the relative level of linkage disequilibrium in the current population, in comparison to its level at the formation of the population. The parameter has the form of a ratio between two correlation coefficients. The numerator is the correlation coefficient computed from the given 2 × 2 table of haplotype frequencies in the population. The denominator is the correlation computed from another 2 × 2 table. The marginal frequencies of this new table are identical to those of the original table. However, one of the entries in the new table is set to be equal to zero. The latter table represents the distribution of haplotypes at the formation of the population (or the formation of the younger of the two SNP), before recombination could add its effect of reducing the correlation between the two SNPs by breaking up haplotypes. (Setting a value for a single entry, and setting the values of the marginal frequencies, uniquely determines the entries of the table. Moreover, in a table with given marginal frequencies, the maximal correlation coefficient is obtained by setting the appropriate cell entry equal to zero.) The exact formula for the computation of D′ can be read off the code of the MATLAB function for its computation.

The parameter D′ is computed from the haplotype frequencies. For example, the value of the D′ coefficient between markers rs737865 and rs165688, based on the data given in Table 3, is 0.8506. This estimate was obtained in a very large sample. Consequently, it is very likely that it is a good approximation of the actual value of this parameter of association in the entire population. However, for our needs we would like to ask how accurately can one expect to estimate the parameter when the sample size is much smaller?

We planned a small simulation, which may help in gaining insight as to the relation between sample size, the actual information, and the accuracy of the statistical tool for inference. In this simulation, the outcome of Table 3 represented the true distribution in the population. A sample of a given size from this population was simulated. The EM algorithm was applied to that sample in order to produce the estimated distribution of haplotypes. The estimated value of D’ was computed from this estimated distribution. This procedure was iterated 10,000 times in order to obtain a distribution for this estimate of D′. We examined the effect of two factors on the distribution of the estimate. One factor was the sample

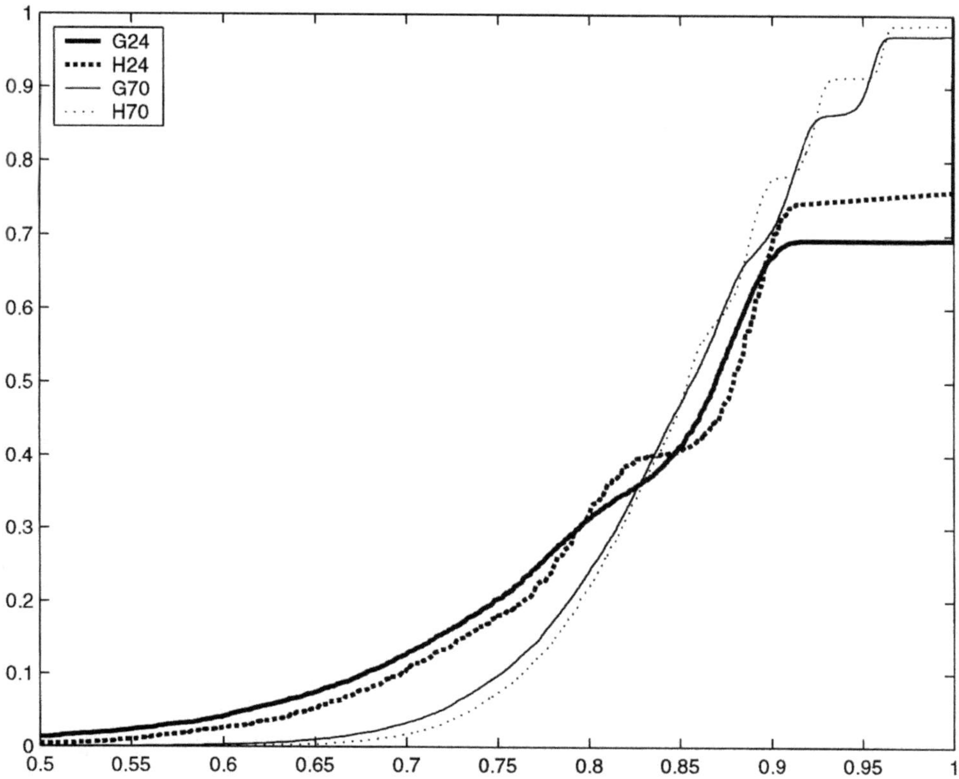

Figure 1. Sampling distribution of D′ in samples of 24 and 70 individuals. Solid lines correspond to application of the EM algorithm. Dashed lines correspond to the case were phase data is available.

size, which was examined by considering a sample of size 70 and a sample of size 24. The other factor was the lack of phase data and the effect of the EM procedure. This factor was examined by simulating haplotypes directly, and computing the parameter from their sampling distribution, without the intermediation of the EM algorithm. The results of the simulation are presented in Figure 1, where the cumulative distribution functions (cdf) of the various estimates are given.

The examination of Figure 1 reveals some interesting phenomena. Consider, for example, estimates based on a sample size of 24 (G24 = application of the EM to genotypic data, and H24 = sampling haplotypes directly). Their distributions tend to converge on given values. This is manifested by the stair-like structure of the cdf's. In particular, there is a definite mass point in 1. The effect of the lack of phase data is expressed by a heavier left tail for the distribution of G24, and a larger probability of the point mass at 1. Apart from the increase in the point mass, the distribution of G24 is somewhat smoother in comparison to the distribution of H24. When one examines the two distributions associated with a sample size of 70 (the finer lines), on the other hand, one may note that the point mass at 1 almost disappeared. The effect of the lack of phase data is much smaller, and the distributions are much more concentrated around the expectation of 0.8506.

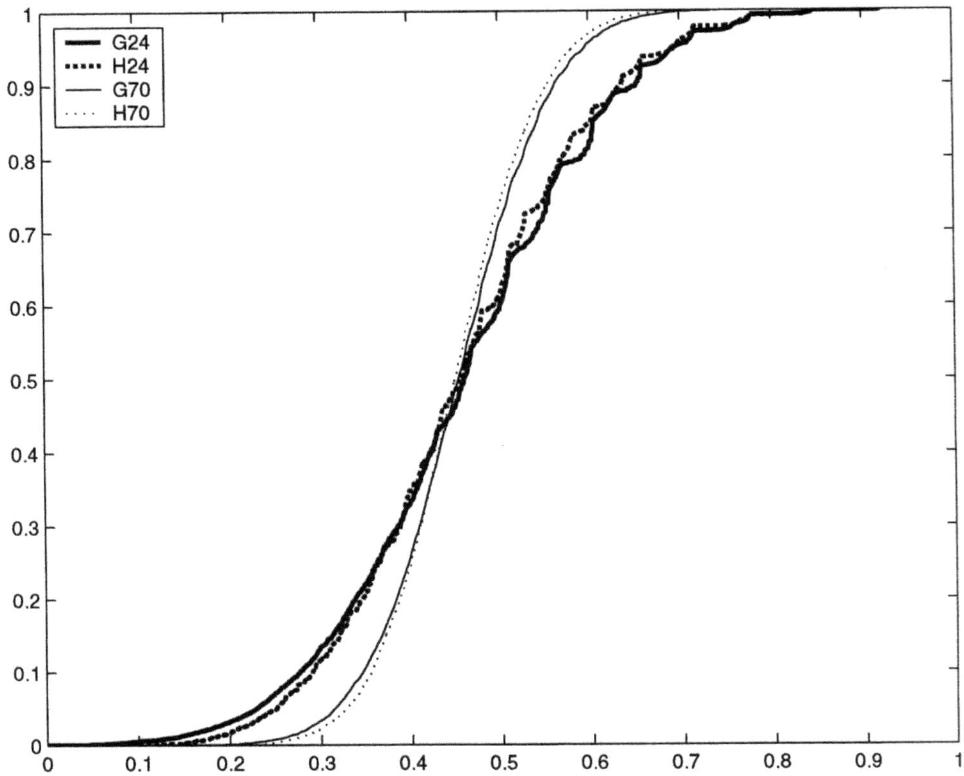

Figure 2. Sampling distribution of r^2 in samples of 24 and 70 individuals. Solid lines correspond to application of the EM algorithm. Dashed lines correspond to the case were phase data is available.

A similar exercise can be carried out in order to assess the accuracy in estimating our target parameter r^2. This time, the parameter obtains the value of 0.4506 in the sample of 2,680 random controls. The distribution of estimates of this parameter with sample sizes of 24 and 70 are presented in Figure 2. Here, the effect of not having phase data is much smaller. (This is evident from the similarity between G24 and H24 and between G70 and H70.) The cdf's are much more similar to the normal cdf, with an increased variance when a smaller sample size is used. The bias of the estimates is negligible. In summary, the estimation of r^2 fits better the standard theory of normal-based statistical inference, where the estimates of D' do not (at least for small sample sizes).

5. Final Conclusions

We examine two of the statistical aspects related to population-based association studies. One aspect is the identification of r^2 as the ratio of the information regarding the association, projected onto the marker. The role of this parameter, in population-based studies, is a parallel to the role of recombination fraction in family-based studies (and in crosses-based experimental genetic studies). However, unlike the recombination fraction, which is determined by the well-studied process of meiosis, the linkage disequilibrium parameter r^2 is a product of the population genetic evolution. Much less is known about the factors that determine this historic evolution. The other aspect we considered is the estimation of linkage disequilibrium. Estimation is possible from genotypic data directly. However, in small samples, estimation may be inaccurate. The statistical tool applied for the estimation, namely the EM algorithm, may increase the uncertainty in estimation.

The findings of the COMT study, and their outcome are summarized hereafter. A total of 7 SNPs were genotyped in pools. Five of these SNPs showed association with schizophrenia. Individual genotyping of these SNPs in a sample of 70 individuals revealed high correlation between two of these SNPs and SNP rs165688 (r^2 values of 0.7712 and 0.9176, respectively, and both D' values equal to 1). The third SNP, rs165599, showed low levels of correlation with rs165688 ($r^2 = 0.0011$ and $D' = 0.0489$). Lastly, as we have seen in the previous section, the SNP rs737865 shows moderate levels of linkage disequilibrium with the non-synonymous SNP rs165688. Based on these finding, individual genotyping was carried out in the full samples for the three SNPs (rs165688, rs165599, and rs737865), for reasons given in the previous sections.

A modest association was found between SNP rs165688 and schizophrenia in men. No effect was detected among women. SNP rs737865 was found to be significantly associated with schizophrenia in men and less so in women. An even stronger association was identified for SNP rs165599, especially among women. The conclusion of these finding strengthen the a-priori assumption of association between the COMT gene and schizophrenia. However, they demonstrate the complexity of the pattern of association. In particular, the Val/Met SNP rs165688 does not seem to be the source of the association. It probably only reflects the association found in the correlated SNP rs737865. Moreover, the interaction between association and gender was a new and unexpected outcome. Such findings demonstrate the great promise provided by population-based association studies for the detection of target pathways.

6. References

Egan MF, Goldberg TE, Kolachana BS, Callicott JH, Mazzanti CM, Straub RE, Goldman D, Weinberger DR. (2001). Effect of COMT Val108/158 Met genotype on frontal lobe function and risk for schizophrenia. Proc. Natl. Acad. Sci. U. S. A.; 98:6917–6922.

Karayiorgou M, Morris MA, Morrow B, Shprintzen RJ, Goldberg R, Borrow J, Gos A, Nestadt G, Wolyniec PS, Lasseter VK, Eisen H, Childs B, Kazazian HH, Kucherlapati R, Antonarakis SE, Pulver AE, Housman DE. (1995). Schizophrenia susceptibility associated with interstitial deletions of chromosome 22q11. Proc Natl. Acad. Sci. U. S. A.; 92:7612–7616.

Lachman HM, Papolos DF, Saito T, Yu YM, Szumlanski CL, Weinshilboum RM. (1996). Human catechol-O-methyltransferase pharmacogenetics: description of a functional polymorphism and its potential application to neuropsychiatric disorders. Pharmacogenetics; 6:243–250.

Murphy KC, Jones LA, Owen MJ. (1999). High rates of schizophrenia in adults with velo-cardio-facial syndrome. Arch. Gen. Psychiatry; 56:940–945.

Shifman S, Bronstein M, Sternfeld M, Pisante-Shalom A, Lev-Lehman E, Weizman A, Reznik I, Spivak B, Grisaru N, Karp L, Schiffer R, Kotler M, Strous RD, Swartz-Vanetik M, Knobler HY, Shinar E, Beckmann JS, Yakir B, Risch N, Zak NB, Darvasi A. (2002). A highly significant association between a COMT haplotype and schizophrenia. Am. J. Hum. Genet; 71(6):1296–302.

Sawa A, Snyder SH. (2002). Schizophrenia: diverse approaches to a complex disease. Science; 296(5568):692–5.

Usiskin SI, Nicolson R, Krasnewich DM, Yan W, Lenane M, Wudarsky M, Hamburger SD, Rapoport JL. (1999). Velocardiofacial syndrome in childhood-onset schizophrenia. J. Am. Acad. Child Adolesc. Psychiatry; 38:1536–1543.

INDEX

INDEX OF AUTHORS

GPSR Compliance
The European Union's (EU) General Product Safety Regulation (GPSR) is a set of rules that requires consumer products to be safe and our obligations to ensure this.

If you have any concerns about our products, you can contact us on

ProductSafety@springernature.com

In case Publisher is established outside the EU, the EU authorized representative is:

Springer Nature Customer Service Center GmbH
Europaplatz 3
69115 Heidelberg, Germany

www.ingramcontent.com/pod-product-compliance
Ingram Content Group UK Ltd.
Pitfield, Milton Keynes, MK11 3LW, UK
UKHW020834270526
471509UK00003B/57
* 9 7 8 1 4 0 2 0 2 6 3 9 3 *